PRACTICAL STATISTICAL SAMPLING FOR AUDITORS

STATISTICS: Textbooks and Monographs

A SERIES EDITED BY

D. B. OWEN, Coordinating Editor
Department of Statistics
Southern Methodist University
Dallas, Texas

Volume 1: The Generalized Jackknife Statistic, *H. L. Gray and W. R. Schucany*

Volume 2: Multivariate Analysis, *Anant M. Kshirsagar*

Volume 3: Statistics and Society, *Walter T. Federer*

Volume 4: Multivariate Analysis: A Selected and Abstracted Bibliography, 1957-1972, *Kocherlakota Subrahmaniam and Kathleen Subrahmaniam* (out of print)

Volume 5: Design of Experiments: A Realistic Approach, *Virgil L. Anderson and Robert A. McLean*

Volume 6: Statistical and Mathematical Aspects of Pollution Problems, *John W. Pratt*

Volume 7: Introduction to Probability and Statistics (in two parts) Part I: Probability; Part II: Statistics, *Narayan C. Giri*

Volume 8: Statistical Theory of the Analysis of Experimental Designs, *J. Ogawa*

Volume 9: Statistical Techniques in Simulation (in two parts), *Jack P. C. Kleijnen*

Volume 10: Data Quality Control and Editing, *Joseph I. Naus*

Volume 11: Cost of Living Index Numbers: Practice, Precision, and Theory, *Kali S. Banerjee*

Volume 12: Weighing Designs: For Chemistry, Medicine, Economics, Operations Research, Statistics, *Kali S. Banerjee*

Volume 13: The Search for Oil: Some Statistical Methods and Techniques, *edited by D. B. Owen*

Volume 14: Sample Size Choice: Charts for Experiments with Linear Models, *Robert E. Odeh and Martin Fox*

Volume 15: Statistical Methods for Engineers and Scientists, *Robert M. Bethea, Benjamin S. Duran, and Thomas L. Boullion*

Volume 16: Statistical Quality Control Methods, *Irving W. Burr*

Volume 17: On the History of Statistics and Probability, *edited by D. B. Owen*

Volume 18: Econometrics, *Peter Schmidt*

Volume 19: Sufficient Statistics: Selected Contributions, *Vasant S. Huzurbazar (edited by Anant M. Kshirsagar)*

Volume 20: Handbook of Statistical Distributions, *Jagdish K. Patel, C. H. Kapadia, and D. B. Owen*

Volume 21: Case Studies in Sample Design, *A. C. Rosander*

Vol. 22: Pocket Book of Statistical Tables, *compiled by R. E. Odeh, D. B. Owen, Z. W. Birnbaum, and L. Fisher*
Vol. 23: The Information in Contingency Tables, *D. V. Gokhale and Solomon Kullback*
Vol. 24: Statistical Analysis of Reliability and Life-Testing Models: Theory and Methods, *Lee J. Bain*
Vol. 25: Elementary Statistical Quality Control, *Irving W. Burr*
Vol. 26: An Introduction to Probability and Statistics Using BASIC, *Richard A. Groeneveld*
Vol. 27: Basic Applied Statistics, *B. L. Raktoe and J. J. Hubert*
Vol. 28: A Primer in Probability, *Kathleen Subrahmaniam*
Vol. 29: Random Processes: A First Look, *R. Syski*
Vol. 30: Regression Methods: A Tool for Data Analysis, *Rudolf J. Freund and Paul D. Minton*
Vol. 31: Randomization Tests, *Eugene S. Edgington*
Vol. 32: Tables for Normal Tolerance Limits, Sampling Plans, and Screening, *Robert E. Odeh and D. B. Owen*
Vol. 33: Statistical Computing, *William J. Kennedy, Jr. and James E. Gentle*
Vol. 34: Regression Analysis and Its Application: A Data-Oriented Approach, *Richard F. Gunst and Robert L. Mason*
Vol. 35: Scientific Strategies to Save Your Life, *I. D. J. Bross*
Vol. 36: Statistics in the Pharmaceutical Industry, *edited by C. Ralph Buncher and Jia-Yeong Tsay*
Vol. 37: Sampling from a Finite Population, *J. Hájek*
Vol. 38: Statistical Modeling Techniques, *S. S. Shapiro*
Vol. 39: Statistical Theory and Inference in Research, *T. A. Bancroft and C.-P. Han*
Vol. 40: Handbook of the Normal Distribution, *Jagdish K. Patel and Campbell B. Read*
Vol. 41: Recent Advances in Regression Methods, *Hrishikesh D. Vinod and Aman Ullah*
Vol. 42: Acceptance Sampling in Quality Control, *Edward G. Schilling*
Vol. 43: The Randomized Clinical Trial and Therapeutic Decisions, *edited by Niels Tygstrup, John M. Lachin, and Erik Juhl*
Vol. 44: Regression Analysis of Survival Data in Cancer Chemotherapy, *Walter H. Carter, Jr., Galen L. Wampler, and Donald M. Stablein*
Vol. 45: A Course in Linear Models, *Anant M. Kshirsagar*
Vol. 46: Clinical Trials: Issues and Approaches, *edited by Stanley H. Shapiro and Thomas H. Louis*
Vol. 47: Statistical Analysis of DNA Sequence Data, *edited by B. S. Weir*
Vol. 48: Nonlinear Regression Modelling: A Unified Practical Approach, *David A. Ratkowsky*
Vol. 49: Attribute Sampling Plans, Tables of Tests and Confidence Limits for Proportions, *Robert E. Odeh and D. B. Owen*
Vol. 50: Experimental Design, Statistical Models, and Genetic Statistics, *edited by Klaus Hinkelmann*
Vol. 51: Statistical Methods for Cancer Studies, *edited by Richard G. Cornell*
Vol. 52: Practical Statistical Sampling for Auditors, *Arthur J. Wilburn*

OTHER VOLUMES IN PREPARATION

PRACTICAL STATISTICAL SAMPLING FOR AUDITORS

Arthur J. Wilburn
A. J. Wilburn Associates
Rockville, Maryland

MARCEL DEKKER, INC. New York and Basel

Library of Congress Cataloging in Publication Data

Wilburn, Arthur J.
Practical statistical sampling for auditors.

(Statistics, textbooks and monographs ; v. 52)
Includes index.
1. Sampling (Statistics) 2. Auditing--Statistical methods. I. Title II. Series.
HA31.2.W54 1984 519.5'2'024657 83-26257
ISBN 0-8247-7124-9

MARCEL DEKKER, INC.
270 Madison Avenue, New York, New York 10016

Current printing (last digit):
10 9 8 7 6 5 4 3 2 1

PRINTED IN THE UNITED STATES OF AMERICA

Preface

This book is written with emphasis on and illustrations from auditing, accounting and management environments. However, essentially the same procedures are useful in other applications where the sample information is not the sole source of evidence and a prior body of knowledge is available.

While writing this book, I endeavored to keep a definite audience in mind and to maintain a mental dialogue with that audience. My first consideration was for the needs of financial managers, auditors, accountants and analysts who design, supervise, apply and evaluate statistical sampling applications. Thus, this book could serve as a text or reference for accounting and business administration courses which are concerned with sampling procedures.

This book is also intended to provide a guide and reference on sampling strategies for managers, auditors and analysts who want to understand statistical sampling methods without necessarily becoming sampling specialists.

Although this book is not being written for statisticians and research scientists, it does provide them with some insight into the unique and peculiar objectives and problems of auditors as contrasted with industrial and scientific applications. It contains the conventional topics as well as newer and unconventional ones, such as flexible sampling and audit appraisal using evaluation matrices.

The topics provide enough guidance to enable auditors to design and execute valid audit samples, to avoid selection and estimation biases, to perform effective audit stratification, and to subjectively appraise as well as statistically evaluate the sample results for audit significance. They should be better able to blend sample results with other audit evidence prior to drawing a conclusion and making an audit decision.

The reader should be able to design a moderate scale audit sample. For complicated and optimum sample designs, a sampling statistician experienced in the particular professional applications should be consulted. In such cases, this book would provide sufficient knowledge to facilitate better communication between auditors and sampling specialists.

The book is oriented towards a working knowledge of practical sampling procedures for use in auditing and related activities. Theoretical statistical discussions and formulas are kept to a minimum, but numerous references to good theoretical source material are provided.

The aim of this book is to adapt statistical methods to assist in the solution of auditing and accounting test-check problems. No attempt is made to modify audit objectives or procedures to fit statistical models.

In writing this book, my original intention was to formalize the lecture notes and applications which have evolved from conducting seminars and workshops for auditors, accountants and analysts since 1959. I was amazed at the volume of statements, discussions, questions-and-answers, and illustrations which were on tapes that had never been reduced to writing. Much of this supplemental material was integrated into seminar material to produce this book.

It is with genuine pleasure that I acknowledge my debt to many friends and colleagues. Of the colleagues who have given me personal support and criticism, I must single out John Neter of the University of Georgia; Leslie Kish of the University of Michigan; Donald Goldschen of the University of the District of Columbia; the late Frederick Stephan of Princeton University; the late Howard Jones, CPA and Professor of Statistics; Raymond Willis of the University of Minnesota; Albert Teitlebaum of McGill University; Donald Roberts of the University of Illinois at Urbana-Champaign; John Sennetti of Texas Tech University; and Donald Leslie, Partner, Clarkson, Gordon & Co. Also I am especially indebted to the late Raymond Snyder, CPA, and Carol Lynch, CPA, who assisted me in interfacing statistical sampling with auditing and with whom I developed case study material used in seminars. Edmund Stover, retired, is the auditor who completed my on-the-job training and also edited my original seminar notes for technical audit procedures. Special acknowledgment goes to Dennis G. Haack, Statistical Consultant, Robert Bickel, Defense Department Auditor, Barbara Morris, Doctorial candidate in Accounting at the University of Georgia and to two anonymous readers, who reviewed the manuscript and offered numerous suggestions. I am indebted to Darrell Oyer, CPA and Manager, Peat, Marwick, Mitchell & Co., for preparing the various computer illustrations and to Earlene Williams, Karen Edris and Kathy Kline for typing and proofing the manuscript.

Stimulation to write this book came from the many students in over 300 seminars and workshops in statistical sampling which I conducted during the past 25 years. These students, who were primarily auditors and accountants, were from all the states and many foreign countries. Their varied experiences, probing questions, suggestions and critiques contributed substantially to this book.

Arthur J. Wilburn

Contents

1

Introduction

I. GENERAL

Many attempts to develop a practical approach to the application of statistical sampling to auditing have not completely satisfied the needs of audit activities. These needs, while they resemble those of some other activities, are quite different in their implications. Thus, it is necessary to discuss the peculiar audit sampling requirements before proceeding to the subject of statistical sampling, per se. These audit implications will be discussed in the first three chapters.

The discussions will cover many aspects of auditing as well as statistical sampling and test checking. Experience reveals that many of the difficulties encountered by auditors in applying statistical sampling are related more to auditing problems than to sampling problems. Hence, for effective application of statistical sampling methods, it is necessary to identify some of the related audit concepts and discuss them in their relationship to sampling. The use of statistical sampling techniques generally requires auditors to exercise more discipline in the determination of specific audit objectives. It may be that the full advantage of statistical sampling methods will not become evident until some of the related audit problems are clarified or solved. This book identifies some of the audit implications and interfaces them with statistical sampling concepts and techniques.

The overall objective of this book is to prepare the reader for efficient, effective and economical use of statistical sampling in auditing and related activities. This objective may be divided into two broad aspects: first, the auditing problem of when statistical sampling is appropriate, and secondly, how to use it efficiently and effectively, which involves the application of certain statistical concepts and techniques in an auditing environment.

The Glossary of Terms in Appendix A should be referred to as necessary to become familiar with new terms and concepts. Some of the terms, as used in statistical sampling, have different meanings from those commonly associated with them. Also, a number of terms have similar meanings and can be used interchangeably. To minimize this source of confusion consistency in terminology will be maintained wherever possible.

II. USE OF AUDIT TESTS

Change usually evolves slowly and one may not recognize that some audit procedures and concepts have been in the process of slow evolution. One of these is test checking. Large enterprises with rapidly expanding masses of financial data forced test checking on the auditor before he really had a chance to fully assess all the implications. Test checking first appeared in auditing towards the 1900s and slowly evolved to its current level. However, mention of statistical sampling in auditing did not appear until the 1930s and did not gain momentum until the 1950s [80].

Whenever an auditor decides that testing is an appropriate audit procedure, statistical sampling also may be appropriate. Furthermore, in some audit situations sampling may be the only practicable approach. In the past, some conditions compelled auditors, using nonstatistical methods, to choose between underauditing and overauditing; there was no happy medium. Underauditing was associated with inadequate selective tests. Overauditing may also result from the use of statistical sampling planning which does not reflect prior audit experience and other sources of reliance. For example, sampling plans and techniques designed for other disciplines and applied to auditing without adequate adaptation may be inefficient and uneconomical.

III. NEED FOR AUDIT TESTING

Due to the enormous volume of financial transactions processed by large organizations, it is virtually impossible, because of limited audit resources to examine all transactions as a means of determining whether the public's or stockholders' interests are being adequately protected. Obviously, some form of testing or sampling is the only practical method for drawing conclusions about the condition of all transactions. It is in this area of limited testing that statistical sampling makes its greatest contribution to auditing. The advantages of statistical sampling techniques over other methods of testing are listed near the end of this chapter. Fortunately, in expressing an *opinion*, the auditor does not require, and it would be impractical for him to obtain, absolute assurance that the auditee's representations or his own findings are accurate. He needs only an acceptable degree of assurance that they are reasonable. For this reason statistical sampling provides another useful audit technique. If absolute assurance were required, neither statistical sampling nor any other testing procedure would be an appropriate method for obtaining evidential matter.

The test check may be defined as any audit step in which only a portion of the documents or transactions of a group are examined. The actual selection of items to be examined may be made in a number of different ways, one of which is the random selection aspect of statistical sampling.

To place testing in its proper perspective in auditing, a brief review of (i) its historical development, (ii) some common nonstatistical methods of selecting items, and (iii) the acceptance of testing by the auditing profession will be helpful. When the general background of audit testing is understood, the logical progression to statistical sampling as a testing procedure is more apparent.

IV. HISTORICAL DEVELOPMENT OF TESTING

Testing is a relatively recent development in auditing history. As mentioned above, the first reference to testing as an audit procedure appeared near the end of the nineteenth century. Even then the use of testing as an audit technique was influenced by several developments. The first was the tremendous growth of many industries during and subsequent to the Industrial Revolution. Auditing of larger companies developed the need for selective tests of the accounts rather than an examination of all relevant transactions.

Another development was the improvement in accounting systems and internal controls of larger companies. Increased size of companies makes the segregation of duties possible and the installation of internal checks and controls. Although accounting literature in the early 1900s recognized the importance of internal controls, the literature preceded actual practice. Meanwhile, auditors continued to expand the use of testing, but decisions concerning the extent of testing were not directly related to an appraisal of internal controls.

A third development was the change in stated audit objective. The stated objective prior to 1900 was the detection of fraud and clerical errors. At the turn of the century, the determination of the fairness of reported financial status appeared as a stated audit objective. This transition was not completed until the late 1930's. However, Government auditing has indicated a return to fraud auditing beginning in 1978 with the establishment of Inspectors General in the Government.

In 1956 the American Institute of Certified Public Accountants (AICPA) appointed a committee to consider the applicability of statistical sampling in auditing. Since the committee was established, there has been increasing interest in statistical sampling by the accounting profession. Much has been done by the AICPA, Canadian Institute of Chartered Accountants (CICA), accounting firms, The Institute of Internal Auditors (IIA) and government auditors to provide a broader education and knowledge of statistical sampling as well as research on its optimal applicability in auditing. AICPA standards on *Audit Sampling* are in SAS No. 39 [3].

V. JUDGMENTAL (NONSTATISTICAL) SELECTION OF ITEMS

Since the adoption of testing by the auditing profession, a number of sample selection procedures have gained general acceptance and are still in use. With the advent of statistical sampling, these earlier test methods have been identified as judgment sampling to distinguish them from random selection methods used in statistical sampling. The terminology is unfortunate since it implies that the auditor does not have to exercise the same degree of judgment in the application of statistical sampling. In fact, the use of statistical sampling probably requires the exercise of a higher degree of judgment due to additional discipline.

A. Cross-Section Testing

Cross-section testing is an attempt to get an across-the-board sample that includes items from all parts of the area being tested. It is common under this type of testing to designate a fixed percentage, such as 10 percent, of items to be included in the test. Many times the selection is made by using a fixed interval, such as every 10th item, for selection. If this method were used with random starts, the sample generally would meet the requirements of the statistical sampling procedure known as systematic selection. This is another illustration of the fact that many traditional testing techniques may be blended with statistical sampling with little modification. However, it is common when one uses the cross-section approach, to go through the records and haphazardly (accidentally) select items until he has the desired quantity of records.

B. Block Testing

Block testing usually includes all items in a given time period or in a given section of the audit area (universe) being tested. For example, the auditor might select one or more months or weeks and examine all items in those blocks, or he might select certain letters or blocks of letters in the alphabet and examine all vendors' invoices in the designated sections. He generally is careful not to select the same or similar blocks each year. The auditee also notices this procedure and relaxes if the block were audited the last time. The obvious weakness of block testing is that it permits the unselected blocks no chance of inclusion. There is an assumption that certain blocks are typical or error prone. Errors may go undetected if they are not associated with the selected blocks. However, if the blocks are randomly selected and there is a sufficient number of them, then one is using a statistical procedure known as cluster sampling.

C. Purposive Testing

Purposive testing is not designed to give a cross section of the entire audit area. Its purpose is to highlight known or suspected problem areas with the least amount of effort. Unfortunately, this approach is used occasionally to distort the representation of the total situation or condition. One concentrates in a limited area, where prior experience indicated major problems, and implies that the whole area (universe) is as bad as the limited area examined. Statistical sampling does not preclude this type of testing provided it is kept separate and no inference is made beyond the area tested. In the final analysis, the consolidation of all evidence must be done subjectively.

D. Convenience Testing

Convenience testing is a type of testing where convenience is the prime consideration in selecting items. This type of testing is generally professionally inadequate and does not reflect good audit judgment. Records that are in storage, in the bottom file drawers, not filed or at another location are excluded when this type of testing is used. Auditors recognize that missing or inaccessible records may be missing or inaccessible because the auditee does not want them reviewed. In both judgment and statistical sampling, missing items should be handled as cues to a system breakdown. Also, in statistical sampling it is invalid to substitute another item for the one that is missing, because the missing item may be very significant to the audit as well as the statistical estimation and evaluation. If missing items exhibit a pattern, an auditor should be alert that the cause of the pattern could be irregular activity including the possibility of destruction of records or fraud.

E. Large Dollar Testing

Large dollar testing places emphasis on the materiality of the items selected. It is an early form of audit stratification (Chapter 3). No examination is made of the lesser dollar value items. In spite of this limitation, inferences are sometimes drawn from the condition of the items examined (usually a significant percentage of the total dollars) to describe the condition of the universe. Obviously, this is not an objective appraisal and could be biased (reflect a systematic error). However, experience has shown that if there are breakdowns in controls and procedures, the breakdown is generally more pronounced in the lower dollar area, since the controls in that area may not be as rigid as for larger dollar transactions.

VI. STATISTICAL SAMPLING AS A MEANS OF TESTING

Statistical sampling is a procedure that conforms to the theory and principles of probability for the selection and statistical evaluation of results obtained from a portion (sample) of a group of items or records to estimate some characteristic of the entire group or as a basis for some action or decision.

When statistical sampling is used the auditor can calculate the risk associated with the use of sample data in lieu of a complete audit of all the records; however, the selection of the sample items is by use of a mechanical device such as a table of random digits or a comparable computer generated program. Random selection allows for the measurement of risk through probability. Without randomness, there is no way to measure this sampling risk.

In using statistical sampling, the auditor can select transactions in a more objective manner. He should recognize that statistical sampling involves the testing process, especially in the random selection of the sample items and in the computation of the statistical evaluation or sampling reliability of the result. Other testing problems may be considered as basically auditing problems with statistical sampling overtones.

This book provides detailed guidance in procedures used for the statistical (random) selection of an audit sample. Such selection methods are necessary if statistical sampling evaluation procedures are to be applied to the results. Some auditors complain that the random process takes too long. This reflects misconceptions about e of rigor required for random audit selection which have resulted in the expenditure of excessive manual effort. Accordingly, various random selection procedures are discussed in some detail so that the auditor can readily adapt more feasible and efficient selection methods to the particular audit situation.

Extensive coverage also is given to estimation procedures and the statistical evaluation of sample results. Some auditors using statistical sampling methods never go beyond making a random selection of items to be tested. If after the subjective audit appraisal of the test results the auditor wishes to generalize about the findings, he should make an estimation (projection) of the universe value or condition and then statistically evaluate the reliability of the estimate [a]. Estimation (projection) techniques have been used by auditors since the introduction of testing. Most of the estimation techniques, as will be seen later, are the same whether applied to judgment tests or to statistical evaluation aspects (Chapter 15) of the sample result. Most statistical sampling literature ignores the subjective appraisal or error analysis aspect, which is very significant in auditing, since it more directly provides support for the audit decision or opinion by revealing the nature, source and impact of findings are discussed in Chapter 13.

VII. RELATIONSHIP OF STATISTICAL SAMPLING TO OTHER AUDITING TECHNIQUES

The auditor's professional judgment is not replaced or diminished by the use of statistical sampling methods. Traditional audit methods are not replaced; they are supplemented. Statistical sampling is a refinement to the auditor's test-checking procedure.

Professionals always strive to use the most efficient and effective methods available. As long as subjective test selection procedures were the best methods available for all test checking, they were used. Now with the more objective selection methods of statistical sampling available, the high standards of the auditing profession require that its applicability to various audit situations be fully explored. Statistical sampling provides the auditor with a procedure for calculating the risk and precision associated with the test results. This does not, however, preclude the subjective selection of any transactions or groups of transactions which the auditor deems material or sensitive. This is accomplished through a process called audit stratification. This concept and its application to auditing will be discussed in detail in various aspects of the sampling procedure, and especially in the subjective stratification of source data described in Chapter 3.

Auditors, accustomed to thinking of risk in abstract terms, may not always be prepared to relate, audit-wise, the more explicit measurements of risk provided by statistical sampling. In this connection, the significance of the measures of "risk" and "precision" will be discussed and illustrated [b]. Guidance will be given in making some of the audit decisions involving the appropriate use of the estimates associated with statistical sampling results. These statistical results, per se, should not be considered the audit opinion or decision. They must be considered in conjunction with all other available sources of evidence [c].

This book provides the reader with a knowledge of basic statistical sampling concepts as they relate to audit activities. The more important problems encountered in the application of statistical sampling to audit situations are discussed in detail. While all possible answers cannot be provided, sufficient guidance is given to assist the auditor in reaching solutions to most of the testing problems. Another important subject, covered in Chapter 13, is the use of probe, or flexible, sampling procedures to conserve audit resources.

The application of statistical sampling is discussed in a practical manner at the level of the staff auditor, yet useful guidance is also provided for the audit manager. For instance, experience reveals that most incredulous costs result from a difference of opinion between the auditor and the auditee concerning the interpretation of criteria or regulations and policies and not from a breakdown in the accounting system. Accordingly, auditors may use statistical sampling, together with other audit techniques, to assure themselves that the particular accounting system is acceptable and that the resulting cost accumulations are reasonable.

VIII. PRIMARY AUDIT OBJECTIVE

All audit effort is directed towards the fulfillment of specific audit objectives. Statistical sampling, when used in audit activities, must aid in achieving these objectives. A full appreciation of the contribution of this relatively new audit technique is dependent upon an understanding of the audit mission and its objectives. Although the objectives of all auditors are not identical with those of the public accountants, they do utilize some of the pronouncements of the public accounting profession which apply to any auditing environment.

The primary objective of auditing is the issuance of a report which reflects an audit opinion. The standard short-form report of the public accountant asserts that the financial statements have been examined in accordance with generally accepted auditing standards. In addition, the public accountant states a professional opinion that such statements present fairly the financial position and results of operations of the client in conformity with generally accepted accounting principles. Other auditors, in their reports also express similar opinions of the fairness of the operations and cost representations. Standards for Governmental audits are issued by the Comptroller General of the United States [d].

IX. LANGUAGE OF AUDIT REPORTS

In reviewing the public accountant's report language, two points are of particular significance in the auditor's use of statistical sampling:

A. Professional Opinion

The first aspect is that the auditor expresses a professional opinion rather than a statement of fact; as such, the auditor is not required to obtain absolute assurance that the statements are accurate. He will have met professional responsibility when there is an acceptable degree of assurance that they are reasonable. The relevance of this aspect to statistical assurance or confidence levels will become more apparent in later discussions.

B. Fairness of Presentation

The second aspect is that the statements present fairly the financial position and results of operations rather than that they are completely accurate in all details. It would be economically impracticable for the auditor to examine every detailed entry behind the financial statements. The auditor has met his responsibility if he has exercised due care in the examination of the statements. The relevance of this aspect to statistical sampling precision or confidence limits also will become more apparent in later discussions.

X. SOURCES OF RELIANCE

In expressing an opinion, the auditor does not place total reliance upon the results of a single test of the transactions. Fortunately, there are many other sources of reliance to support the judgment. In performing an audit of financial statements, the auditor will have made a system survey and will have obtained and examined a considerable amount of relevant material and information. Based upon the review and appraisal of all the available evidence, an opinion will be expressed on the statements and operating conditions.

Some of the sources that the auditor relies on in forming an informed professional opinion are discussed in subsequent sections of this and other chapters. The auditor should always be aware of other sources of reliances, since they influence some of the decisions made in applying statistical sampling. In this regard, the auditor's position is different from samplers in many other professions. Other samplers normally have only the information developed from the sample with which to make a decision. By contrast, the auditor usually will have available, from other sources, additional information about the area being tested. Therefore, the auditor will not be placing sole reliance on the sample results. For this reason, some of the higher values of statistical sampling assurance, normally considered necessary in other professions, will not in most instances be necessary in auditing. With the other sources of reliance, the auditor may consider lower values of assurance more appropriate in many circumstances. If other sources of reliance are ignored, as was the case in earlier applications of statistical sampling to auditing, the auditor may overaudit since the sample size is the sole source of the auditor's assurance.

XI. RELIANCE ON INTERNAL CONTROL

One source of the auditor's reliance is on the system of internal control. The AICPA second standard of field work in Generally Accepted Auditing Standards requires an evaluation and appraisal of internal control as a basis for determining the extent of reliance to be placed on the system. The evaluation and appraisal will normally involve two phases.

The first phase involves acquiring a knowledge and understanding of the procedures and methods prescribed. Ordinarily, this is obtained through inquiry and by reference to written instructions. Based on this information a preliminary evaluation of the effectiveness of the procedures is made, assuming for the moment that compliance with them is satisfactory. The use of testing is generally not applicable in this phase of internal control appraisal.

The second phase is to evaluate and appraise the extent of compliance with the procedures. For those procedures with an audit trail, in the form of documentary evidence, testing methods may be used. This evidence may consist of signatures, initials, stamps and the like, which indicates preparation, checking or approval of documents.

At the completion of the second phase the auditor will have determined the extent to which he can rely upon the system of internal control. This is an important source of reliance and influences the extent of other tests, which will reflect the risk one is willing to take.

The American Institute of Certified Public Accountants has provided specific guidance concerning the auditor's reliance on internal accounting control and resulting reliability levels for substantive tests. The AICPA states that, ". . . the extent of substantive tests required to obtain sufficient evidential matter . . . should vary inversely with the auditor's reliance on internal accounting control." AICPA Statement on Audit Standards (SAS) No. 39, *Audit Sampling*, indicates that the risk level for substantive tests of account balances and classes of transactions is not an isolated decision, but is a consequence of the auditor's evaluation of reliance on internal accounting control, analytical review procedures and other relevant substantive tests [3]. A schedule similar to the following was once disseminated by the AICPA to illustrate this impact:

Auditor's assessment concerning reliance to be assigned to internal accounting control and analytical review procedures	Resulting assurance level for substantive tests
90%	50%
70%	83%
50%	90%
30%	93%

In other words, the more subjective reliance that an auditor can assign to internal control and other sources of reliance, the lower the corresponding level of statistical reliability or assurance for substantive tests. Earlier applications of statistical sampling to auditing did not reflect this concept, and consequently resulted in overauditing in many instances.

XII. RELIANCE ON EVIDENTIAL MATTER

Another source of the auditor's reliance is on the evidential matter he accumulates. Evidential matter supporting the financial statements has been defined as consisting of the underlying accounting data and all corroborating information available to the auditor.

Another auditing standard indicates that sufficient competent evidential matter should be obtained. It is therefore appropriate to explore the sufficiency of evidential matter in testing.

Little specific guidance is provided in making a decision concerning the sufficiency of evidential matter. The auditor is given a general framework of things to consider in reaching a decision. However, the amount and kinds of evidential matter required to support an informed opinion are matters for the auditor to determine in the exercise of his professional judgment. This is accomplished after a careful analysis and appraisal of the circumstances in a particular case [e].

XIII. PROFESSIONAL AUDITOR JUDGMENT

The dominant consideration in the planning, execution, appraisal and evaluation of any audit is the exercise of sound professional judgment.

In discussing the standards of audit field work, the term judgment is preceded by the word "professional." This implies that the judgment is to be made, only after a careful consideration of all available evidence, by a person having adequate technical training and proficiency as an auditor.

Judgment varies among individuals and will vary with the same person from time to time. A person's judgment reflects training, audit experience, location in the organization, and mental attitude and physical condition at the time judgment is used to make a decision. Changes in these factors as well as others, can alter a person's judgment. Some other factors affecting judgment are ability, educational background, experience in the particular audit environment, maturity, imagination, initiative, attitude and outside pressures.

All these factors and others affect a person's judgment and, in turn, the decisions that are made. An evaluation and appraisal of the decision determines whether the judgment was good. Good judgment might be defined as that which results in the proper decision based on available information under particular circumstances.

The competence of the evidence and its acceptability are also matters of auditor judgment. The various types of evidence vary considerably in their influence on the auditor as the opinion is developed. The pertinence of the evidence, its objectivity, its timeliness, and the existence of other evidence corroborating the conclusions to which it leads reflects its competence.

The auditor obtains evidence by employing various techniques. Some basic audit techniques are (i) physical examination and count, (ii) confirmation, (iii) examination of authoritative documents and comparison with the basic records, (iv) recomputation, (v) tracing bookkeeping procedures, including ADP systems, (vi) scanning, (vii) inquiry and floor checking, (viii) examination of subsidiary records, and (ix) correlation with other related information. The use of testing methods might be considered appropriate in the application of a number of these techniques.

A better understanding of the auditor's reliance on evidential matter may be gained by briefly reviewing some of the more significant types of audit evidence. These will be covered under two broad groupings of internal sources and external sources of evidence.

XIV. INTERNAL SOURCES OF EVIDENCE

A. Physical Evidence

Physical evidence is obtained through the audit techniques of observation, inspection and enumeration. It might refer to events or to tangible items and is regarded as high quality evidence. Examples of physical evidence might include such things as:

1. Actual observance of work being performed such as making a floor check or witnessing the performance of an operation or task.
2. Inspection of facilities, such as buildings, furnishings and equipment.
3. Counting of assets such as inventory items.

Among the objectives of physical observation are to ascertain that: (i) the item actually exists, (ii) it is being properly protected, (iii) it is being economically used for the intended purpose, and (iv) relevant work is actually being performed.

B. Documentary Evidence

Documentary evidence is probably relied upon more often than any other type. This form of evidence includes vendors' invoices, purchase orders, receiving reports, checks, correspondence, contracts, subcontracts and many other similar documents. The value of the documents as evidence varies considerably and depends in part upon whether it was created within the organization or came from an outside source.

C. Ledgers and Journals

Ledgers and journals, books of original entry, are themselves evidence. They will evidence that expenditures are supported by entries in the accounts; account totals correctly summarize the detailed entries; and the entries are proper accounting interpretation of transactions. The reliability of ledgers and journals as evidence is dependent on the validity of the principles and policies upon which they were developed and upon the adequacy of internal control during the preparation and review of the records.

D. Comparisons, Trends and Ratios

Comparisons, trends and ratios provide a procedure for spotting significant changes. Unusual changes should be investigated to ascertain the reasons. Examples of these techniques are comparisons of expense accounts from year to year, a study of trends of labor rates, and the analysis of ratios of indirect to direct employees.

E. Computations

Computations, like comparisons, trends and ratios, are made to prove the accuracy of records. These might be recomputation of figures, alternate methods of calculation, or a reconciliation of two amounts.

F. Verbal Evidence

Verbal evidence, involves answers to questions the auditor asks in the course of an examination. Generally this type of evidence is useful in disclosing situations that require audit examination or may confirm other types of information already obtained. Generally, verbal evidence should not be the sole support of an audit conclusion or opinion.

XV. EXTERNAL SOURCES OF EVIDENCE

Other sources that the auditor may rely on in addition to the system of internal control and evidence he accumulates himself include the following:

A. Prior Audits

Prior audits indicate potential problem areas and thereby concentrate attention on these areas. The initial audit of an operation will normally be in greater depth and detail than subsequent audits. Audit effort may be subsequently reduced for operations that have been found reasonable and acceptable in the past.

B. Industry Information, Studies and Practices

Industry information, studies and practices within a particular area may be important sources of reliance to an auditor. This is particularly true concerning reasonableness. In determining whether salaries are reasonable, a comparison might be made with salaries in similar organizations. The subsistence and location allowances paid in certain remote locations or foreign countries might be compared with payments by other companies. The useful life of certain types of equipment might be determined from studies made by industry or government.

XVI. RELATIONSHIP OF STATISTICAL SAMPLING EVIDENCE TO OTHER SOURCES OF RELIANCE

Whenever possible, the auditor should consider all of the sources of reliance described above. Then there will be a much higher degree of assurance that the findings, decisions and opinions are reasonable. Furthermore, if all other sources of reliance indicate favorable conditions, the auditor should limit his tests of transactions to the minimum number which will support an informed opinion for a reasonable degree of risk. For example, when a survey indicates that the controls are strong and operating effectively and the sampling of the records discloses no exceptions, the auditor has a much stronger conviction concerning the reliability of the records than he could have from the test of the records alone. Correspondingly, the amount of testing will be much less when the test is used to confirm a preliminary favorable conclusion from an analysis of a system than it would be if sole dependence were on a test of records. Furthermore, little testing would be required to confirm that costs are still acceptable when data analysis indicates that the current cost representations are consistent with those previously audited and found to be acceptable. More specific guidance is presented in the following paragraphs on the interrelationships of various sources of reliance and the weight to be accorded each in developing an informed opinion.

One of the principal advantages of statistical sampling is that it provides an objective means of calculating the sampling risk. This may be a difficult advantage for the auditor to grasp if he does not know what risk is appropriate in the particular audit area he is testing. He naturally desires a high degree of precision with low risk. Yet, he is surprised at the large sample size required for the reliability he thought was appropriate. The reaction often is to revert to judgment sampling. Part of the problem is that sample sizes shown in tables do not recognize the impact of the auditor's other sources of reliance. Hence, the problem is relating the reliance the auditor has from other sources with the information provided by statistical sampling.

It has already been mentioned that the auditor should consider all sources of reliance in forming an opinion on the acceptability of operations and cost representations. The question often arises, however, as to how this is accomplished. Unfortunately, literature on statistical sampling for accounting and auditing has provided little or no guidance in resolving this problem. This is evident from the fact that many sampling procedures and tables do not contain evaluation data for assurance or confidence levels below 80 percent. In other words, these procedures have accorded little weight to other sources of reliance available to the auditor. Flexible sampling provides a means for reducing the amount of testing when other sources indicate favorable conditions.

For example, suppose an auditor wishes to conduct a review of purchases during an audit of research grants and contracts at a large university. Also, assume that the following facts and conditions prevail: (i) a recently completed survey indicates that the university's procedures and controls are strong, (ii) past reviews of this type have never disclosed any significant findings, (iii) the level of purchases is consistent with that of prior years and the current year's volume, and (iv) knowledge of the auditee's systems and management stability suggests a high degree of integrity.

Under the above circumstances to what minimum extent should the transactions be tested in order to determine procedural compliance? A statistical sample of 50 transactions (assuming no error is found) from a universe of over 1,000 transactions provides the auditor with a 78 percent assurance (or 22 percent risk) that the error rate in this universe of transactions is less than 3 percent. How to determine and evaluate a minimum sample is discussed in detail in the chapter on flexible sampling. From an audit viewpoint the above risk can be construed as tolerable, particularly in light of the favorable conditions indicated by all the other sources of reliance. Obviously, any finding in the sample should be investigated for nature, cause and impact. It is important to note that, whenever conclusions are drawn, it is essential that the justification be clearly documented in workpaper files and that the rationale be explicitly stated along with criteria and assumptions. Guidance on presentation of results is given in Chapter 2.

XVII. ADVANTAGES OF STATISTICAL SAMPLING IN AUDITING

The advantages of statistical sampling may be summarized as follows:

1. The amount of testing (sample size) does not increase in proportion to the increase in the size of the area (universe) tested [f].
2. The sample selection is more objective and thereby more defensible.
3. The method provides a means of estimating the minimum sample size associated with a specified risk and precision.
4. It provides a means for deriving a "calculated risk" and corresponding precision (sampling error), i.e., the probable difference in result due to the use of a sample in lieu of examining all the records in the group (universe), using the same auditors and procedures.

5. It may provide a better description of a large mass of data than a complete examination of all the data, since nonsampling errors such as processing and clerical mistakes are not as large.
6. It may provide better audit coverage for a similar audit effort.
7. With proper documentation of the sampling plan, it can be executed by different auditors at various locations with a high degree of consistency.
8. Objective statistical evaluation of the reliability associated with the results is possible and hence probably more defensible than judgment sampling.
9. It is an efficient, effective and economical procedure for accepting a system and related cost accumulations when no exception is revealed in a properly planned and executed sampling program such as flexible sampling.
10. Statistical sampling does not preclude the use of judgment sampling to assure adequate coverage of high dollar and sensitive items through audit stratification or separation.

Under some audit circumstances, statistical sampling methods may not be appropriate. The auditor should not attempt to use statistical sampling when another approach is either necessary or will provide satisfactory information in less time or with less effort. Some of these circumstances may be summarized as follows [g]:

1. Sampling should not be used when *exact data* or complete accuracy are required. These usually reflect legal requirements such as a complete census of the United States every 10 years. It is paradoxical that no member of Congress wishes to risk the loss of his congressional seat because of a declining district population reflected by a sample; however, funds are provided annually to employ sampling methods to improve the quality and accuracy of the 100 percent census.
2. When searching for very *rare occurrences* sampling tends to be inefficient. Sampling for the rare occurrence is synonymous to looking for the "needle in a haystack." A very large sample would be required because with randomized selection there would be a very low probability of revealing a rare occurrence.
3. Sampling is not desirable when fairly precise *data* are needed *for* minor subdivisions or *individual units*. A reasonable size sample usually provides sufficiently precise results for the total and for major breakdowns. However, when precise results are needed for minor subdivisions, a relatively larger sample is required because each subdivision is evaluated separately.

4. Sampling should not be used whenever a *complete examination* is *cheaper*. Sampling operations require certain special costs which are not required for a complete examination. In designing the sample there are costs associated with determining what kind of sampling to use, obtaining an appropriate sampling frame, identifying the sample, drawing the sample, and summarizing, analyzing and evaluating the results. Usually there are cost savings which result from examining only a portion of the universe. However, for a *small universe* this may not be true, since the reliability of a sample is primarily dependent on its absolute rather than relative size. There is only a slight dependence on the relative size of the sample to the universe size. Thus, when a universe is small, the sample necessary to achieve the desired reliability may be relatively large in comparison to the universe. In rare cases, the supplemental costs of sampling may make a complete examination more economical. Whenever the audit sample size would be over one-half of the universe, sampling may be uneconomical.
5. Whenever the audit objective is to highlight *known problem areas* or to cover only high dollar or sensitive transactions, judgment sampling would be more appropriate since the auditor knows or suspects the location of the "needles in the haystock."

XVIII. SUMMARY

Although the inception of test checking occurred in the latter part of the nineteenth century, the transition from the practice of a detailed audit was slow and it was not until the early 1930's that testing became the rule.

The suggestion that statistical sampling might be considered as an audit technique first appeared in the 1930's. However, it was not until the 1950's that interest in its possibilities in auditing became sufficiently widespread to merit official consideration within the accounting profession.

The principal reason for reviewing and illustrating the auditor's sources of reliance is to emphasize that there are many sources of evidence in addition to the information from the sample. Seldom will the auditor have to make a decision or render an opinion on the basis of information from only one source. The extent of testing in an area should depend upon how much reliance is available from other relevant sources. When statistical sampling is used, the risk levels used in sampling applications in other professions are not necessarily relevant in determining appropriate risk levels for auditing. The auditor's reliance upon sampling is augmented by many other sources of reliance which usually are not available to other professions.

The ultimate objective of auditing is the preparation of a report expressing a professional informed opinion on certain representations and operations. This opinion is based upon an audit performed in accordance with generally accepted auditing standards, including the evaluation and appraisal of internal control and the review of sufficient competent evidence.

XIX. NOTES

a. Sampling reliability, assurance, and confidence level are often used interchangeably in statistical sampling to represent the complement of sampling risk, that is, it equals one minus the sampling risk.

b. The AICPA SAS No. 39 defines ultimate risk as ". . . a combination of the risk that material errors will occur in the accounting process used to develop the financial statements and the risk that any material errors that occur will not be detected by an auditor . . ." It further states that the auditor may rely on the internal accounting control to reduce the first risk and on analytical review procedures and substantive tests of transactions and balances to reduce the second. Thus, ultimate risk includes both sampling risk and nonsampling risk. In this book, unless otherwise indicated, "risk" for both compliance and substantive tests is restricted to sampling risk. This is the risk that audit procedures which are limited to a sample of details of transactions and balances might produce different conclusions from those reached by the auditor if the procedures had been applied to all items or transactions in the same manner and with the same care. This aspect of the audit risk can be controlled and estimated when statistical sampling is used. The nonsampling risk is associated with the nature of the procedures, the timing of their application, the system being audited, and the skill, knowledge, maturity and perserverance of the auditor. Those readers interested in a detailed discussion of the nonsampling audit risk should see Roberts [57].

c. There is no problem of later consolidating the test results because auditors are accustomed to this kind of procedure. The reports reveal which transactions and account balances were tested in detail and which were sampled. The auditor appraises and evaluates all these results together with other relevant evidence in the formulation of an informed opinion.

d. The official document for such pronouncements is "Standards for Audit of Governmental Organizations, Programs, Activities and Functions," which is issued by the U.S. General Accounting Office (GAO). The latest revision was issued in 1981.

e. The AICPA third standard of field work states that sufficient evidential matter is to be obtained by inspection, inquiries, observation, and confirmations to provide a reasonable basis to support an opinion about the financial statements and records being audited.
f. Under certain monetary-unit or probability-proportional-to-size (PPS) sampling procedures, using the concept of an average sampling interval, the sample size will vary with the total value of the universe.
g. Whenever it is possible with computerized systems and procedures to review an entire universe for certain attributes and characteristics, any sampling procedure may be inappropriate.

2
Considerations in Developing an Audit Sample

I. GENERAL

When an auditor decides to sample items in a particular area, there are a number of factors which should be addressed prior to commencing the operation. These factors or elements are discussed in a somewhat logical order in this chapter. These factors are not unique to statistical sampling, hence precise and consistent definitions should be made even when subjective (judgment) sampling is used, because they have an impact on the audit usefulness of the sample result.

Auditors, applying statistical sampling methods for the first time, soon discover the need for precise and consistent definitions and criteria. The use of judgment sampling may not require precise definitions and consequently, auditors have become less exact in some sampling definitions and criteria [a]. Not only must the individual elements be defined precisely, but the definitions and criteria must be consistent with each other and the audit objectives.

An advantage of statistical sampling is that it imposes an additional discipline which requires the auditor to make clear, consistent audit definitions and criteria. When these are not clear and consistent, problems often arise in the audit process. Some auditors tend to blame statistical sampling for these problems rather than realizing that the basic audit definitions and criteria are at fault. Therefore, much care should be devoted to defining the objectives before any other factors are considered.

The initial consideration in the development of a sampling program is clear and precise statements of the audit objectives to be achieved by the testing. Vague statements of objectives can result in sample products which are inconsistent with the objectives, and hence, may be useless. Improper objectives can result in studying the wrong problem or precluding the consideration of relevant sources and data.

Different audit objectives may require different sampling procedures. With clear statements of objectives, effective and efficient sampling procedures can be planned to satisfy these objectives.

The audit objectives may vary among (i) verifying the acceptability of a system's operation (occurrence rates), (ii) verifying the reasonableness of the value of a set of records (inventory), or (iii) a search for rare occurrences such as fraud.

Obviously different auditors, under the same circumstances, may formulate different audit objectives. Whether one set of audit objectives is better than another would have to be decided in terms of auditing standards and criteria. The purpose for discussing audit objectives is to emphasize their impact on the sampling approach.

Ijiri and Kaplan [25] describe four separate and distinct objectives of audit testing. A brief discussion of these objectives follows.

A. Estimation Objective

The objective may be to select a sample which represents the universe as closely as possible, somewhat of a snapshot of the universe. The authors mention that representative sampling in a broad sense also includes judgment sampling since the auditor's experience and knowledge are used to achieve a fair representation of the universe. Since there is no support that a representative sample has been achieved, it is best to consider the statistical sampling aspects of this objective under estimation sampling, where the aim is to estimate a universe characteristic (dollar amount or occurrence rate) from a sample within calculable limits for a specified confidence level. In some instances this objective may be to estimate overstatement errors in a particular account balance. While at other times, this objective may be to estimate the rate of occurrence of insufficient documentation.

B. Correction Objective

If the objective is to locate the maximum number of occurrences or amount of overstated cost, the auditor is applying corrective auditing.

This objective is supported by the practice of endeavoring to select samples from areas where mistakes are more likely to occur. Based on knowledge of the operation, an auditor can select judgmentally a sample which is expected to contain more errors than a sample selected randomly from the entire universe. This is accomplished through audit stratification where the universe is delimited by the items judgmentally selected. In other words, the auditor separates or stratifies out from the universe those areas (strata) which his judgment indicates are most likely to contain the highest proportion of deviations and/or the highest amount of unsupported costs. This process is referred to as audit stratification since the items from those strata are handled separately from the randomly selected items. This process permits the auditor to interface statistical and judgment sampling which optimizes the use of knowledge and experience. Detailed discussions of audit stratification are covered in Chapter 3.

In any audit sampling procedure, it is necessary to analyze each error or occurrence to determine whether it is a systematic or a random occurrence. Detecting a systematic occurrence, whether by random or judgment selection, requires corrective auditing. The analysis of findings is discussed in detail in the Error Analysis and Audit Appraisal section of Chapter 13.

C. Protection Objective

The objective of protective sampling is to maximize the total dollar value of selected sample items. In other words, it assures that all large dollar value items are tested. Often judgment dictates the selection of all items over a stated dollar amount, such as $5,000 as well as specified sensitive smaller valued items. The auditor accomplishes this objective by allocating disproportionately more items to be tested in the high dollar stratum. This is consistent with the disproportionate allocation concept of stratification. However (in Chapter 8) items over a specified dollar amount are usually tested 100 percent, and considered separately from the sampled items. In this manner there is no sampling error associated with the test of large dollar items. Also, in many audit situations a relatively few items contain the majority of the dollars.

Protective sampling reflects the auditor's recognition of the difficulty of detecting discrepancies and irregularities which may occur in only small pockets of the universe. He feels more secure if at least a relatively large proportion of the total dollars has been examined.

Monetary-unit sampling, discussed in Chapter 12, may be regarded as a form of protective sampling, since the procedure provides for the selection of all transactions equal to or greater than a specified value.

D. Prevention Objective

The prevention objective is to create an atmosphere of uncertainty as to which transactions are likely to be selected during future reviews. From an audit point of view, random future testing periods are desirable, but generally advanced notice of audit is necessary to minimize disruption of the auditee's operation. In addition to the random sample, one may wish to choose some transactions based on a seasoned judgment, to ensure that certain types of transactions do not escape audit. Transactions judgmentally selected should be handled separately from those randomly selected. It is possible to consider the prevention objective in designing the sampling plan, especially in subjective stratification for audit purposes.

In preventive sampling—which may be regarded as a form of psychological sampling—the auditor endeavors not to follow any pattern in the test items he examines. Thus, random selection procedures are very appropriate for this purpose.

E. Detection Objective

In addition to the four objectives mentioned by Ijiri and Kaplan [25], detection or exploratory sampling should be added. The aim of this objective is to ascertain whether a certain type of occurrence exceeds a specified maximum relative frequency, such as failure to adhere to basic internal control procedures or evidence of irregularities. A sample size is chosen so as to provide a specified probability of obtaining at least one example if the occurrence rate is greater than the specified frequency. For a more detailed discussion of detection or exploratory sampling see Chapter 10.

II. ELEMENTS AFFECTING SAMPLE PLANNING

The author was involved in the preparation of both the exploratory sampling tables (Appendix C) and the flexible sampling tables (Appendix D). He also developed the flexible sampling procedure which utilizes stop-or-go sampling [79]. Although both tables are based on the same mathematical function: the hypergeometric distribution, which provides probabilities when sampling without replacing the sample items, the flexible approach, described in Chapter 13, embodies the advantages of detection sampling while avoiding its limitations. The flexible procedure is much more responsive and efficient because it emphasizes a minimum sample and it does not necessarily stop with a single finding. By contrast, the detection sampling approach is like looking for a "needle in a haystack" provided one makes an assumption about how many needles are in the haystack. In the flexible procedure this assumption is not required. Also, many audit advocates of detection sampling are actually proclaiming the virtues of flexible sampling. Under a detection sampling plan the testing operation ceases with the discovery of a single occurrence of the specified characteristic, while this is only one of many options when using flexible sampling. Thus, whenever a detection objective is involved, the flexible sampling procedure is recommended for its economy and efficiency.

In selecting a sample, the auditor would have all the test objectives in mind. While most of the literature on audit sampling has emphasized the estimation objective, the correction and protection objectives may be more important. In any case, the following illustration of the interaction of the distinct testing objectives may be useful.

During an audit of a neighborhood health center, the pharmacist gave the auditor what he described as a complete and accurate listing of controlled substances (prescription drugs). It was a manual listing, 510 pages long, with 40 items to a page, and totals $90,000. Line items to be examined were selected from a daily status report, which constituted the physical representation of the universe (sampling frame). Other audit procedures revealed no adverse condition. The next step was to ascertain whether unit prices, extensions and footings were reasonable.

For the protection objective all items with recorded extensions greater than $100 were reviewed for accuracy of unit prices and extensions. This process is one aspect of stratification for audit purposes. Since there is a 100 percent review of this area (stratum), no sampling error is associated with the results.

Unit prices and recorded extensions under $100 were statistically sampled. There were no findings in this case; however, if there had been some monetary errors in the sample, these could have been projected to the universe to estimate total monetary errors for items with extensions under $100; thus, satisfying the estimation objective.

If there had been any findings in the sample, the auditor could have used the flexible procedure described in Chapter 13 to detect the nature, cause and impact of the findings. If the detection procedures had revealed a clustering of findings, such as for a particular kind of prescription, the auditor could have accomplished the correction objective by scanning the entire listing for entries relating to this drug. Also, if the manual listing were prepared by several employees, the prevention objective might be satisfied by augmenting the statistical sample by purposively reviewing portions of each employee's entries.

A. Area to be Sampled (Universe)

The determination of the area to be sampled should be consistent with the audit objectives. Theoretically, it should include all items necessary to meet the objectives and exclude all items not necessary to the objectives. For example, if one audit objective were to verify the allowability of inventory issues on cost reimbursable contracts, the universe should include only inventory issues to this type of contract. Inventory issues to other types of contracts should be excluded. However, if an objective were to determine consistency of costing between types of contracts, the universe would be expanded to include all inventory issues. Unfortunately, the exact universe is not always correctly defined. This may lead to strange definitions of universes, such as the universe of all undocumented entertainment costs, when in fact the real universe is total expense vouchers for the applicable time period.

Ordinarily one could imagine a file of vouchers, each of which contains a desired value for one of the universe items. In auditing, however, the value or characteristic may not be common to each voucher. For example, cash discounts may not be applicable to all vouchers. A lot of time would be saved and a better estimate would result if there were some way for the auditor to direct his effort to only those vouchers with available discounts, but this is not statistically possible. He must rely upon experience and his knowledge to effectively stratify the vouchers as a means of directing his efforts to the most productive areas. However, in auditing, those items judgmentally separated or stratified for audit purposes or more detail or purposive examination should be excluded from the delimited universe, since these items involve no sampling error.

B. Sampling Unit

Another important consideration is the sampling unit. A sampling unit is a selected item from which information is sought. The total of these units is the universe. The number of sampling units will determine the size of the universe.

It is often possible for the sampling unit to be defined a number of different ways in the same universe. For example, in the audit of travel for a year, the sampling unit could be defined as each individual travel voucher, each employee's travel folder for the year, line items on a machine run (one travel voucher may charge two different work orders or contracts) or the journal vouchers which summarize all travel vouchers for a week.

Another example is the audit of accounts payable distribution. The sampling unit might be each line item on the accounts payable distribution run. However, the auditor may find that many line items represent the combining of a number of invoices or that there are several line items on a single invoice.

The proper identification of the sampling unit has long been a problem for auditors. More oversampling can be attributed to this source than perhaps any other single cause. Consider, for example, the case of an auditor who had the task of developing a sampling plan to estimate the total undocumented cost in a travel expense account. The sampling unit was identified as employees who incurred any travel expense in the year being audited. Each employee who traveled submitted from one to twenty individual expense vouchers during the year. By defining the sampling unit as an employee and then reviewing all vouchers submitted by each of the 80 randomly selected employees, the auditor would review only 80 employees, but 426 individual vouchers.* Had the sampling unit been defined as individual expense vouchers, the sample could have been limited to a much smaller number of randomly selected expense vouchers with the same level of risk and precision. The reason for this is that the groupings of employee vouchers form clusters, and in this type of statistical sampling each cluster is weighted by its relative size but evaluated as if it were a single unit.

The audit of an inventory listing may involve similar problems of choosing an appropriate sampling unit. It may be a line item on a listing. Yet, behind each item listed there may be a large number of detailed records supporting the one item selected. Sometimes this is referred to as the "Christmas tree effect."

C. Sampling for Attributes or Variables

When the auditor tests to determine the rate of occurrence or any other qualitative characteristic in a set of records or an operating activity, it is called sampling for attributes. On the other hand, when the objective is to estimate dollars, or any other quantitative characteristic, the procedure is called sampling for variables.

Fortunately, one can sample for attributes and variables simultaneously, because the characteristics of interest may be either quantitative or qualitative. For example, in reviewing direct material costs, the auditor may desire to estimate the percentage of purchases made without competition. One has the flexibility of switching from one type of sampling to another based on an examination of the findings from a preliminary sample. For instance, the auditor may initially select a

---

*The 426 is the total of the varying number of vouchers for each employee.

sample large enough to support his acceptance of an account if no error is disclosed. If the preliminary sample reveals errors with dollar impact, he may then modify his sampling procedure to include an estimation of undocumented costs. On the other hand, suppose his initial sample were designed to estimate improper dollar charges against an account and his sample reveals no undocumented charges. Then there is no basis, from the sample, to estimate any undocumented dollar amount; however, the result can be statistically evaluated for a probable upper error limit. This procedure is explained in detail in Chapter 13, Flexible Sampling. Also, if a dollar-unit sampling scheme had been used to select the initial sample, it could be evaluated for a probable upper monetary error limit. Nevertheless, there is still no basis to estimate an undocumented dollar amount from any sample which discloses no dollar errors, but one can estimate probable monetary limits.

D. Data to be Collected

The auditor must be specific in determining what information is desired. Obviously, the information collected should be relevant to the audit objectives. Also, no essential information should be omitted. On the other hand, there is a tendency to collect more detailed data than are necessary to achieve the audit objectives. This tends to lower the overall quality since one is prone to include data not properly analyzed. How the sample information is to be used and its relative impact on the audit opinion are the significant considerations in determining what to include in the sample. Some questions to be answered in this planning phase are (1) Is the sample for an initial or a subsequent audit? (2) Will all the needed data come from the sample? (3) Will the sample data affect a nationwide or a local audit decision?

After deciding what information is desired, the next step is to ascertain the optimal available listing from which to draw the basic data.

E. Sampling Frame

The sampling frame from which the sample is actually selected consists of the physical lists and the procedures that can account for all the sampling units without actually listing them. The sampling units consist of the individual elements from which the sample is drawn or, more specifically, those elements in the universe whose characteristics are to be estimated, appraised and evaluated from the sample result. Thus, the ideal sampling frame would be a listing of all the relevant sampling units and only those units. The auditor may have such a frame in the form of an electronic machine listing, a trial balance, or a tape produced for some specific purpose. However, a perfect frame, one in which every relevant unit appears separately, and only once and no other elements appear, is very rare.

Thus, one of the major problems in statistical sampling is finding an appropriate list or sampling frame. This problem involves blank and ineligible items, duplications and omissions from the frames.

An auditor may have as an objective the examination of travel vouchers for undocumented claims, however travel vouchers may be intermingled with other expense vouchers in the sampling frame. Any vouchers not related to travel would be considered ineligible items. Inappropriate subtotals also would be regarded as ineligibles. A blank occurs when there is no item listed on a line, due to spacing or the end of subaccount or a time period.

The importance of a proper definition of the sampling frame is illustrated by the sad experience of a former magazine. This magazine endeavored to predict the outcome of a presidential election. To accomplish that objective, a large sample was selected from telephone directories (sampling frame). On the basis of responses a spectacularly inaccurate prediction was made.

It was a large sample and included all parts of the country, but it was drawn from only a segment of voters—those who were listed telephone subscribers—and inferences were made regarding the voting preferences of all voters. Precluded voters, those who were not listed telephone subscribers, obviously had a significantly different preference concerning the candidates. Also, the time frame was in the 1930s when there were fewer telephones than today. The point to be emphasized is that the sampling frame selected must be consistent with the objective and adequately reflect the universe of interest. For more details, the reader should see Haack [21].

The sampling frame chosen should be one that fulfills the audit objectives with the least amount of effort. It is not always possible to choose the sampling frame that would be most desirable due to the manner in which data are recorded or filed. This is often the situation and the auditor will have to choose the best alternative among available sampling frames. If there are deficiencies in the sampling frame, adjustments should be made in the sampling design to compensate for them. Hence, it is important that proper inspection and investigation be made of the frame to anticipate problems before the random selection procedure is started.

If the frame is small, the auditor may consider removing all ineligible items, but in most cases the frame is too large for this approach to be feasible. Where an ineligible item occurs in the selection, it is inappropriate to take the very next applicable item. This would give greater probability of selection to items that immediately follow ineligible items.

In the selection process, both blank and ineligibles are ignored. The sample size or selection interval, however, should be adjusted to provide for the expected number of exclusions. An estimate of the proportion of exclusions may be based on prior experience or from a preliminary sample.

In simple random sampling each item of the universe has an equal chance of selection. Thus if an item appears on a frame more than once (duplication), some biases may be introduced; therefore, to keep the sample unbiased, the results should be weighted by the universe of the probabilities of selection. Items that are listed more than once would be weighted by the reciprocal of the number of times each is listed.

The most serious problem associated with a sampling frame is the omission of some significant aspect of the universe. Ascertaining the completeness and accuracy of a listing is more of an auditing problem and should be dealt with by use of other audit techniques. However, there are some other alternatives for handling missing sampling units. One way is to add a supplement to include missing elements in a separate stratum, such as a supplementary payroll listing or newly hired employees who are not yet on the regular payroll. Where a separate stratum is too costly, a half-open interval procedure may be used. This involves listings which are defined to include an interval up to, but not including, the next listing. The missing elements are fitted into the interval in a clear and practical manner. For instance, on the payroll listing, the interval could begin after the employee listed last in each work unit. Kish [32] discusses in detail and illustrates solutions to various sampling frame problems in a manner comprehensive to nonstatisticians.

The objectives of the audit and other sources of reliance determine the best course of action in dealing with faulty frames. Usually, the best available listing (frame) is used and supplemented, as necessary, with other statistical and auditing procedures. Some additional guidance is given in the following paragraphs.

Another way to increase sampling efficiency is to combine similar accounts or transactions where consistent with the audit objectives. For example, rather than treating each class of expenditure as a separate frame, such as travel, space costs or equipment, the audit objective might permit the combining of these items into one frame. If the sample results show an unsatisfactory condition in only one class of expenditures, the frame for additional testing might be redefined to include only that class, but with the understanding that the additional testing is not applicable to the other classes of expenditures.

The auditor in sampling for undocumented costs actually has two universes. There is a strong possibility that the distribution of undocumented costs in the sampling area may be considerably different from the distribution of all costs in the same area. Since the auditor is primarily concerned about the total undocumented costs and will be attempting to draw inferences from the sample regarding this amount, the sample should be designed with this objective in mind. He should use any knowledge he has (from prior audits and surveys) of the distribution of undocumented costs to reduce the variability through stratification and 100 percent examination of certain types of transactions.

Another consideration in defining the sampling frame is to decide which items are to be examined 100 percent and thus for audit purposes precluded from the sampling frame. These might be high value items and any other items in which the auditor has a special interest. A similar consideration is whether the sampling frame is to be further stratified and some types of items given a higher probability for selection. If this kind of plan is followed it would result in a stratified random sampling procedure.

In defining the sampling frame, sampling efficiency and the ultimate effectiveness of sample findings can be increased by making the sampling frame as broad as possible consistent with the audit objective. One way to accomplish this is to lengthen the time period. Rather than using a quarter of transactions as the sampling frame, the year may be considered the sampling frame. The sample size required for any given sample reliability for a universe covering a year is considerably less than the total of the samples required if each of the 12 individual months were treated separately. This is one of the major advantages of statistical sampling cited earlier.

F. Definition of Occurrence or "Error"

Another element to consider is the definition of an occurrence or "error." The specific characteristics or attributes that will be tested on each selected item must be clearly defined. Each characteristic will have a separate objective which contributes to the overall audit objective. Consequently, the characteristics or attributes tested must be compatible with the overall audit objective. Effort should not be wasted on testing characteristics that do not contribute to achieving this overall objective.

The term "occurrence" is more comprehensive than "error" or "finding" and hence preferable. Often initial errors or differences are reconciled by adjusting entries elsewhere in the records. Also, an attribute may not constitute an error, such as the percentage of overdue accounts receivable, or the proportion of college students with government-insured loans. Furthermore, the use of attributes to develop statistics such as overhead rates or operating ratios do not involve errors. Another example would be the proportion of men, women and children participating in a certain program.

An example of audit objectives can be illustrated in an audit of Foster Care under Title IV of the Social Security Act. An overall audit objective might be to determine participant eligibility. The characteristics to be tested as separate objectives would include the following:

1. The child was removed from the family home and placed in foster care by judicial determination.
2. The child was receiving aid to families with dependent children assistance at the time of removal or was eligible for such assistance had application been made.
3. The child was placed with an eligible foster care provider.

The definition of an occurrence or error and the listing of characteristics to be tested should also be compatible with the sampling unit designation. For example, in a Medicaid audit, the pharmaceutical claim might be designated as the sampling unit. Some of the characteristics to be tested could apply to prescriptions rather than to claims. The situation could easily arise where the same prescriptions may support two or more sampling units (claims). The universe of prescriptions may contain far fewer items than the universe of pharmaceutical claims. It is possible to test characteristics applicable both to pharmaceutical claims and to prescriptions in the same sample. However, the auditor should be aware of problems that can arise and design his sampling procedure accordingly.

Another consideration is the audit significance of an occurrence. Although this concept is treated more thoroughly in a later section, its importance suggests a brief discussion at this point. The concept can best be illustrated by the case of an auditor who reviewed a randomly selected group of transactions and found only one discrepancy. He then evaluated the sample as one concerned with merely checking controls and procedures (attributes) and concluded that the records were acceptable. However, the one error disclosed by his sample had a potential effect of $100,000. The sampling plan should never obscure the audit objectives. Regardless of how acceptable the overall rate of error disclosed by the sample, all occurrences or findings must be evaluated and appraised for their audit significance and for any possibility of a pattern, trend, or clusters of similar findings, indicating the "tip of an iceberg."

In auditing, the concept of "an acceptable error rate," per se, tends to be meaningless since each error or finding must first be investigated to determine its audit significance and whether it tends to be localized.

As many attributes or characteristics as desired may be tested. However, each characteristic and individual source of error must first be considered separately since the evaluation and appraisal should be made by individual characteristic and not in total. Generally, the wording of the characteristics or attributes should be designed so that the examination of the sample item will result in a "yes" answer to indicate a satisfactory condition and a "no" to indicate an unsatisfactory condition. In addition, the dollar effect of the unsatisfactory condition should also be recorded for "no" answers, if applicable, even though the original sampling plan may have been designed to test attributes.

G. Judgmental Stratification for Audit Purposes

Prior to the random selection, the auditor should set aside (stratify) for separate (possibly complete) examination all important, material, sensitive, nonrecurring and rare transactions or groups of items, so as to obtain more detailed information about them. This is a subjective process based on the expertise of the auditor and is referred to as audit stratification. This process may be performed before, during or after the sample selection, but often it is not completed until the post-stratification phase because some of the significant conditions, trends and patterns are unknown prior to the analysis of the sample data.* Also, if these kinds of clusters of transactions are systematically and uniformly precluded from a modified or delimited universe, the precision or quality of the sample results will be improved. The separated items are still important in the formulation of the audit opinion and decision.

Precision of results depends on the degree of variability of the sampling units, extent of stratification and the size of the sample. However, variability and effective stratification are far more significant than sample size if the sample is not small. The more similar or homogeneous the items, the more precise the results. Thus, it is desirable to separate the modified or delimited universe into relatively more homogeneous groups (strata), and to randomly sample independently within each group. This procedure produces improved precision separately for each group, as well as for the combined weighted results. This is called stratified sampling as contrasted with stratification for audit purposes. Under the latter procedure, the universe is redefined to preclude the items set aside or stratified for audit purposes, whereas in stratified sampling the universe remains unchanged, just separated into strata.

More detailed discussions of audit stratification are presented in Chapter 3.

H. Sampling Error

Sample results are subject to uncertainty because (i) only a part of the total has been examined and (ii) there are usually observation and measurement errors. The latter errors are called human or nonsampling errors and are not reflected in the computation of the sample precision. They are controlled by careful execution of the sampling plan and adequate audit supervision.

---

*Audit post-stratification should not be confused with statistical post-statification. In audit post-stratification, a procedure somewhat similar the evaluation matrix shown in Figure 2 is used to assist in locating clusters of findings.

The extent to which the difference between the sample result and the corresponding universe value is not controlled is expressed as the sampling risk. Estimating the probable difference involves two interrelated factors: the tolerance or precision and the assurance, or confidence, level.

The tolerance or precision is a probabilistic measure of the closeness of a sample estimate to the corresponding universe characteristic. Precision is somewhat similar to the tolerance factor associated with any measurement. For example, an estimate may be made to the nearest $1,000, to the nearest $100, or to finer tolerances. Precision also may be expressed as a maximum or upper limit, such as "less than $100" or "less than 3 percent error." In most cases, the primary consideration influencing the selection of an acceptable precision is the potential effect of the sample result on the audit decision. The tolerance, or confidence, interval is obtained by extending the precision amount on both sides of the sample estimate. For instance, $500 plus and minus $50, or the corresponding interval of $450 to $550.

The confidence level is the complement of the risk associated with a specific precision, or confidence, interval. It indicates how frequently the estimation procedure used will yield differences between the sample value and the universe value as small, or smaller, than the computed precision. For example, from a sample to determine the average value of accounts receivable, the auditor may conclude that there is a 90 percent probability, or confidence, that the average value of accounts receivable is within $30, either way, of the sample average of $525. Correspondingly, there is a 10 percent risk that the average value is greater than $555 (525 + 30) or less than $495 ($525 − 30).

The selection of a confidence level involves consideration of other sources of reliance as discussed in Chapter 1. Thus, in most cases, the confidence level is a subjective determination which reflects the risk of being wrong versus the cost of obtaining more reliable sample information.

Procedures for computing sample precision are discussed and illustrated in the chapters on Monetary-Unit Sampling, Flexible Sampling and Estimation and Evaluation.

I. Sample Size

The size of the sample, per se, is no indication of its precision. Also, an auditor cannot offset poor auditing or sampling procedures by increasing the sample [c]. Stratification and the choice of sampling units are more significant. After these are properly handled, then an increase in the sample improves the precision but the point of dimishing returns is reached rapidly.

The size of the universe is of no practical importance in determining the size of the sample. In fact, the universe size is usually ignored in calculating the sample precision, except when the sample is a significant part of the universe. The sample size for a universe of 10,000 may be about the same as for a universe of 1,000,000 if the number of strata and the degree of variability of the primary variables or characteristics are the same. It is the absolute size of the sample and not its proportion of the universe that determines its precision. This concept is illustrated in Figure 1.

For example there are no basic distinctions between a national and a local sample. Thus, a statistical sample of an institution, city, county, or state requires similar procedures, and probably the same size sample, as a national sample. It is therefore more efficient and economical to direct the sampling activities to the highest organizational or geographic level consistent with the audit objectives. Obviously, the results of a nationwide sample are not applicable to individual states unless stratified by states with sufficient sampling for each strata.

The auditor cannot delegate to the statistical sample the responsibility for specifying the kind of information needed, the criteria and methods of eliciting the information, the segregation or stratification of material and sensitive items, nor the rationale of the audit opinion which integrates the sample results with other audit evidence. The statistical sample only provides a range of probability precisions. The auditor must translate these into meaningful audit opinions and decisions.

The precision of the sample results is often thought to include errors of observation or measurement. If the methods of examination or of obtaining information or of analyzing and appraising findings are not satisfactory, no sample, not even a 100 percent examination, will provide useful audit information.

Many auditors who have been exposed to classical statistical methods may remember mathematical formulas or tables for determining sample sizes if the desired precision and confidence level are specified. Such formulas and tables, however, merely shift the problem from specifying sample size to one of specifying the sample precision.

The most critical factors in determining sample size for audit purposes is the value of the information and the availability of evidence from other sources of reliance. The value of information depends on how likely it is to influence a decision. Given auditing's present state of the art, the value of the information is made subjectively by the auditor.

Also, there is a basic difference between classical and nonclassical statistical methods. Nonclassical procedures incorporate prior judgments, other information and subjective probability into the decision process. The procedures, discussed in Chapter 4, enable the auditor

As in matching a piece of dress fabric, it is the absolute not the relative size of the sample that is important.

When daughter decided that she wanted a dress just like Mother's - how much of a sample did Mother need for a precise matching of the pattern, at the store?

Obviously, what was needed was a sample big enough to cover all the elements of the pattern; not so small as to show only part of the pattern, and not so large as to include any element of the pattern more than once.

In the illustration above, the size of sample has no relationship to the size of Mother's dress or of daughter's dress.

Similarly, in scientific sampling it is the absolute size of sample that is important. The percentage of the universe covered is of no importance, except when the sample is a substantial part of the universe.

Figure 1. Absolute sample size is important. Figure 1 originally appeared in an AT & T publication. It is reproduced, by permission, from AT & T.

to use his judgment is a logical and explicit manner. Also see Kraft [33] and Tracy [76].

In classical statistics, risk is expressed as the standard error associated with the sample results and estimation procedure. However, the auditor's risk should not be based solely on sample evidence, since he has other sources of reliance and must exercise maturity and audit judgment.

Guidance for determining minimum and optimum sample sizes are given in the chapter on Flexible Sampling.

J. Upper Occurrence and Monetary Error Limits

Obviously, if no monetary misstatements are detected in the sample no dollar estimation of universe characteristic can be computed from the sample; however, an upper occurrence limit could still be estimated under flexible sampling. Also, if the sampling unit were a monetary unit such as a dollar, an upper monetary error limit could be estimated under dollar-unit sampling even when there is no finding in the sample. Thus, in essence, flexible and dollar-unit sampling may be considered parallel procedures when there is no finding in the sample. The difference is in the sampling unit; in the former it is an item, while in the latter it is a monetary unit. However, if there are findings, these two evaluation procedures, as well as others, still may be used.

K. Random Selection of Samples

Statistical sampling is dependent upon the principle of random selection, which eliminates personal bias and subjective considerations. However, audit stratification makes provisions for handling subjective considerations without invalidating the statistical sampling portion of the audit.

Before a random sample can be drawn, the sampling frame or listing of sample items, must be established. The frame should be an adequate representation of the universe, since the statistical evaluation of the results applies only to items in the frame.

The random selection of a sample from 50 vouchers may be accomplished by (i) recording the identification number of each of the 50 vouchers on a separate slip of paper, (ii) placing the folded slips of paper in a container and thoroughly mixing, and (iii) withdrawing the required number for the sample, thoroughly mixing after each selection. This procedure may be feasible when the universe is very small, but difficulties are apparent when the universe consists of thousands of items. Random digit tables and computer selection routines provide means to overcome such difficulties. Such devices provide numbers which are thoroughly mixed, i.e., randomly arranged. Considerable time can usually be saved if a computer selection program is used instead of manually selecting from a random digit table; however, it is important that the auditor understand the random selection process.

There are two basic random selection methods: unrestricted or simple random selection, where each item is drawn at random from the sampling frame; and systematic or interval random selection, in which, after a random start, items are selected from uniform intervals. Other statistical sampling procedures involve the use of one or both of these two basic methods, or a modification or a combination of them.

L. Sample Design

The sample design is the scheme of a sampling process; it covers the procedures for selecting a random sample from a specific frame.

Although there are a variety of techniques for selecting a random sample, feasibility and economy are the prime considerations in deciding which scheme to use in a particular situation. For any audit situation there are usually several possible sampling schemes, all valid, but of varying degrees of difficulty with some more appropriate while others may be more economical. The strategy is to choose the sampling scheme that will provide the required information with a reasonable assurance at the lowest possible cost and within the least time consistent with resource constraints.

Feasibility refers to procedures which are practical, simple, reasonable and easy to execute. Simple procedures reduce the risk of making human or nonsampling errors. Thus, simplicity is desirable even if there is some sacrifice in theoretical efficiency, because the reduction in nonsampling errors may provide better overall sampling precision. Each sample design should be considered an adaptation of sampling theory and techniques to the particular audit objectives modified by the environment and resource limitation.

Economy concerns the accomplishment of the audit objectives with minimum effort for a reasonable precision.

In addition to feasibility and economy, the following considerations influence the optimal selection techniques: (i) complexity of the audit objectives, (ii) complexity of frames for the most critical characteristics to be tested, (iii) desired sampling precision, (iv) availability and quality of other evidential material, (v) availability of computer programs to assist in the sample selection and evaluation of results, and (vi) time limitations.

Based on the above considerations, a random selection scheme is chosen. It should be controlled through adequate supervision. The ideal procedure would be feasible, economical, efficient and effective. For detailed discussions of random selection procedures see Chapters 10 and 11.

M. Sampling Plan

A sampling plan should be an integral part of the audit program and, where possible, should be prepared in a similar format, showing in detail the information and procedural steps necessary to stratify, randomly select and statistically evaluate the results. Although it may not be possible to outline the subjective audit appraisal of the results since they are not known until after the examination of the sample, provisions should still be made to include this analysis. The plan defines the sampling frame or listing to be used, the characteristics to be appraised and evaluated, the types of factual determinations to be made about the characteristics, and the criteria and judgments to be exercised. It also indicates the method of randomly selecting the sample, the sampling assurance desired, and the minimum and optimum sample sizes deemed necessary for the flexible features of the sampling operation. See Chapter 13 for details. It may be necessary to stop sampling because of unacceptable results and resort to other techniques prior to the completion of the sample. Thus, the plan should provide for stopping at any point in the process.

A sampling plan is necessary to coordinate the requirements for the various test-checks which may appear in different parts of the audit program; and to assure that no pertinent facts or steps have been omitted in the preliminary planning of the sampling process or in conducting the sampling operation. Failure to adequately plan the entire sampling operation may result in considerable rework and often a complete revision of the sampling procedure after some sample items have been examined.

Another principal value of a sampling plan, aside from its control function, is the support it provides to the audit findings and opinions by revealing the factors and conditions considered in the planning and the execution of the sampling operation. This support is important, especially when the audit findings and workpapers are subjected to review and critical analysis by auditees, by higher authority, and by the judiciary.

The sampling plan may be separate from the audit program, an integral part of it, or in a few cases a part of the footnotes to the audit workpapers. A separate sampling plan may be advisable when the audit program is not in sufficient detail to permit proper coordination of the sampling requirements. The audit program usually and properly limits the test-check instructions to a general statement on the type and purpose of the required tests. Frequently a single sample may be used for multipurpose test-checks required in different parts of the audit program. Consequently, if the sampling procedures were made an integral part of the audit program, the details of the sampling procedures would often be widely scattered. Flexibility in both the sampling and audit operations, and the coordination of the divergent elements in the mechanics of sampling and auditing, may be more readily achieved if separate but coordinated programs are used.

Although it is usually better to have a separate sampling program, there are occasions when a separate one is unnecessary. For instance, if only one test-check were to be made from a sample or if the sample affects only a single portion of an audit program and, if the requirements for implementing and controlling the sampling operation were not too detailed, the sampling instructions could be incorporated into the audit program. Also, in simple sampling situations, when the details of the sampling plan are brief and routine, the sampling procedure could be stated as a footnote to the workpapers.

The sampling program should make optimal use of applicable computer programs. All available software is not needed for each sampling application. The use of unneeded software tends to cloud the basic issues and audit appraisal. The auditor, however, should exercise great care that he does not lose his independence to computer assistance. There should be means of validating computer results with basic records.

N. Sampling and Nonsampling Errors

The three principal sources of error in sampling are (i) sampling variability, generally referred to as sampling error or precision, which depends on the sample design and, to a much lesser extent, on the sample size; (ii) sample selection biases which reflect how accurately the sample design is carried out; and (iii) effects of defective frames, erroneous criteria, and mistakes in collecting, processing, analyzing and interpreting data which cause differences between what is observed and actual conditions or characteristics. The latter two types of error are called nonsampling errors and are not covered in the statistical evaluation of the sample results.

One procedure for controlling nonsampling errors is for two or more auditors in widely scattered locations to collaborate. The results obtained by combining efforts are substantially better than the sum of separate audits. Also, experience indicates that tighter control can be maintained over smaller groups of auditors in several locations than over a larger group in one location, although it may require more supervisory effort.

Another method for increasing sample credibility is to compare the current audit results with those of similar prior audits. If the results replicate those of earlier audits, both the prior and current audits gain in credibility, even if the auditing and sampling procedures differ. If, however, the results of a current audit contradict the results of similar earlier audits, the auditor should investigate to determine whether the differences are caused by sample differences, different criteria of observation and measurement, or something else, such as organizational, operational or system changes. Any of these changes could effect the results.

O. Subjective Audit Appraisal of Sample Results

The audit examination and appraisal of a statistically selected sample are the same in all essentials as the examination and appraisal of a sample selected by any other method. The same characteristics and variables are examined; criteria of acceptable and reasonable transactions and operations are the same; and the same level of judgment should be exercised in appraising the findings for audit significance. Also, the methods of analysis and the precautions to avoid incorrect conclusions are the same. However, for statistically selected samples, the auditor has an additional capability of estimating the reliability of his factual findings. However, these findings should be in consonance with other evidence. The consideration of all of these interacting factors is called audit appraisal. This process involves separately investigating the nature, cause and audit significance of each individual finding. The auditor should be alert to ascertain whether the findings are systematic or in clusters, or whether they seem to follow no pattern, or nonrecurring.

Thus, it is essential that all findings, both factual and judgmental, be correctly and adequately recorded. This should be accomplished for any audit, but when the reliability of the results is not involved, as in the case of judgment sampling, there is a tendency to be less disciplined in the recording and analysis of findings.

The evaluation matrix, provides a simple means of tabulating the audit findings for each characteristic examined in the sample. Figure 2 is illustrative of an evaluation matrix. Down the left column is a listing of the characteristics tested and across the top is a list of sources of errors. In this illustration, the audit included a review of an auditee's direct material charges. The vendors' invoices supporting the direct material charges were sampled for the indicated characteristics; see Figure 2. The vendors' names are shown across the top of the matrix. The check marks in the body of the table reveal the sources of each finding or error. This evaluation matrix depicts the results of a sample of 100 randomly selected vendors' voices supporting the direct material charges.

It is obvious from the matrix that freight charges and Vendor Brown are problem areas. As a consequence, the auditor isolated or stratified all of Vendor Brown's invoices for a separate, more detailed audit. Also, a larger sample of freight charges was made to obtain additional information relating to causes and financial impact.

Characteristics Tested	ABLE	AVON	BAILEY	BANKS	BROWN	CADY	CAGE	CARTER	CROSS	FORD	GREEN	JONES	KING	ROYAL	WILSON
1. Account Coding															
2. Invoice extensions															
3. Cash discounts taken															
4. Invoice quantities and price agree with purchase order															
5. Invoice quantities and receiving reports are in agreement															
6. Correctness of freight charges					✓✓✓ ✓✓✓			✓							
7. Purchase properly approved															
8. Allowability of cost															
9. Allocability of cost															
10. Reasonableness															
11. Invoice approved for payment															
12. Payment of invoice															

Figure 2. Evaluation matrix for sample results from vendor's invoices.

A further investigation revealed that Vendor Brown, a small contractor, is apparently reliable and honest but has difficulty interpreting the freight rate structure. All of his freight charges have been erroneous but the errors are not systematic. In some cases he overcharged the auditee and in other cases he undercharged. The auditee is handling this problem by providing assistance to Vendor Brown in interpreting freight rate schedules and in making necessary computations.

Of course, the cluster of findings against Vendor Brown cannot be projected against the other vendors nor to the universe as a whole. This is another reason for handling the Brown problem separately, once it is disclosed. Likewise, the findings associated with freight charges cannot be projected across-the-board. Except for freight charges, the characteristics reveal no deficiencies, and thus, they are deemed to be acceptable. This illustrates the fact that an overall error rate is usually of little value in audit activities.

An investigation of the freight rate errors associated with other vendors revealed only minor clerical mistakes with no discernible pattern.

For a more detailed discussion of audit appraisal and the use of the evaluation matrix, see Chapter 13.

P. Statistical Evaluation of Sample Results

The sample result is objectively (mathematically) evaluated by the computation of the standard error for the estimate, which, in turn, may be used to compute the precision associated with any desired probability, or confidence, level.

Briefly, the process involves (i) computing a point estimate from the sample, (ii) computing the estimated standard error for the estimate, and (iii) then computing from the standard error the precision for the desired confidence level.

One estimates a total amount by "blowing up" or projecting the sample result to the universe. This single value obtained from this process is called a point estimate. Such an estimate may be computed in a number of different ways, which are called estimators or estimation techniques. Examples of estimators are ratios, means and regression coefficients.

For the same estimator, the point estimate is computed in an identical manner for either a judgment sample or a statistical sample. In both cases if the auditor were using the ratio of audited dollars in a sample to the dollars examined to estimate total audited costs, the computed ratio would be multiplied by the book value to obtain a universe estimate. For example, suppose $9,500 were the audited amount in vouchers totaling $10,000. Then the ratio of audited to recorded amounts would be 0.95. If the total book value of vouchers were \$500,000, the ratio point estimate of audited costs would be \$475,000 (500,000 × 0.95).

The major advantage of statistical sampling over judgment sampling is that the calculated risk associated with the results can be estimated. Statistical sampling evaluation provides an estimated probability associated with the actual universe value being in a range of values (confidence interval) around a point estimate.

For example, the statistical evaluation might indicate that there is an 80 percent probability that the universe total is \$200,000, plus and minus \$15,000, or between \$185,000 and \$215,000. In other words, if repeated samples of the same size were selected from the universe, using the same care and procedures, the computed probability intervals of about 80 out of 100 of the samples would be expected to include the actual universe value of interest.

Chapter 15, Estimation and Evaluation, describes in detail how to make and use the various computations involved in this process. Here, as in the case of random selection, computer programs provide invaluable assistance. They free the auditor of the drudgery of the mathematical computations. Nevertheless, he should understand and appreciate the process so that he can properly exercise judgment and flexibility in designing and modifying sampling plans.

Q. Presentation of Sample Results

There is a lack of sufficient guidance on what recorded information is necessary when statistical sampling is used in auditing. Actually the scope of recorded information is similar to that required for any other type of audit test-check.

Good audit reports reflect certain standards and qualities. Hence, there should be no significant change in the quality nor tone of reports which are based partly on the results of statistical sampling. Where the results of selective tests have been previously cited in audit reports, the reference would be to the use of statistical sampling. The report, per se, should not contain any more test information or technical details than previously.

The details associated with statistical sampling applications properly belong in the audit workpapers and not in the report. However, if it is obligatory that a sampling reliability statement be included in the audit report, the statement should be limited to citing the standard error for the estimation (projection) procedure used. In this manner, the auditor is not sanctioning any particular confidence level, which reflects a judgment.

It is important that an explicit opinion be expressed in the report. The mere presentation of data derived from a sample is insufficient. Also, the rationale for the opinion should be clear and adequately supported using all available evidence. Criteria used as a basis for conclusions and recommendations should be clearly identified. Additionally, assumptions should be identified and explained, because they imply a limitation or a judgment. Finally, care should be exercised to avoid treating uncertain quantities as if they were facts, such as projected workloads or overhead rates.

The workpapers, and not the audit or survey report, should convey full information on the technical aspects of the sampling procedure, so that reviewers can appraise the procedure and the statistical conclusions. The sample information should constitute only a part of the total evidence upon which an audit opinion is formulated. Even when it is obligatory to include technical sampling details in a report, it is best to keep the coverage to a minimum, since the useful evidence is not uniform in quality and the auditor's decision making process may be subtle. For example, if the statistical sample primarily provided more clues for further audit stratification and contributed little to the audit decision, technical details of the sample would contribute little to the report. On the other hand, if the sample is the sole source of evidence (and this should be rare in auditing), then more technical information would be useful.

A suggested list of technical sampling details for the workpapers may include the following:

1. A statement of the aim of the test-check.
2. A description of the sampling unit.
3. A description of the sampling frame and the rules for using it.
4. A description of any sampling frame deficiencies and actions taken to correct them.
5. The rules or criteria for determining material and sensitive items (if any) that require segregation (audit stratification) and special separate treatment.
6. The procedure for randomly selecting sample items.
7. The kind of estimators, such as ratios, used in the computation of the point estimates.
8. The projections (estimations) calculated from the sample.
9. The standard errors of the more significant estimations. These are shown on the printouts from applicable computer programs. It should be remembered, however, that a small standard error does not mean that the estimate is necessarily accurate or useful, since accuracy also involves nonsampling errors, and usefulness reflects professional judgment applied to all available evidence. A small standard error implies that the results of a 100 percent examination would have been the same as the sample within a small margin of difference, if the 100 percent examination had been carried out with the same

criteria, same auditors, sharing the workload proportionately, and with the same care that they used on the sample. Furthermore, a small standard error does not indicate successful coverage; it does not indicate that the audit work was good. The standard error does not measure the biases of wrong information, nor of other errors that may affect both 100 percent coverages and samples. It is possible for a sample result to be useful and still possess a large standard error. A result obtained by definition, criteria, and techniques that have been drawn up with great care, based on experience of mature auditors, may have a large standard error because the sample was small; yet such a result may be preferable to one obtained with a larger sample, with a smaller standard error, but whose definitions, criteria and techniques were not in consonance with the best audit practice.

10. Comparisons of sample data against known information and prior audits. Some of these analyses may be part of the survey or a flexible sample. Flexible sampling methods are useful to reveal misunderstanding in procedures, sampling frame deficiencies, inadequate criteria, and to provide clues to a need for audit stratification or subsidiary sampling of certain strata. An evaluation matrix and the results of further investigations, if there were any findings or adverse conditions revealed by an examination of the sample, would be useful.

III. SUMMARY

In the application of statistical sampling, the first consideration is to define clearly the audit objectives, the universe of records and the scope of the examination. Assumptions, criteria, rationale, and definitions should be precise, including the definition of the sampling unit and the characteristics (attributes and variables) to be examined. Usually these are an integral part of the audit steps. Next, the universe is represented by an available sampling frame or listing.

The data to be collected and the collection method must be decided. The method of collecting items randomly, generally, is a modification or a combination of simple random selection and systematic selection, involving several random starts. For example, the sampling plan may require a 100 percent testing of high dollar items; a simple random selection of middle valued items; and a systematic selection of low dollar items.

Any meticulous, absolutely exact or time consuming procedure with precisely predetermined sample size is neither justifiable nor desirable in most audits. Predetermined sample sizes are generally based on assumptions which may not be applicable to the audit circumstances.

Randomization is achieved by use of a random digit table or a computerized random generator. It ensures statistical validity of the results by eliminating biases (systematic errors) due to personal preferences of the auditor. However, personal preferences relating to such considerations as dollar amounts or particular kinds of transactions can be accommodated through audit stratification. Naturally, the universe would be delimited to exclude items stratified for audit purposes. Feasible and economical selection methods which are reasonably random and achieve the audit objectives are preferable. The particular selection method will reflect the degree of professional judgment and extent of imagination and ingenuity exercised by the auditor in designing the sampling plan.

The sampling design and the related plan must be considered. Both resource cost and precision of the sample results depend on the sampling design or scheme used. The auditor, however, should guard against the uneconomic practice of attaining a higher sampling precision than is required for the audit objective. Unlike most other professions, in auditing there are many other sources of reliance in addition to the sample, such as knowledge of internal control and audit survey results. Thus, if sampling plans and desired precision do not reflect information from other sources, critical audit resources may be wasted.

The sampling plan must reflect simple, clear and adequate instructions and resources. Short instructions that maximize understanding should be used. Instructions for diverse tasks may need to be separated. Also, there should be guidance on how to cope with anticipated problems.

The subjective audit appraisal of the sample results to discern the nature, cause and audit significance of each finding is an essential step in the development of conclusions and rationale to support audit opinions. The evaluation matrix is a device which assists in the appraisal by tabulating the findings by characteristic and source to indicate clusters.

When statistical sampling is employed, the auditor is provided a means of calculating the risk associated with the sample results.

In the final analysis, the use of statistical sampling depends primarily upon the exercise of good audit judgment, effective audit stratification, consideration of all other sources of reliance, and some understanding of the technical considerations involved. It should be remembered that the most precise procedure applied to the wrong problem or objective is valueless; whereas, a less precise technique applied to the correct problem or objective does provide some information.

3
Stratification for Audit Purposes

I. GENERAL

It is generally agreed that audit effort should be concentrated in those areas representing the most significant cost levels, regardless of whether the method of testing is judgment or statistical sampling.

Cost submissions and related accounting data may contain undocumented costs of material significance to the audit. Most undocumented costs, however, are usually found in relatively few accounts or in a limited segment of the data and are not distributed throughout an audit area (universe). For example, a typical overhead cost representation by a large institution for any given year will usually contain relatively few accounts with undocumented costs, whereas the majority of the accounts are acceptable and therefore do not warrant extensive review.

Generally, financial data exhibit a wide variation between the smallest and the largest individual dollar amounts, with the bulk of the amounts being relatively small and a few very large ones. Hence, there is a tendency towards a high degree of variability and for the extremely high values to be concentrated in a few limited transactions or accounts. An example of this situation is the account for consultant services. In such circumstances, the precision of dollar estimates from simple random sampling could be low because the extreme values would increase the variability. As previously mentioned, the auditor, based on judgment, experience and other information, should separate from the total universe, for more thorough testing or detailed review (i) all significantly high value transactions, (ii) all unusual charges and credits of substantial value, and (iii) any other transactions or segments of significance to the audit. He should then examine the remaining transactions on a sampling basis, being alert for causes, patterns, trends and clusters among the findings. The original universe or frame is redefined or delimited to exclude those transactions or segments separated for audit signficance.

II. AUDIT STRATIFICATION

This process of separating audit areas into different segments (such as high value items, sensitive accounts, and the area to be statistically sampled) for separate and varying degrees of examination is known as stratification for audit purposes. The aim is to isolate for separate handling the major problem items or accounts thereby reducing the risk of missing significant undocumented transactions. Furthermore, by separating the high value and sensitive items from the universe or frame, the auditor decreases the variability among the items to be sampled, thereby reducing the required amount of testing and increasing the precision of the subsequent sampling results. The items and accounts separated for audit significance should be appraised separately and not mixed with items randomly selected, since they involve no sampling error. In other words, the results from items or accounts which are reviewed in detail are not diluted by sampling uncertainties.

This stratification procedure for audit purposes is equally valid with statistical sampling procedures as it is for more traditional audit test methods. In fact, preliminary stratification of the audit area is essential to the economical use of statistical sampling in auditing. Of course, the criteria for stratification may vary according to the audit objectives and the area under review. For instance, in the audit of direct labor charges there is usually no particular reason for stratifying by dollar value, since there is relatively little variability in labor rates. In this case, stratification would usually be directed towards isolating or separating certain types of employees, departments or functional areas based on experience or results of a preliminary sample. In the audit of direct equipment expenditures, however, the primary criterion for stratification will often be dollar values, since by examining all transactions exceeding a certain dollar value (or, if not examining all at least performing a more thorough review of high dollar transactions) the auditor obtains more dollar coverage with perhaps less effort. Stratification by dollar value does not preclude further stratification by sensitive employees, departments, accounts or functions.

The above discussions have emphasized that there are actually two separate audit steps associated with the selection phase of statistical sampling: first, the preliminary judgment stratification of the universe or frame for audit purposes and, secondly, the statistical selection of sample items from the remainder of the delimited universe. Of course the delimited universe may be further stratified statistically to improve the precision of the sample.

It should be noted that the examination of only the high value items in an audit area may not give a true picture of the conditions in the total area, even though the high value items are examined in detail (100 percent). Such a review provides information only about the specific type of items examined. On the other hand, if the auditor relies solely on a statistical sampling procedure without audit stratification, he will usually need a much larger sample and even then, when the audit area contains a few "localized" significant items, the sample results may be misleading. For example, a sample of 200 reveals only 6 errors but all the errors are attributable to one employee. Hence an overall error rate of 3 percent would be misleading. Effective auditing and efficient use of statistical sampling require that the auditor in addition to audit stratification, also stratify statistically selected samples to improve the precision, i.e., reduce the sampling error. Statistical stratified sampling is discussed in Chapter 8.

A. A Misconception

In the initial application of statistical sampling methods, many auditors are disturbed because the random selection procedure fails to select some transactions they wish to examine. With this apparent deficiency in the random selection process, auditors are prone to consider discarding statistical sampling and revert to judgment selection. They tend to feel that subjective methods of selection provide better samples for their purposes. This attitude reflects a common misconception. Except for possibly setting aside high value items, many auditors believe that they are prohibited from using judgment in the selection of any of the remaining items. In the process of emphasizing the random selection of sample items, the opportunities for exercising audit judgment are often obscured. One has unlimited opportunities to exercise judgment in the planning before selection, and by stratification during and after selection. There is the opportunity to decide (i) whether high value or any other transactions deemed sensitive will be set aside for separate or detailed review, and (ii) whether the remaining sampling area will be further stratified to give certain types of transactions a greater chance for selection. There are two basic types of stratification available to auditors: (i) stratification for audit purposes to control sensitive and material items, and (ii) statistical stratified sampling to improve the efficiency of the sampling operation.

B. Complete or Special Examinations

The auditor is aware that he may set aside the high value items for a more detailed or complete examination. Often by examining a relatively few high value items, a significant portion of the total cost will be reviewed. This procedure provides sufficient dollar coverage which is an important aspect of auditing. However, the auditor may not be aware that there is also the opportunity to set aside any other items or group of items he wishes to examine, provided they are appraised separately from the randomly selected items. This flexibility should dispel the contention that the statistical sampling process does not afford an opportunity to select certain sensitive and lower value items that are of audit interest. The auditor may have other information available that indicates there are problem areas which should be handled separately. For example, universities frequently transfer labor charges between grants, subsequent to the grant period. From experience, auditors have learned that these transfers are often made, not because work was or was not performed on a particular grant, but because certain grant funds were over or under expended. Consequently, after-the-fact labor transfer charges are separately identified for a detailed examination.

Another reason for examining certain items might be that the auditor, through a survey, flexible sampling or by inspecting the listing or frame, may find that certain items appear to be out of line. Even though many of these items may be proper charges, they should be examined. For instance, it is observed that animals are being charged to a research contract for electronic equipment. Another example is the finding that cases of ping pong balls are being charged to a study contract where the auditor knew the employees principal lunch-time activity was playing ping pong. However, in both instances there were valid reasons and needs for these items to test radar equipment being procured under the respective contracts.

Another reason for selecting particular items for examination is that the auditor by scanning the listing sometimes finds items that are obviously in error. For example, items being charged on one contract that obviously belong on another contract. Another example would be the finding of cost estimates on proposals for items which are to be provided as Government Furnished Equipment.

In the examples cited above, the auditor does not have to rely upon statistical sampling methods to provide such items before they can be examined. He can simply separate or exclude these items from the list used to select the sample. In fact, this may be accomplished before the sample selection, during the course of the selection or after the selection has been completed, provided the stratification rules are applied in a consistent manner. However, any occurrences or errors found or undocumented costs on such items should not be projected to the entire universe because of their peculiar nature. Nevertheless, these items may provide cues or patterns or other information about problem areas which require further audit effort.

Superficially, it may appear that audit stratification "loads" the audit results in the auditor's favor. He is taking out the obviously erroneous items from those which he judgmentally identified, and is randomly sampling from the remaining items for estimation or projection purposes. Actually, if the auditor were adequately informed so that all areas of procedural errors or undocumented costs could be set aside for 100 percent examination, the statistical sample would disclose no error or undocumented cost. The only errors or undocumented costs would be specific items found in the areas of 100 percent examination. There would be no basis for making an estimation or projection from the sample. By handling all judgmentally selected items separately, the statistical sample results are not invalidated. Thus, the audit practice of screening out possible problem areas for 100 percent examination is highly desirable. This also avoids some of the difficulties encountered in handling nonrecurring or rare findings which are not typical of the universe. Error analysis and appraisal are discussed in Chapter 13.

C. Desire for Unequal Chance of Selection

In some audit situations, it is desirable that some items, other than those examined 100 percent, be given a greater chance for selection. The auditor in many instances will find some areas in which there is more interest than others, yet the items in these areas are too numerous for a complete examination. Also, there is a desire to examine more of these items than other types. Consequently, he would like to give them a greater chance for inclusion in the sample than items which are probably acceptable. This is accomplished by dividing the area to be sampled into strata and drawing a proportionately larger sample from the strata of interest than from others. This is a type of statistical stratification known as disproportionate allocation. This procedure is permissible using statistical sampling provided it is executed properly. The auditor may divide the sampling frame into as many different strata as he desires, although having too many strata may become cumbersome and provide only trivial proportional improvement in the precision. Also, an increase in the number of strata involves extra costs in planning and drawing the sample. General guidance indicates that few strata, perhaps between two and six, yield most of the possible gain from a single variable such as dollars. In addition to stratification by dollar amounts of items, it may be by periods of time, by departments or any other consistent criteria.

D. Computer Assistance

One of the advantages of packaged computer programs is the quick classification of data. This relates to the problem that often confronts an auditor in determining where and to what extent to stratify. Initially, strata limits may be determined on an arbitrary basis because the auditor does not have sufficient information about the nature of the universe. For example, in performing comprehensive audits of purchases, it has been a common practice to identify, for a complete review (usually in a manual operation), all charges in excess of perhaps $10,000 or $25,000 or some other arbitrary value, and then randomly select a sample of the charges less than that amount. Even though the strata limits are arbitrary, stratification generally improves the precision. With the availability of data from the initial sample and computerized assistance, much more information can be provided about the composition of any account or total. Also, an entire account or subaccount may be requested to be listed if it is of interest to the auditor. He may scan the account, review it in detail or draw a subsidiary sample from it. With this initial information, the auditor can make better decisions about further stratification. In other words, in a computerized environment, the auditor has more flexibility to experiment and refine both the audit approach and sampling procedure to provide optimum audit coverage for available resources.

Ideally, stratification should reduce the variability within each subgroup for all the characteristics being tested. When this is impossible, stratification should be focused on the most critical characteristic. The objective of stratification is to have the items within each subgroup (stratum) as much alike as possible and as different as possible from items in other strata. If the auditor does not have any knowledge of the sensitivity or distribution of characteristics he is testing, a preliminary sample will often furnish some useful information needed for more effective stratification. For example, consider the audit of inventory issues where there are a number of separate stockrooms making the issues. If the auditor has no knowledge of whether mischarging is more prevalent at one stockroom than another, stratification by stockrooms or clerks may not be effective. However, on the basis of a flexible sample he may discover which stockrooms or clerks are responsible for most of the errors and thus stratify to isolate the problem areas for separate and more detailed review.

Where manual methods of statistical stratification are employed, approximate optimal dollar stratification is achieved by setting strata boundaries such that all strata contain about the same total dollar amount. Obviously, this scheme would exclude those sensitive, nonrecurring and material items previously stratified for separate audit consideration.

In the final analysis, however, effective stratification reflects mature judgment and knowledge of the audit environment. Prior audit information, intuition, the judgment of the auditor, or good guesses can all be used effectively in setting up strata. Also, if judgment is exercised in determining the strata, the sample results will not be biased by the action, and the precision may be improved if the judgment is good.

III. SUMMARY

Aside from the general statistical guidance of endeavoring to achieve reduced variability within stratum and increased variability among strata, mature judgment is the primary basis for deciding where and to what extent to stratify in auditing. In an initial audit it may be a trial-and-error process. Although judgment should be exercised in the determination of strata and sensitive transactions, the actual selection of items (except those for 100 percent review) must not be influenced by judgment. The items which are to be statistically evaluated must be selected by random methods. When there is some knowledge about the variability of a characteristic within each stratum, statistical guidance can be given concerning the allocation of the sample among the strata for that characteristic. It should be remembered that any logical audit stratification, no matter how it was determined will improve the sample precision. Guidance on statistical stratification is summarized in Chapter 8.

4

Sample Size Considerations

I. GENERAL

Most of the topics discussed in Chapter 2 have an impact on sample size determination. This chapter extends some of those discussions, provides additional topics, and furnishes general guidance on sample size considerations. Specific guidance on sample sizes is discussed in chapters devoted to specific sampling methods. For example, sample sizes for dollar-unit sampling are discussed in the chapter on Monetary-Unit Sampling.

The determination of an adequate audit sample is more sophisticated and less straightforward than auditors have been lead to believe. Often statisticians have advised auditors to assume the worst possible condition or situation in determining sample size. This procedure, however, usually results in oversampling, and thus, an unnecessary expenditure of critical resources, which has contributed to the slow acceptance of statistical sampling among some auditors and accountants.

When data are selected by statistical sampling, the universe to be sampled is divided into sampling units. The sample data are collected by selecting two or more of these sampling units. A selected sampling unit is often referred to as an observation.

The number of sampling units into which the universe is divided is called the universe size. The number of sampling units selected is called the sample size.

II. SUBJECTIVE CONSIDERATIONS

One cannot determine sample size merely by looking at a listing of data. Before sample size can be determined, an auditor must decide how the universe is divided for sampling purposes, i.e., what constitutes the sampling unit. For instance, in testing accounts receivable the universe may be divided so that a sampling unit consists of (i) a single amount such as a debit, credit or balance, or (ii) the amounts on a single line of a ledger, or (iii) the amounts for one account, or (iv) the amounts for a particular customer, or (v) the amounts on one of several ledgers, or (vi) the amounts in some group of columns on a page of a computer listing, or (vii) the amounts on a single page or card, or (viii) the amounts on cards or other records in a single file cabinet drawer. Additionally, there may be "empty" sampling units corresponding to pages or lines containing no relevant audit data.

The division of a universe into sampling units is dictated primarily by feasibility and convenience. In order to easily identify the sampling units, the division is made so that different units are associated with different individuals or items such as accounts, pages, cards, or lines.

The size of a sample is no criterion of its reliability. Hence, an auditor cannot offset poor testing procedures by selecting a larger sample. Stratification and the choice of sampling units are more important. After these are properly handled, then an increase in the sample improves the reliability, but the point of diminishing returns is reached rapidly.

Also, the auditor cannot delegate to the sample the responsibility for specifying (i) the kind of information needed, (ii) the criteria and methods of eliciting the information, (iii) the segregation of material and sensitive items, nor (iv) the rationale of the audit opinion, which integrates the sample results with other audit evidence. A statistical sample only provides a range of probable tolerances. The auditor must translate these into meaningful audit opinions or actions.

After the auditor has delimited the original universe to exclude material, sensitive, nonrecurring, and any other items or groups of items he feels warrants special or detailed treatment, he is then confronted with the determination of sample size.

There is no single way to approach the problem of sample size, which is difficult to determine precisely. There is a tendency to use sample sizes which have been used by other auditors under similar circumstances.

The idea that a precisely predetermined sample size is always possible is probably the most widespread misconception about statistical sampling methods among auditors and accountants. Articles appearing in some journals have emphasized this premise. The author of one such article endeavored to compare the results from what he called a judgment sample with a statistical sample. Interestingly, he concluded that there were no significant differences between the results of the two samples from the same frame. Unfortunately, the author did not recognize the essential difference between a statistical and a judgment sample. In fact, both of his samples were drawn by random selection procedures, and therefore qualified as statistical samples. He thought, however, that one was a judgment sample, since sample size and its allocation to strata were not determined by mathematical formulas, but were based on the expert knowledge of accountants who were experienced in the area under examination. The author stated that the random selection did not make it a statistical sample, since the sample size and its allocation to strata were determined judgmentally; whereas, in the other sample an optimum allocation formula was used. The only conclusion which could have been drawn from a comparison of the two samples is the statistical concept that optimum allocation is generally more efficient than proportionate allocation of sample size among strata.

III. NEED FOR A MINIMUM SAMPLE

The use of samples to estimate characteristics of universes from which they are drawn is based on the premise that some sample items will overestimate the actual universe value and other sample items will underestimate the value, but when these values are combined into an overall sample the tendency is to approximate the value of the universe characteristics. For this tendency to operate effectively, the sample size must be sufficiently large. Fortunately, through audit stratification of material, extreme, nonrecurring and sensitive items, the minimum for the remaining sample items is often smaller than in other disciplines, because items in the delimited universe are much more similar than those of the original universe. Also, there are other possible sources of audit evidence to support the decision. Nevertheless, the minimum sample must be large enough to allow any undesirable condition to manifest itself.

A sample is too small if its results are not precise enough to contribute to an audit opinion or decision. Conversely, a sample is too large if its results are more precise than is warranted by its intended use or if nonsampling errors caused by human mistakes overwhelm the sampling precision. This suggests some trade-off between the cost of designing and executing the sample and the value of the results in the formulation of an audit opinion, especially in the light of additional information from other sources of evidence and audit experience.

IV. SAMPLING COST VERSUS PRECISION

Too often in audit situations, both the cost of sampling in terms of resources and sample tolerance or precision are specified in advance. Since these are competing factors, usually one or both will have to be modified during the course of the audit.

Also, predetermined sample tolerance for a specified confidence level may be vague when the audit has several objectives, with conflicting demands on the allocation of the sample. Furthermore, the overall accuracy is usually less than the computed precision because of the adverse effects of nonsampling (human) errors, about which knowledge is generally inadequate.

When considering the sample size necessary to achieve a specified precision or tolerance, the auditor desires to have a satisfactory degree of certainty that the sample point estimate does not differ from the value being estimated in the universe by more than a specified amount or a specified proportion, which may be considered the permissible tolerance.

It is difficult to fix the sample size and the sampling rate, or fraction, if the critical factors of unit variance and unit cost of collecting data are subject to considerable uncertainties. Even when a predetermined sample size is specified, the sampling rate, or fraction, cannot be specified if the size of the universe is unknown.

If an arbitrary sampling rate, or fraction, is used, the selection process should not be terminated when the desired sample size is reached. This is especially true with systematic sampling procedures, where the items are selected at or within uniform intervals after random starts. Such truncations would preclude the selection of some sample items.

V. MINIMUM AUDIT SAMPLES

Any arbitrary sampling rate, or fraction, should be a sensible minimum based on a reasonable expectation of unit variance. Experience reveals that the absolute minimum audit sample for attributes is 50 when controls and conditions are deemed to be excellent; and at least 100 when estimating dollar amounts. If errors are found, these minimums are generally insufficient to provide a reasonable precision for the confidence levels desired by most auditors. Obviously, in situations where the sample data are the sole basis for an opinion, the auditor would want more precision and less risk;

thereby, requiring a larger minimum sample. However, effective stratification is usually more efficient than merely drawing a larger sample. The determination of larger minimum samples is discussed in the chapter on Flexible Sampling.

VI. REPLICATED SAMPLING

In replicated sampling the total sample n is selected in k independent samples of equal size so that each of the k samples is a separate sample covering the entire universe or sampling frame. Usually the use of replication does not change the sample size; however, when used in conjunction with flexible sampling, which is described briefly in the next section, the overall sample size may be reduced.

Replication provides a simple and direct way of calculating point estimates and their standard errors (precisions). Replication can also be used to avoid the complexities associated with multistage sampling. For example, one replicated sample may be applied to the total universe, while another replicate is applied to an area or stratum of special interest. Additionally, one or more replicates may be used for detection or as a probe. This is the basic procedure of flexible sampling, where the size of each replicate constitutes a minimum sample assuming no finding.

Although the number of independent subsamples may range from 2 to 20, most often there are 2, 4, 5 or 10. It is possible to replicate any sample design merely by dividing the sample size in half and repeating the same sampling procedure twice to make two replications. Deming [12] has a detailed discussion of replicated sampling designs and applications. Also see Chapter 15 on the statistical evaluation of replicated samples.

VII. FLEXIBLE SAMPLING

In considering sample size it would seem that the auditor is more concerned with the *minimum* sample, since he has many other sources of information and evidence. His primary use of sampling is to reduce the risk associated with undetected adverse conditions and material amounts of undocumented costs. Thus, if the sample reveals a single undesirable condition that he was not aware of, his audit strategy should change. He should immediately become concerned with the nature, source and potential impact of the finding. The minimum sample feature of flexible sampling is consistent with this concept, and also it prevents the wasting of resources through oversampling. Briefly, under flexible sampling, a minimum sample provides a specified probability, like 95 percent, that the occurrence rate in the unaudited records will not exceed a specified percentage, like 3 percent, provided no error is detected in this minimum sample. For example, if 100 transactions are selected by simple random sampling and no error is found, the auditor has a 95 percent assurance that the error rate in the remaining transactions is less than 3 percent. If, however, there are findings, the concept of a minimum sample no longer applies. At this point, the auditor has a number of options, among which are sample replication, or terminating sampling and resorting to other acceptable audit techniques. Probably the most appealing aspect of flexible sampling compared with classical approaches is that if it is not necessary to estimate or postulate any expected universe characteristics. For detailed discussions on the application of flexible sampling to audit activities, see Chapter 13.

Supplemental sampling should be planned in advance, otherwise the risks associated with the sample results may be greater than calculated. (See next section.) Often a large initial sample can be selected, then separated into subsamples to be used as supplements after an initial subsample is examined. Replicated sampling, because of its ease of application and estimation of standard errors, may be a better procedure. The overall large sample should represent a reasonable maximum, assuming the highest estimate of unit variance. The size of the initial replicate may correspond to a minimum sample, assuming there are no findings. The replicates, after the first two, could be held in reserve to be used if necessary.

Experience indicates that for most audit activities a sample three or four times the size of the minimum sample, assuming no errors are found, may be considered a maximum. In other words, if an audit decision has not been reached by that time, it is usually best to cease sampling and resort to other acceptable audit techniques. However, if the frame is highly skewed with a low error rate that exhibits a high monetary tainting, a larger sample may be necessary to adequately support an audit opinion.

VIII. SAMPLE ENLARGEMENT INCREASES RISK

If an auditor concludes that more verification is required and he enlarges a preliminary sample, the sampling risk from the enlarged sample is greater than if the larger sample were selected in the first place. Since there is a tendency among auditors to expand preliminary samples which are too small for the intended purpose, it is significant to understand that this problem can be avoided by using replication.

IX. CLASSICAL DETERMINATION OF SAMPLE SIZE

A. Mathematical Formulas

Classical sampling theory provides formulas for computing sample sizes for various sample designs. To use these formulas, however, information is required which may not be known. For example, precision for the estimate must be specified. This procedure merely shifts the problem from one of specifying sample size to specifying precision and assurance. However, neither the size nor the precision of a sample can be determined accurately until after the sample data have been examined, analyzed and evaluated.

For a discussion of classical determinations of sample sizes, see Roberts [57]. The following discussions cover the considerations in a less technical manner.

B. Predetermined Sample Sizes and Uncertainty

There are mathematical formulas which relate sample sizes to the permissible tolerance and assurance of a particular sampling scheme. However, the computed sample sizes may not be precisely accurate, since the necessary parameters in the formulas are subject to uncertainty. Thus, any predetermined sample size reflects the uncertainties of the particular parameters used in the computation. Nevertheless, one of the advantages of statistical sampling is that the underlying theory enables the construction of these formulas. On the other hand, it is somewhat paradoxical that the mathematical formulas used to determine sample sizes generally are the same formulas originally derived to compute the precision of the results after the samples have been examined. Hence, as has been mentioned, the formulas require the estimation or postulation of some information which the samples are intended to provide. Thus, exact predetermined sample sizes are not needed as the formulas may lead one to believe. If the estimations or postulations are not sustained by the sample results, the predetermined sample size is said to reflect poor planning. Obviously, undersampling can be handled, but oversampling has already wasted critical resources. One way to avoid oversampling is to use a flexible sampling plan with equal replications, where each replicate constitutes a minimum sample if there were no audit findings. If there are any findings, the prescribed minimum sample would be insufficient. Thus, the sampling scheme should provide for at least two, and preferably four, replicates of the same or approximately the same size.

C. Sample Precision

The precision of a sample estimate is measured and gauged by the standard error of the estimate, which is computed by mathematical formulas derived from sampling theory. The smaller the standard error, the greater is the mathematical precision of the results. The standard error is associated with the 68 percent probability or confidence level. Other confidence levels are merely multiples of the standard error. The standard error varies from zero if the sample consists of the entire universe to that associated with a sample of two, which is the minimum size for computing variability.

If simple random sampling is used and the sample size is less than 5 percent of the universe, the standard error is computed by dividing the universe standard deviation (which is the square root of the universe variance) by the square root of the sample size. As a verbal equation, this may be stated as follows:

$$\text{standard error} = \frac{\text{universe standard deviation}}{\text{square root of sample size}}$$

Thus, the standard error or precision of the sample results varies inversely with the square root of the sample size. For example, suppose the standard deviation of the average undocumented cost in an account is estimated to be $10 on the basis of a prior audit, then the precision (standard error) of the result will vary with the sample size as shown in Table 4.1.

One observes from Table 4.1 that the standard error or precision does not decrease in proportion to the increase in the sample size. That is, doubling the sample size does not reduce the standard error by one-half. This is due to the fact that the denomination in the formula is the square root of the sample size. Actually, there is a rapidly diminishing return in improved precision for increased sample size. For instance, to reduce the standard error by one-half, from $1.00 to 50¢ in the above example, the sample has to be quadrupled from 100 to 400. Also, to reduce the standard error from $1.00 to 25¢, the sample has to be increased sixteen-fold from 100 to 1,600. Thus, it is generally more efficient to resort to further stratification in lieu of increasing the sample size when better precision or tighter tolerance is desired.

It should be remembered that the single value (point) estimate with which the tolerance or precision (standard error) is associated is expected to remain somewhat stable for increased sample sizes. Thus, the value of the standard error is relative to the value of the point estimate and to the significance of tighter tolerance in support of the audit opinion. For instance, if the point estimate associated with the computed precisions in Table 4.1 were $3.00, then the

Table 4.1. Relationship of Sample Size to Precision*

Standard deviation	Sample size	Formula computation	Precision (standard error)
\$10	100	$10 \div \sqrt{100}$	\$1.00
10	100	$10 \div \sqrt{200}$	0.71
10	400	$10 \div \sqrt{400}$	0.50
10	500	$10 \div \sqrt{500}$	0.45
10	1,000	$10 \div \sqrt{1,000}$	0.32
10	1,600	$10 \div \sqrt{1,600}$	0.25
10	5,000	$10 \div \sqrt{5,000}$	0.14
10	10,000	$10 \div \sqrt{10,000}$	0.10

*For simple random sampling where the sample is less than 5 percent of the universe.

corresponding tolerance or confidence interval for the sample of 100 would be \$3.00 plus and minus \$1.00, or \$2.00 to \$4.00 at the 68 percent confidence level (since the standard error denotes that level). On the other hand, if the point estimate were \$20.00, then the sample of 100 would have given a tolerance or confidence interval of \$20 plus and minus \$1.00, or \$19 to \$21 for the same confidence level. Although the value of the standard error (of \$1) is the same in both instances, the relative impact is obviously different.

D. Components of a Sample Size Equation

If the above verbal equation for computing the standard error for simple random sampling is solved for sample size, the following equation results:

$$\text{sample size} = \frac{\text{universe standard deviation squared}}{\text{standard error squared}}$$

In practice, this equation is further modified to provide for confidence levels different from the 68 percent level of the standard error, and a permissible tolerance is substituted for the standard error, resulting in the following:

$$\text{sample size} = \frac{(\text{estimate of universe standard deviation})^2 \times (\text{confidence coefficient})^2}{(\text{permissible tolerance})^2}$$

If the sample is greater than 5 percent of the universe this formula would be further modified by a finite universe correction factor, which is not necessary for the present discussion where the emphasis is on the interrelationship of the various factors, but will be discussed in detail in the chapter on sample evaluation. Also, different sample size formulas are used for random selection schemes other than simple random sampling.

In the evaluation of sample results, the confidence coefficient used depends on the size of the sample; however, in the computation of sample size, the confidence factor used does not change with sample size, since size is the variable being estimated.

E. Estimating Universe Variance

The formula requires an estimate of the universe variance to compute sample size. Unfortunately, in practice, the universe variance is generally not known, and hence, must be postulated or estimated from experience or from a preliminary sample. Obviously, the computed sample size will be affected by the postulation or estimation of the universe variance, which is one of the most significant factors in determining sample size. In other words, the formula for computing sample size depends on some characteristic of the universe which is to be estimated through sampling.

Despite the above uncertainties, some ways of estimating the universe variance are by (i) using information from an audit survey, (ii) using results from previous audits in the same or similar areas, (iii) taking a preliminary sample and using its variance as an estimate of the universe variance, and (iv) using audit experience and judgment.

The audit survey method of obtaining information is well known among auditors. If statistical sampling is used in the audit survey, the sample and the results may be treated as a subsample of the larger audit sample. Often, however, the survey is limited to a special part of the universe that is convenient or that is expected to reveal certain problems or conditions. Thus, adjustments should be made for any limiting or selective aspects of the survey coverage. For example, if the survey is restricted to a few areas or clusters, the variance within a cluster may be a gross underestimate of the overall universe variance.

Using values from previous audits in the same area is another way of estimating the expected universe variance. However, necessary data may be limited to the most critical characteristic. If similar prior audits are available, the data from them may require adjustments for time and procedural changes before they are useful.

Finally, a usable estimate of the universe variance can sometimes be made by the auditor based on his knowledge of the operation together with experience and judgment.

F. Tolerable Error and Allowance for Sampling Risk

After an estimate of the universe variance (or standard deviation) has been made, the next consideration is the amount of error tolerance for some specified confidence level. Auditors often specify a higher level of confidence than that associated with the standard error (68 percent confidence level), thereby disregarding other relevant audit evidence in many situations. According to the AICPA guidance in SAS No. 39 [3], if an auditor can assign sufficient reliance on internal accounting control and analytical review procedures, the standard error may provide sufficient assurance for substantive tests.

Also, allowable tolerance or precision may be expressed either as an interval or as an upper or lower limit. It may be expressed in terms of a percentage or proportion, such as an occurrence rate, or it may be in terms of an amount, such as dollars.

If the auditor is sampling to estimate a percentage or a rate, he may specify a range of allowable percentage tolerance for the estimate of the actual occurrence rate at a given confidence (probability) level, say 90 percent. In other words, if repeated samples of the same size were taken from the universe, using the same audit and sampling procedures and care, the computed confidence (probability) intervals of about 90 out of 100 of the samples would be expected to include the actual occurrence percentage.

If the audit objective can be achieved with a one-sided confidence interval, such as a maximum occurrence rate, a smaller sample will result. This is the strategy of monetary-unit sampling in Chapter 12 and flexible sampling in Chapter 13. For instance, the sample evaluation may indicate that there is a 95 percent assurance or confidence that the universe occurrence rate for some characteristic does not exceed 3 percent.

Similar statements are associated with sampling involving variables, such as dollars. For an interval estimate the sample evaluation may indicate that there is 90 percent assurance that undocumented cost for a particular program is within $5,000 of an estimated $80,000. That is, if repeated samples of the same size were taken, the confidence intervals of 90 out of 100 would be expected to include the actual universe value of interest.

G. Confidence Level or Degree of Assurance

The confidence level indicates the degree of assurance (probability) that the results of a sample are reasonable estimates of specific universe characteristics. Whenever one takes a sample he is taking a risk that the sample results may not be reasonable. By specifying the confidence level, the auditor has a means of measuring his risk, which is the converse of the confidence level.

The selection of a confidence level depends upon how much risk one is willing to take of having the universe value lay outside the computed tolerance range or confidence interval. When one speaks of a 90 percent assurance, it is expected that if repeated samples were selected, the actual value would fall within the confidence intervals of about 90 out of 100.

Confidence levels are usually expressed in percentages such as 68 percent, 80 percent, 90 percent, or 95 percent, or in proportions, 0.68, 0.80, 0.90, or 0.95. However, in the formulas for computing sample size, equivalent Z factors are used. These factors or coefficients are discussed in Chapter 14.

H. Illustration of Sample Size Computation

If the maximum allowable tolerance, the desired confidence level, and an estimate of the universe standard deviation are specified, an estimate of sample size for simple random sampling can be computed by applying the above sample size formula.

To illustrate the computation of sample size, suppose the universe standard deviation is estimated to be $10.00, the maximum allowable error is prescribed to be $2.00 and a 95 percent confidence is desired. In Chapter 14, the Z confidence factor for 95 percent is 1.96. Substituting these values in the sample size formula, the results are:

$$\text{sample size} = \frac{(10)^2\,(1.96)^2}{2^2}$$

$$= \frac{100 \times 3.8416}{4} = \frac{384}{4} = 96$$

Thus, based on the specified information, a simple random sample of 100, rounding of 96, would be drawn. See Roberts [57] for other formulas.

There is no need to manually compute sample sizes. There are computer programs and published tables for use in determining sample size.

X. BAYESIAN APPROACH TO SAMPLE SIZE

Bayesian methods take their name from Thomas Bayes, an eighteenth century English cleric, who derived a formula for dealing with prior probabilities.

Bayesian statistical methods, which are controversial among statisticians, incorporate prior information from experience and subjective probability [34] which reflects judgment into the decision-making process. These methods are based on the notion that the theory of probability does not replace audit judgment and experience.

Thus, in the Bayesian decision approach, judgment and experience are permitted in assigning probabilities when dealing with uncertainties. This approach appeals to auditors because their decisions regarding uncertainties are not based merely on sample data, but on other sources of evidence, experience and judgment. In other words, Bayesian methods enable the auditor to use his judgment in a logical manner.

The Bayesian approach to sample size determination emphasizes the relative value of new sample information in the light of evidence already available from other sources of reliance. The value of new information is regarded as the most significant and critical element in determining the sample size. Also, the value of the new information depends on how likely it is to influence an action or an opinion. Obviously, the value of any information is personal and depends on the knowledge, maturity and beliefs of the individual auditor.

In many audit situations it may be difficult to justify a sample that is not expected to reflect a financial impact or to aid in sustaining a significant decision. The use of a sample to estimate an overall error rate which merely reinforces information already gained from an audit survey or other evidence may not be justified, especially if there is little or no financial impact. In other words, the cost of conducting an audit sample must be weighed against the value of the elicited information in influencing an audit opinion or decision.

If an auditor has concluded from an audit survey that an operation is grossly unacceptable, then there is little value in additional information from a sample. On the other hand, if he is not sure about an opinion or a decision, the additional information could be useful, such as assisting in stratification. Usually examples of poor internal control can best be located judgmentally during an audit survey. Also, if the cost of sampling exceeds the value of the reduction in uncertainty about a decision, an auditor should seriously consider the use of other audit techniques. Thus, if an auditor has a strong opinion about an activity or an account, the relative decrease in uncertainty due to additional sampling may be small, and therefore, may not be justified.

Inasmuch as Bayesian statistics are controversial, it is useful to summarize the pros and cons of the method.

Bayesian inference is the optimal statistical method if there is adequate known prior information. Compared with classical methods, the Bayesian approach often yields more credible point estimates and smaller confidence intervals. Bayesian methods are particularly useful in applications where sample sizes are small.

The major criticism of Bayesian methods is that they are highly subjective. Often there is inadequate prior relevant information and assumptions are necessary. However, classical methods also involve assumptions which are often implicit in the techniques.

The AICPA guidance on reliance on internal accounting controls, discussed in SAS No. 39, reflects aspects of the Bayesian approach. Schlaifer [61] provides background, technical and illustrative information on Bayesian methods. Articles by Kraft [33] and Tracy [76] should be of particular interest to auditors and accountants because they discuss Bayesian methods as used in accounting environments. Also, computer programs are available to assist in the formulation of applications and in the evaluation of sample results. Roberts [57] discusses some computer programs. The significant audit aspects of the Bayesian approach are summarized in the following paragraphs.

Auditors often have to make decisions whether to certify a balance or to make further verification to eliminate or reduce the monetary differences between the audited amounts and the recorded book amounts. There are two schools of thought among professional statisticians regarding the best way to use statistical sampling data in the decision-making process. However, notions and ideas from both schools are useful to auditors.

Statisticians of the classical school consider their science to be exact and objective. To avoid comprising this objectivity, classical statisticians may require someone else to specify the desired or tolerable precision and allowable sampling risk, since these are subjective judgment decisions.

By contrast, statisticians of the Bayesian school endeavor to solve decision problems in such a manner as to minimize the cost of making an incorrect decision. If sufficient data are available, the best decision may be determined by a Bayesian analysis, based on the mathematical theory of probability. There is no difference of opinion concerning the mathematics involved in such an analysis; however, adequate data and information for making such an analysis are often unavailable. When this is the case, subjective judgmental information and opinions are quantified for use in the mathematical formulas, on the assumption that these subjective estimates are reasonable and sufficiently adequate for the formulation of a decision.

The classical statistician objects to the Bayesian approach because it is subjective; the Bayesian statistician objects to the classical approach because it ignores other relevant information that may be available in addition to the data from the sample.

XI. SUMMARY

The size of a sample is no criterion of its reliability, and a large sample does not compensate for poor audit procedures. The choice of sampling units and their effective stratification are more significant.

A statistical sample provides a range of probable precisions. The auditor, based on experience and his knowledge, must integrate sample data with other evidence to formulate an audit opinion.

Sample sizes are determined usually by (i) following examples of others, (ii) administrative decisions, (iii) use of formulas or computer programs, (iv) flexible sampling using at least two replicates where each constitutes a minimum sample assuring no findings, and (b) Bayesian methods where the relative value of new sample information is emphasized.

When sample sizes are determined by some mathematical process, the formulas used are generally those originally derived to estimate the precision of the sample results, which is like putting the cart before the horse. Thus, the necessary parameters must be estimated or postulated, and hence, the accuracy of predetermined sample sizes is subject to the validity of these estimates. Information from audit surveys, prior experience, and preliminary sample results can assist in resolving uncertainties of these estimates.

Auditors should be careful in enlarging initial samples, since the mathematical risks are greater than those of a simultaneous sample of the same size with the same findings. It is better to use replicated sampling if one is not sure of the ultimate sample size.

The factors or elements affecting audit sample sizes may be summarized as follows:

1. The degree of reliance on internal control and other sources of evidence.
2. The value of the sample information in achieving the audit objectives, including materiality and sensitivity, and whether attributes, or variables, or both are involved.
3. The variability or rarity of the most critical characteristic or event under investigation.

4. The nature of the sampling unit, whether individual element or clusters.
5. The desired precision or amount of sampling variation to be tolerated.
6. The desired degree of assurance that large dollar "errors" (undocumented costs) are not missed.
7. The risk associated with the desired precision; or conversely, the confidence level involved.
8. The distribution of the items in the sampling frame and the effectiveness of stratification.
9. The order or arrangement of items in the sampling frame.
10. The accessibility of sampling units.
11. The amount and scope of detail data required from the sample.
12. The random selection method, such as simple random sampling, optimal stratification, etc.
13. Kind of estimation procedure used to project or estimate from sample data, such as ratio, difference, etc.
14. The scope and control of nonsampling errors, such as clerical mistakes associated with the sampling process.
15. The rapidity of detecting an undesirable condition or activity.
16. The relative significance of the sample information in the formulation of an audit opinion or action.
17. Possibly the most critical factor is the time and resources available for planning, executing, evaluating and appraising the sample.

When time and resources will not accommodate the computed sample size, the auditor is faced with a number of alternatives, among which are: (i) proceed with a smaller sample with a probable reduction in precision; (ii) resort to a more efficient sampling scheme which may be more complicated; (iii) perform further audit stratification to reduce variability; or (iv) use a flexible sampling plan with at least two replications, each one-half of a minimum sample for the desired precision and confidence, assuming no finding.

The determination of an optimum sample size may be difficult, requiring estimations of sampling costs in terms of time, resources and funds versus the audit value of new information, not merely additional similar information. The value of new information is judgmental and depends on (i) its significance to the audit opinion, (ii) what is already known from other sources of evidence, and (iii) the relative effect of uncertainty on the ultimate audit opinion or decision.

Generally, the audit sample sizes are determined by allocating available resources to competing activities and projects. Thus, the idea of a minimum sample has gained many supporters among auditors. For a detailed discussion of how to determine minimum audit samples, see the chapter on Flexible Sampling.

5
How to Achieve Randomness

1. GENERAL

The manner in which one selects samples from the universe is crucial in statistical sampling, since the sampling process determines the manner in which chance factors affect the relevant statistic, and consequently, affects the sampling distribution of the statistic. In order to derive the sampling distribution of a statistic, one must know the manner in which chance factors affect the choice of items from the universe. In other words, one must draw samples on a random basis in such a manner that one knows from the procedure that chance factors are operating.

A probability, or statistical, sample is one in which chance factors determine which items from the universe will be included in the sample, and the selection of items is made in such a manner that it is theoretically possible to calculate the probability that any specific universe item becomes an item in the sample.

Nonprobability (judgment) samples are selected in such a manner that the probability of an item of the universe being included in the sample cannot be calculated. The use of statistical inference on nonprobability samples is not legitimate if the intended purpose is to generalize to the universe. Some common sampling procedures result in nonprobability samples, such as the selection of individuals from a "stream" of people passing a particular location. Also, nonresponses can result in a technically nonprobability sample, since nonresponses make it impossible to calculate the probability of each item in the universe becoming an item in the sample.

II. RANDOMIZATION

Randomization means that a chance mechanism is used in the selection process. It ensures the statistical validity of the results by eliminating selection biases (systematic selection errors) due to personal preferences of the auditor. However, personal preferences relating to such considerations as dollar amounts or particular kinds of transactions can be accommodated through stratification. Naturally, the audit universe would be delimited to exclude the peculiar items, since they would be handled separately. For example, the auditor may desire to see all transactions of $5,000 or more and those showing a credit balance. Then the delimited universe would be all debit transactions under $5,000.

In random selection it is assumed that there is independence between observations. The concept of independence means that the choice of one sample item has no influence on the choice of any other items for inclusion in the sample.

To illustrate the concept of independence, suppose one desires a sample of married persons but uses a list which contains only the names of married men; thus, when a man is randomly selected from the list, his wife is also included in the sample. However, the concept of independence is violated, since the women in the sample are there only if their husbands are selected. If the selections were independent, each person has a probability of 1/N, where N is the total of all persons, of being included in the sample. However, the probability of a woman being selected, if her husband is not selected is zero, and the probability of a woman being selected if her husband is selected is unity or 100 percent. Hence, neither of these conditional probabilities is equal to the unconditional probability of 1/N. Therefore, this selection process is not independent.

A random sample may be selected by assigning a different serial number (or other identification symbol) to each item in the universe, recording each identification number on a separate slip of paper, placing the slips of paper in a container, and withdrawing the required number of slips from the container, mixing thoroughly before each drawing. This method might work when the universe is small, but it is obviously impractical with a large universe.

Theoretically, a number of mechanical devices could be used to ensure equal probability of selection. Early methods of producing allegedly random digits used dice and shuffled cards. The dice, each labeled 0 to 9 twice, were rolled to produce random digits. Obviously, if this method were used in the presence of an auditee, it might convey the impression that the audit is a big gamble. Also, experience has shown that a reasonably random shuffle of cards may not be easy to accomplish.

Actually the auditor need not go through such an involved process. A more convenient and efficient way to select a random sample is through the use of an approved random digit table. Random digit tables have been published since 1927 including those by Fisher and Yates [17], Tippett [74], Kendall and Smith [29], Interstate Commission [65], and the RAND Corporation [53]. All of these tables have been subjected to various statistical tests and are deemed sufficiently random for selecting audit samples, even though some are better than others. One of the statistical tests ensures that there is approximately an equal number of each of the digits 0 through 9 in the table. Unfortunately, some alleged random number tables found in some publications have not been adequately tested to ensure an acceptable level of randomness. One of the most tested random digit tables is the RAND Corporation, *One Million Random Digits* [53]. It is the largest and probably the most reliable of the published tables.

Sequences of random digits that are produced by a deterministic process are called pseudorandom digits. Pseudorandom digits and numbers can be generated by a computer, thus eliminating the problem of storing a large number of random digits. However, the auditor should assure himself that the computer routine has been adequately tested for randomness. Also an acceptable random digit table may be stored in a computer, but the access routine may be deficient, in limiting the number of digits in the random start number, such as two digits. The auditor should have assurance that the access routine to the computer has no limitations.

The RAND table [53] was generated electronically instead of by a deterministic process, and it is available in punched-card format to facilitate its use in computer programs; however, the random digits would have to be stored which could create an additional expense.

Another widely used random digit table is the Interstate Commerce Commission, *Table of 105,000 Random Decimal Digits* [65]. This table, which appears in Appendix B, will be described in the next section.

Regardless of the method used to generate random digits, only tables or procedures which have been subjected to statistical tests should be used. For descriptions of some statistical test methods see Feller [15] and Kendall and Smith [29]. The RAND table has been thoroughly tested and therefore need not be tested further. Such sources as telephone directories and serial numbers on currency should not be used because they have not been adequately tested for randomness.

In auditing, if a random identification number appears more than once, any subsequent selection of identical numbers is disregarded. This is called sampling without replacement. Aside from being a practical approach, the effect reduces the sampling error, that is, improves the precision of the results. As the sample becomes a larger fraction of the universe, the sampling error is reduced. Obviously, there is no sampling error if the total universe is examined.

Since auditors usually sample without replacement, the assumption of independence is not quite satisfied unless a minor adjustment is made. However, when the universe is relatively large compared to the sample, the auditor can neglect the adjustment. On each successive drawing in sampling without replacement, the probability of an item being selected is slightly increased because there will be fewer and fewer items left unselected after each drawing. However, there is independence from one drawing to the next except that no item can be selected more than once. As a rule of thumb, if the sample size is over 5 percent of the universe, an adjustment, known as a finite correction factor, should be applied to compensate for the lack of replacement. This factor is an integral part of sample evaluation procedures and related computer programs, and thus, need not be considered separately. The impact of this factor is discussed in Chapter 15.

III. HOW TO USE A RANDOM DIGIT TABLE

An inspection of the *Table of 105,000 Random Decimal Digits* (in Appendix B) reveals that the digits are grouped in blocks of five with a space after every fifth row. The table is also divided into 14 columns of five-digit groupings, and has 1,500 rows of digits, resulting in 21,000 groupings of five digits each. However, columns and rows are numbered merely to provide a simple means of identifying any particular point (digit) in the table. Of course, this also facilitates the location of starting and stopping points (digits).

The arrangement of the table and the groupings of digits are solely for convenience and better readability and has no other significance. The table does not consist of 105,000 groups of five digits; the 105,000 refers to the number of individual digits and not the groups. There are only 21,000 groups of five digits (21,000 groups times five digits each equals 105,000 individual digits).

It is improper to always use the numbers in the table formed by the convenient groupings of five digits. Using the table in this manner will preclude many five-digit numbers from ever being selected. For example, if the auditor is using the table to select a random sample from a universe containing 99,000 records, he would need to select five-digit random numbers from the table since the largest number 99,000 has five digits. If he always uses the convenient groupings of of digits as they appear in the table, there would be approximately 78,000 items in universe that would have no chance or opportunity of being selected, since there are 99,000 items in the universe and only 21,000 groups of five digits in the table, or a difference of 78,000, thus precluding any chance of selection for most of the 99,000 records. Also, if the auditor is using the table in this manner and it is known, certain documents could be numbered or arranged so that they would have no chance of being included in any sample. Therefore, it is the starting digit of the necessary grouping which should be randomly selected as will be shown in the chapter on simple, or unrestricted, random sampling.

Random selection, alone, may not be sufficient in the audit environment. The psychological aspect of an audit must be fully appreciated. For example, the surprise element in test checks is often ignored. When making a floor check or selecting sample items for future examination in a concurrent audit, if auditors "telegraph" their intentions or follow a set pattern of operation, as exemplified by their previous activities, the human element aspect of the random process may be missing. The workers in the floor-check area may be on their guard or the auditee may be able to control or modify the sample coverage or content based on his knowledge of the relative audit significance which has been placed on various activities and kinds of transactions.

IV. FEASIBILITY IN RANDOM SELECTION PROCESS

One of the most common complaints from auditors when statistical sampling is used is the amount of time and effort involved in selecting a random sample. The application of any newly initiated procedures usually involves more effort. This is referred to as "start-up time." However, experience gained and more effective use of stratification should significantly reduce subsequent efforts. If the auditor is using a considerable amount of time to select the sample, he is probably not using the most feasible sampling frame or selection method. Also, many auditors are continuing to manually select random samples, although computer programs are available for randomly identifying and selecting samples. When computer programs are used, the auditor should ensure that the sample is actually selected from the relevant universe and in accordance with an approved sample design. Otherwise, he will lose his independence, since the computer could be programmed to preclude the selection of certain records.

One false premise that many beginning samplers have is that the items in the universe must be numbered so that a random digit table can be used. There have been instances in which auditors have gone through a large group of documents they wished to sample and numbered every document. The sequential numbering of items to be sampled is not necessary. There is a method of random selection that does not require the numbering of items. It is described in the chapter on systematic or interval sampling.

V. BASIC RANDOM SELECTION PROCEDURES

There are two basic random selection procedures. One is known as simple, or unrestricted, random selection in which each item is drawn by reference to a random digit table or by use of a computer generator. In the other procedure, called systematic, or interval, selection, the items are selected at or within a fixed or uniform interval after a random start. Systematic selection is based on the premise that the items to be selected are randomly arranged; that is, they do not form a pattern which coincides with the interval size. Ways of avoiding this potential problem are discussed in the chapter on systematic, or interval, selection. All random selection procedures involve the use of one or both of these two basic methods or a modification or a combination of them. The auditor should select the most feasible method for the particular circumstances. Sometimes this may consist of combining manual and computer stratification and selection procedures.

VI. SUMMARY

Statistical sampling is not the expert selection of representative or typical items, cases, areas or periods of time. Instead, it is the selection of sample items by means of an approved random selection device. While expert knowledge, judgment, sincerity and honesty are necessary in the selection of a statistical sample, they alone are not sufficient to assure an appropriate random selection. Although these professional traits are necessary supplements to random selection, they cannot be substituted for it; nor is there any substitute for the random or chance selection of sample items. In fact, the essential difference between judgment sampling and statistical sampling is the random selection process.

Appropriate random sampling procedures not only free auditors from selection biases but assign equal chance of selection to samples of the same size, from the same universe, permitting the formulation of a sampling distribution of estimators, thus permitting the statistical evaluation of the sample results. By contrast, the sampling distribution associated with an arbitrary or judgmental selection procedure is unknown.

Computer programs are available that provide random numbers as well as randomly identify relevant transactions and documents. These programs can conserve considerable audit effort. Some printouts from computer applications will be illustrated in subsequent chapters.

6
Simple or Unrestricted Random Sampling

I. GENERAL

A sample may be any group of items selected from a larger group which represents the totality of all conceivable relevant items. The larger group is often called a universe, which is equivalent to the universal set, implying applicability to every item.

When sampling from an infinite universe, or sampling with replacement, a simple random sample is defined as one selected in such a manner that every item in the universe has an equal and independent chance of being selected.

Also, if the universe is finite and one samples with replacement, the sample is said to be a simple random sample if it is selected in such a manner that every possible sample of the same size has an equal chance of being the sample selected.

If an auditor wishes to draw a simple random sample, there is a need for a complete list of the items in the universe. Then he needs a means of selecting the sample from the list that ensures independence and equal probability in the choice of sample items. The requirement of independence and equal probability are usually achieved by using an approved table of random digits, or some other device which provides random numbers. These identification numbers are then used to select items from the list (sampling frame), for example, a computerized payroll listing. However, a major problem in selecting a simple random sample is the availability of a relevant and complete list of all the items in the universe of interest. In most cases the list is likely to be incomplete, contain obsolete items, or contain duplicates. Sampling frame problems and solutions are discussed in Chapter 2. For more detailed discussions see Kish [32].

In any case, the auditor must be reasonably sure that the frame is an adequate representation of the universe, since the statistical evaluation of the sample applies only to items in the frame. Often the number of items in the target universe exceeds the number of items in the frame. However, any extrapolation from a grossly incomplete frame to the target universe is judgmental and, therefore, has no statistical support.

This chapter will discuss and illustrate one of the two basic random selection procedures, namely, simple, or unrestricted, random sampling. The other basic selection procedure, systematic, or interval, sampling, is described in Chapter 7.

II. SIMPLE RANDOM SELECTION PROCEDURE

Under the simple random selection procedure, each item in the sample is determined by reference to a table of random digits or other tested random selection devices. The procedure for making a manual selection using a table of random digits is as follows:

1. Determine the beginning and ending numbers of the group or universe to be sampled. The largest number determines the number of random digits needed in the selection process. When identifying serial numbers have not been assigned or when the same series of numbers have been assigned to more than one type of document or time period, it may be better to use the systematic selection procedure described in Chapter 7.
2. Turn to any one of the pages in the *Table of 105,000 Random Decimal Digits* (see Appendix B), and looking away from the page, place a pencil point on the page. If the digit on which the pencil point falls and the following digit combine to form a number from 01 to 30 (corresponding to pages in the table) turn to the page in the table identified by this number and proceed as described in the next paragraph. If not, continue pointing at the page (looking away) until an applicable number is obtained. If another table, such as the RAND Corporation table [53] were used, the selection would correspond to the number of pages in the table.
3. Turn to the randomly selected page number between 01 and 30.

4. Select the first digit of the initial random number in the sample by pointing, eyes averted. The other digits of the first random number may be selected in any consistent manner; it is not necessary to select sequential digits, and as previously indicated in Chapter 5, the convenient groupings of five-digits should not always be used. For example, if the items to be audited have three-digit numbers, the two adjacent digits in any direction from the one originally selected may be used to form the first random identification number. Also, every other digit from the original digit may be selected, or some other consistent scheme may be used. Actually the scheme does not need to be consistent, since random digits are being combined to produce random numbers. A consistent scheme, however, is easy to describe and evaluate.
5. Decide in advance the direction to proceed in selecting digits from the table; that is, across the lines, either to the left or to the right, downward or upward in the columns, or diagonally. Any direction is equally valid provided the established criterion is maintained.
6. Following the established direction and grouping, proceed to select consecutively listed numbers in the table. These random numbers will be used for identifying the sample items; however, in selecting random numbers from the table, disregard those which:
   (i) Fall outside the range of the group (universe) being sampled.
   (ii) Have already been selected for the sample (duplicates), or are not applicable to the universe or subgroup (stratum).
7. Continue this selection process until the desired quantity of random numbers is identified. Usually a few extra numbers are identified to provide for inapplicable items.
8. Rearrange the selected numbers in sequential order to facilitate locating the items or transactions in the files, but also retain the original random listing in case it is desirable to reduce or terminate the sample. Any reduction in sample size should be in the inverse order of random selection.
9. If there is a possibility that the entire sample may not be examined, then the documents should be reviewed in the original random order as selected from the table; that is, in the original random order before they were placed in numerical sequence.

III. ILLUSTRATION OF SIMPLE RANDOM SELECTION

An example of simple random selection is presented to illustrate the steps involved. The illustrative universe is reduced to a manageable size. In actual practice, the auditor will be working with much larger universes and sample sizes.

A. Audit Objective. The auditor desires to verify the residual inventory on a terminated contract.

B. Audit Approach. The audit approach is to take a simple random sample of 25 stock record cards from the 600 inventory items on the contract. The audit tests will be made on the 25 records selected.

C. Simple Random Sampling Approach. The documents are numbered from 1 through 600. Since the largest, and hence the controlling number has three digits, the random numbers selected also will have three digits.

   1. It is decided that in selecting random numbers from the random digit table, the selection will proceed downward to the bottom of the page and then continue in the corresponding position on the top of the next page. If the last page in the table is reached, then continue on page 1 since the table is circular.
   2. After a random page is selected, e.g., page 9, then the starting point will be selected by placing a pencil point on page 9 without looking. Suppose the starting point is on line 411 of column 4, second digit, reading to the right. On page 9 of Appendix B, this digit is a one, and the two adjacent digits to the right are 51. Combining these three digits produce the random number 151. The selection will be the first 25 random numbers between 1 and 600 with duplicates, if any, being replaced with another randomly selected number. In this example the random three-digit numbers from the starting point and their disposition are as follows:

Random number	Disposition	Random number	Disposition
151	Selected	190	Selected
501	Selected	050	Selected

(Continued)

Random number	Disposition
641	Not selected (over 600)
974	Not selected
807	Not selected
612	Not selected
250	Selected
226	Selected
587	Selected
408	Selected
528	Selected
930	Not selected
230	Selected
483	Selected
330	Selected
794	Not selected
366	Selected
101	Selected

Random number	Disposition
687	Not selected
407	Selected
720	Not selected
752	Not selected
264	Selected
217	Selected
310	Selected
506	Selected
197	Selected
471	Selected
078	Selected
805	Not selected
472	Selected
450	Selected
812	Not selected
233	Selected

IV. ORDERING RANDOM NUMBERS

It is not necessary to rearrange the selected random numbers in numerical sequence unless it would facilitate the audit. However, the advantage in examining the items selected in their original random order, is that, at any interim point, there is a complete although smaller random sample available. This allows the auditor to make an interim appraisal and, if he feels he has sufficient evidence, to terminate his sampling and resort to other acceptable audit techniques. If the interim appraisal reveals a pattern or is consistent with other sources of reliance and prior experience, continued sampling may not be efficient. On the other hand, if the items are examined in numerical

	000	100	200	300	400	500	600	700	800	900
0000	0018							0718		
1000		1162		1369						
2000	2001		2240	2313						
3000		3146						3737		
4000										
5000		5112	5255							
6000	6040 6070									
7000	7012									
8000										
9000				9346						

Listing of Selected Random Numbers
7012
3146
0718
2240
2313
1162
6040
2001
0018
5112
6070
3737
9346
1369
5255

Figure 3. Summary of numbers selected at random.

```
IF YOU WANT INSTRUCTIONS, TYPE '1'; IF NOT, TYPE 'Ø'.
1 ØR Ø? 1

THIS PROGRAM PRODUCES UP TO 1ØØØ DIFFERENT RANDOM NUMBERS
WITHIN A SPECIFIED RANGE AND PRINTS THEM IN ASCENDING ORDER

IF YOU WANT THE NUMBERS PRINTED FIRST IN RANDOM ORDER AS THEY
ARE GENERATED, TYPE '2'; IF YOU WANT THEM PRINTED IN ASCENDING
ORDER ONLY, TYPE 'Ø'.
2 OR Ø? 2

PLEASE TYPE THE QUANTITY (Q) OF RANDOM NUMBERS YOU WANT, THE
LOWEST NUMBER (L) IN YOUR RANGE, THE HIGHEST NUMBER (H), AND A
FIVE-DIGIT RANDOM NUMBER (R) FROM A TABLE OF RANDOM DIGITS.
Q,L,H,R=? 1ØØ,1,1ØØØØ,761Ø2
```

NUMBERS LISTED IN RANDOM GENERATED ORDER

69Ø1	218	5642	9692	9888	5Ø32	7Ø78	276Ø
9Ø75	51Ø8	6646	8Ø45	9698	1444	2935	7343
4133	972Ø	9722	6738	9862	3884	2589	3944
4942	7251	3637	3149	979	1Ø96	2213	7362
3588	745	5123	8251	369	917Ø	2852	3491
6374	8Ø86	3882	16Ø8	7Ø71	6539	6415	4319
8822	7ØØ4	7586	6379	6841	8978	184	2926
673	72	6874	7155	46Ø7	19Ø8	8487	888
97Ø7	353	772	22Ø6	8892	48Ø4	1271	2645
3171	96Ø3	968Ø	79Ø9	9638	2559	764Ø	8342
7566	4168	8577	6894	1999	4328	5Ø5Ø	4594
4843	6563	3591	3514	3467	9735	7Ø51	7238
6792	7271	4443	8835				

SAME NUMBERS LISTED IN NUMERICAL ORDER

72	184	218	353	369	673	745	772
888	979	1Ø96	1271	1444	16Ø8	19Ø8	1999
22Ø6	2213	2559	2645	2589	276Ø	2852	2926
2935	3149	3171	3467	3491	3514	3588	3591
3637	3882	3884	3944	4133	4168	4319	4328
4443	4594	46Ø7	48Ø4	4843	4942	5Ø32	5Ø5Ø
51Ø8	5123	5642	6374	6379	6415	6539	6563
6646	6738	6792	6841	6874	6894	69Ø1	7ØØ4
7Ø51	7Ø71	7Ø78	7155	7238	7251	7271	7343
7362	7566	7586	764Ø	79Ø9	8Ø45	8Ø86	8251
8342	8487	8577	8822	8835	8892	8978	9Ø75
917Ø	96Ø3	9638	968Ø	9692	9698	97Ø7	972Ø
9722	9735	9862	9888				

Figure 4. Printout of computer generation of 100 random numbers.

order it is necessary to look at all the sample items before making an appraisal and evaluation. Sampling may be terminated at any point when a very unsatisfactory condition is found; however, if the examination has been in sequential order, sample results cannot be inferred to the universe unless the entire sample is examined. If the numbers are rearranged in sequential order and if the sampling is terminated at an interim point, items in the latter part of the universe would have no chance of being included in the sample. It is possible that the condition of the items not examined, through early sample termination, will differ from those tested, because of seasonal or other factors related to the arrangement or location of items.

If it is desired to put the sample items in sequential order, three ways are suggested. One is to write the random numbers on individual cards or slips of paper and arrange them in sequential order. If they also are numbered in the random selection sequence, it will be possible to restore the cards or slips to their original random order if needed. A second method is to record the numbers on a workpaper similar to the one shown in Figure 3. A third method is to use a computer program. Figure 4 is illustrative of a printout from a random generator from one of many available computer programs.

One of the reasons for arranging the random numbers in sequential order is to facilitate the location of the documents in the files. Even when this procedure is followed, it is not necessary that the documents be examined in sequential order. If interim sample appraisals are desirable, the documents must be arranged in the original random order prior to examination.

Another procedure which permits interim audit appraisals without placing the documents back into their original random order is to examine the sample items in subgroups (subsamples). Take the first 50 numbers in their random order, place them in sequential order, and then select and examine the documents. Then do likewise with the next 50 numbers. Appraisals, made after the examination of each subsample, will be valid when applied to the universe. Each subsample is randomly selected from the entire universe, and if all subgroups are of equal size, and there is a separate random start for each group, the process is known as replicated sampling. This procedure is discussed in Chapter 13.

V. PRACTICAL EXAMPLES IN AUDITING

A. Broken Number Series

There are few situations where the auditor encounters a universe in which the items are numbered consecutively beginning with one and continuing through to the last document. Examples of situations that are usually encounted are where (i) there are gaps in the number series and/or (ii) the numbering of documents in the period under audit does not begin with the number one. If there are gaps in the series, any random number selected that falls in a gap is disregarded. For example, stock requisition forms are assigned to two departments in a university conducting federally sponsored research. Each department is given a block of numbers and when one department has completely utilized an assigned block, it is given another block. There will be a gap in the series of numbers which have not been used. In this case, if there were a gap between numbers 7830 and 8000, then all random numbers selected between 7830 and 8000 would be disregarded.

Where the numbering series begins higher than one, the random numbers selected lower than the number of the first document and higher than the number of the last document in the audit period will be disregarded. For example, stock requisition forms are prenumbered and the numbers in the period of audit are from 2,367 through 8,952. All random numbers selected lower than 2,367 and higher than 8,952 will be disregarded.

B. Intermingled Inapplicable Documents

There may be items of no audit interest intermingled with those that are being sampled. For instance, there may be items pertaining to another department or cost center which is not being audited. The simple solution in such a case is to disregard the random numbers which fall outside the limits of the audit universe. The difficulty is that when the auditor reaches the end of his list of selected random numbers, he may not have sufficient sample items since he has disregarded certain random numbers. A better solution is to estimate from experience the proportion of random numbers that may not be useable and to increase the total random numbers selected from the table by this proportion. For instance, in the preceding example on stock requisition forms, assume that it is estimated that one-third of the requisitions apply to overhead accounts which the department is not claiming and consequently are outside the area of audit interest. This means that an average of three of every four items selected would be useable. If a sample of 90 is desired, there would be about 60 useable items in the original list of 90 random numbers. This means that initially 50 percent more random numbers than the desired sample size should be selected. Thus, with a list of 135 random numbers, one-third, or 45, could be disregarded and there still would be 90 applicable numbers left. It should be noted that it is a better procedure to "flag" the disregarded numbers since the auditor may desire to test the unclaimed items for the purpose of obtaining specific details on the nature of the items the department is not claiming as clues to the disposition of other similar costs the department has claimed in overhead. This technique has been used effectively in audit practice.

C. Two-stage Random Selection

Sometimes in accounting records, a new numbering series beginning with one is started at the beginning of each month. For a fiscal period there would be twelve series of numbered documents beginning with one. The sampling approach in this situation could be the addition of two extra digits to the random numbers selected. The first two digits of the random number would represent the month and the remainder of the number would represent the item in that month. On the other hand, if the monthly transactions are kept separate, each month could be regarded as a subuniverse or stratum and a simple random subsample could be selected from each month. This process is known as stratified sampling and is discussed in detail in Chapter 8.

Another method consists of selecting the numbers in two stages. For example, in the first stage a list of random numbers is selected to represent the months. The second stage is the selection of a separate list of random numbers to represent the items. The two lists are combined to indicate the items to be selected.

There are, however, a few rules that must be followed when random numbers are selected in two separate stages and then combined. These rules may be summarized as follows:

1. Select a list of random numbers to represent the months. The procedure is the same whether the first stage is a month, a page, a file cabinet or whatever represents the first stage of selection. The numbers selected are left in the original random order. Duplicates are not eliminated, because the same numbered item may be selected from more than one month.
2. Select another list of random numbers to represent the particular item to be selected from a month. The month having the largest number of items will determine the upper limit of the random numbers selected. For example, if June has 950 items, which is the largest number in any month during that year, random numbers would be selected between 1 and 950. The numbers selected are left in random order, and duplicates are not eliminated.
3. The two lists of numbers, in random order, are placed in correspondence. At this point duplicates of the combined numbers are eliminated; that is, both the month and the corresponding item in the month are eliminated. Also, unusable

numbers are eliminated. For example, the month of February, which may have only 556 items, might be paired with the number 639. Since there is no item number 639 in February, both the month (February) and the item number (639) are disregarded and the next applicable combination of random numbers is used.

D. Identification Numbers Containing Alphabets

Many accounting records have an alphabet as part of the item identification number. When the numerical portion includes no duplications or if there is only one alphabet in the series being sampled, the alphabet can be ignored. However, if there are two or more alphabetical designations and the numerical portion of the identification run in the same series, there may be a problem. One solution is a two-stage selection approach similar to the one discussed above when the numbering series is recycled each month. The first stage is the selection of a list of random numbers to represent the alphabets. The second stage is the selection of another list of random numbers to represent the items. The procedure and rules are the same as indicated where the numbering series is recycled each month.

On the other hand, if the alphabetical designations identify unique departments, activities, locations, or time periods, the alphabet can be ignored in the selection of random numbers. Each alphabet could be used to identify a subuniverse or stratum. Then a simple random subsample could be selected from each stratum. For example, A could be used to identify records in one department or period and B could be used for another department or period.

E. Large Identification Numbers

Often, auditees will maintain a single numbering series for a particular form over several years. Consequently, for the period being audited, there may be large identification numbers within a fiscal period. This would make the manual selection of applicable random numbers laborious since so many simple randomly selected numbers would be disregarded. One solution is to reduce the applicable portion of the number series to a manageable size by eliminating digits that are insignificant to the identification of documents. For example, the same series of numbers has been used on stock requisitions for a number of years. In the period under audit the document numbers range from 76,328 through 84,920. The numbers have five digits and, if five-digit random numbers were selected, the manual selection process would be infeasible since so many numbers would be disregarded. A solution is to eliminate the initial digits, since they are not needed, and select four-digit random numbers. Random numbers selected between 6,328 and 9,999 would automatically be identified as being preceded by the "7" digit and random numbers between 0000 and 4,920 would be identified as being preceded by the "8" digit. The random number selection process would be faster since only four-digit random numbers between 4,920 and 6,328 would not be useable. Obviously, a computer selection procedure is much more desirable under these circumstances.

F. Unnumbered Items

In some sampling areas (universes), the individual documents or items are not numbered. This makes it more difficult to manually apply simple random selection methods. However, if the universe is not too large, so that the average interval between random numbers is small, it is possible to put the random numbers in numerical order and count from one number to the next. If there is a uniform number of items on a page, the average interval can be larger because this will reduce the necessity for counting each individual item, since the "count" may be estimated by multiplying the number of items per page times the number of pages. Also, if there is a uniform number of items on a page, the numerical location of any item can be ascertained. For example, suppose the area to be sampled consists of 50,000 unnumbered items. These are listed on a computer printout containing 1,000 pages with 50 items per page. Thus, item number 23,225 would be the 25th item on page 465. (23,225 divided by 50 equals 464.5 pages or the 25th item on page 465.) This procedure leads to the other basic random selection method known as systematic, or interval, selection which is described in Chapter 7.

VI. SUMMARY

Simple random sampling is a selection scheme which satisfies the assumption of independence as well as giving each item in the universe an equal chance (probability) of being included in the sample. Another way of expressing the same notion is that simple random sampling gives each combination of items of the same size from the same universe an equal chance of selection. This distinguishes simple random sampling from other popular methods of random sampling such as stratified sampling.

The simple random selection procedure involves the use of a table or random digits or some other approved random device, such as a computer selection program, in the identification of each sample item. This selection procedure is often used when identifying serial numbers have been assigned to records.

An auditor is fortunate if the universe he is auditing and his data collection methods permit the use of simple random sampling procedures, since the sample selection, data analysis, and sample precision computations are easy and relatively inexpensive, and can be facilitated by using computer programs.

It is important to realize that in all types of statistical sampling there must be randomization and a fairly complete listing of universe items. In auditing, the complete listing problem is not unique to statistical sampling. This assurance is needed for any kind of audit test-check.

As one will ascertain in the discussions of other random sampling methods, the nature of the required lists (sampling frames) may differ from one sample design to another, with some being much simpler to obtain. The auditor should examine his sampling frames carefully and ascertain how they were assembled and the nature of any deficiencies so that he may make necessary adjustments.

One of the major advantages of simple random sampling is that public relations are easier because of its apparent fairness of selection which is appreciated by the auditee. This method, however, fails to use prior audit information and other sources of evidence in the optimum allocation of audit effort. On the other hand, audit stratification, where the stratified items are considered separately, would still permit the use of simple random selection for the delimited universe. Nevertheless, it is important to recognize when simple random sampling is not appropriate, such as when there is no feasible way to number the universe items. Also, some more complex selection methods are usually much more efficient and effective.

7

Systematic or Interval Sampling

I. GENERAL

In many situations where auditors desire to use statistical sampling, the items in the universe are either not numbered or the numbering system is such that it is not feasible to use simple random sampling, since each item would require an identification number. Also, if the universe is very large, it is laborious to manually use simple random selection methods. However, there is a method of obtaining a random sample under such circumstances. This method is known as systematic sampling because the selection consists of using fixed or systematic intervals after random starts.

Most auditors are familiar with this type of sampling since they have traditionally drawn a fixed-percentage sample, but without the random starts. An example is a sample in which every tenth item is selected. The use of random starting points is the feature that makes this approach a random method of selection, provided the sampling frame is arranged in random order with respect to the universe characteristics being examined. For example, if accounts receivable amounts are being examined, systematic sampling may be used, although the accounts are filed alphabetically, provided the amounts have no relation to the names on the accounts. In other words, if the ordering used in compiling the sampling frame (listing) can be considered random with respect to the characteristics being examined, a systematic sample will be equivalent to a simple random sample. However, to prevent selection bias several random starts are recommended. Each random start identifies a separate subsample, and the combined subsamples produce the overall sample originally planned. In any case, at least two random starts are required.

In systematic sampling, initial sample items are randomly selected and the remaining items are selected by applying a fixed or systematic interval, k. The interval between selected sample items is obtained by dividing the estimated universe size, N, by the sample size, n, and multiplying by the number of separate random starts (rounded down). In other words, in systematic sampling, instead of using a table of random digits or an equivalent computer program to select each item, one simply goes through a list taking every kth item, after a random start.

For a simple random sample greater than one, an unbiased estimate of the sampling precision can be calculated. Since a systematic sample can be regarded as a sample of one independent unit, the above property does not apply. However, an unbiased estimate of the sampling precision can be obtained if two systematic samples, with different random starts and an interval 2k, are selected within each zone (interval).

In most of the examples, small sample and universe sizes are used for illustrative purposes only; it is usually impractical to sample audit universes less than 200 items. Under such circumstances, scanning or some other audit technique may be more efficient. Also, audit samples less than 50 are not recommended because of the nature of distributions of accounting records. If dollars or other variables are being estimated from the sample, an overall minimum of 100 sampling units is advisable. Even then, large dollar and sensitive items should be stratified from the original universe for separate and more detailed examination, resulting in an overall test of more than the minimum sample. Obviously, if the universe is very small and there are other sources of evidence to support the audit opinion and decision, these guidelines may be modified.

For illustrative purposes, suppose a random sample of seven items is to be selected by systematic sampling from the universe of 25 expense vouchers in Figure 5, which have been sorted by expense voucher number. Using two random starts, the selection interval would be 6 ($25 \div 7 = 3.57$, rounded down to 3 and multiplied by $2 = 6$). To determine the starting points, a random digit table such as Appendix B is used.

Using that table, looking away from the page, point a pencil to establish the starting point; the third digit on line 57 of column 5, and reading down, the first digits between 1 and 6 are 3 and 5.

Name	Voucher Number	Amount
Wolfe, L.	01	$1,050.30
Thompson, T.	02	612.50
Gross, J.	03	335.65
Rogers, W.	04	681.25
Davis, T.	05	350.00
Ford, H.	06	1,350.68
Page, A.	07	768.17
Hanson, F.	08	700.50
Oden, C.	09	1,196.84
Williams, E.	10	250.65
Lee, S.	11	1,279.75
West, J.	12	875.63
Rowe, G.	13	306.00
Wallace, R.	14	1,040.20
Smith, D.	15	234.96
Bonds, J.	16	1,350.50
McCarthy, D.	17	234.08
Starr, C.	18	1,050.53
Polk, D.	19	746.34
Litton, S.	20	292.56
Fox, P.	21	350.75
Dyson, A.	22	1,150.37
Howe, R.	23	933.09
Carter, W.	24	1,933.72
Newman, G.	25	350.33

Figure 5. Twenty-five expense vouchers.

Therefore, the first voucher in the first subsample would be voucher number 03. The next voucher selected for this subsample would be number 09. The items selected for the two subsamples would be:

	Voucher number	Amount
1st subsample	03	$ 335.65
	09	1,196.84
	15	234.96
	21	350.75
2nd subsample	05	350.00
	11	1,279.75
	17	234.08
	23	933.09

Eight vouchers have been selected rather than the desired sample of seven. Often a slightly increased sample size results when the selection interval is rounded down to an integer. However, one should not stop the selection process when the desired sample size is obtained unless the entire universe has been covered. Otherwise, the tail end of the universe will have no chance of being included in the sample. It is better from an audit point of view to use the eight vouchers rather than endeavoring to reduce them to seven.

II. ESTIMATING UNIVERSE SIZE AND SELECTION INTERVAL

In using systematic sampling it is often necessary to estimate the universe size in order to determine the sampling interval. An estimate, however, is unnecessary if there is a total or exact machine-count of the universe. The estimate may be made by counting the line items on a few pages in a printout to determine the average number of items on a page and multiplying this average by the number of pages in the universe of transactions.

In simple random sampling, it is conservative to overestimate the universe in determining sample size. However, for systematic sampling it is better to underestimate the universe size since the selection interval is determined by dividing the actual, or estimated, universe size by the desired sample size. Underestimating the universe will result in a larger sample, but it is easier to reduce a systematic sample than to increase it. For the same reason, any sampling interval fraction is always dropped. This results in a slightly smaller selection interval and consequently the number of sample items identified is slightly larger. In many audit applications it is better to use the few extra sample items than go to the trouble of properly eliminating them. For example, it is estimated there are approximately 11,500 supply requisitions during the year. A sample of 120 items is desired. Dividing the estimated universe size of 11,500 by the sample size of 120 results in an interval of 95 5/6. The 5/6 is dropped and an interval of 95 is used. This would result in about 120 (if 11,500 is accurate, it would be 121) sample items. In this case, use the 121 because 120 is not that "pure," since sample results are evaluated after the items are examined and analyzed.

III. CHANGING THE SAMPLE SIZE

There may be a desire to examine a different number of items than those originally selected or a significant error may have been made in estimating the universe size. It is somewhat more difficult to increase the sample size than to reduce it when using systematic sampling. It is better to provide for the selection of slightly larger number of sample items in the sampling plan.

If the sample selected is too small, it is not necessary to discard the sample items already selected and draw a completely new sample. The number of additional items is determined and a new systematic sample can be selected using the number of additional items to supplement the original sample. This means determining a new interval and random start; do not expand the original sample because it will become a sequential sample with different probabilities. Also, additional items may be selected by using a random digit table, particularly if there are only a few and the records are numbered. It should be noted that it is easy to estimate the universe size after an initial systematic sample has been identified. This is accomplished by multiplying the resultant sample size by the size of the selection interval. Any error in estimating the universe size using this technique would not exceed the size of the selection interval. This method of estimating the universe size is another advantage of systematic sampling. For example, from a universe of 1,000 time-and-effort reports a systematic sample of 100 has been selected and it is desired to increase this to 200 by selecting another 100 reports. The approach is to take a new systematic sample of 100 and combine it with the 100 reports already selected. The selection interval would be nine and a random start between one and nine would be selected, since only 900 cards are left after the initial 100 are selected. A better procedure is to initially plan for a larger sample in case it is needed. Also, more than one random start is desirable as one method of avoiding the potential problem of a systematic arrangement of records coinciding with the selection interval. Two or more independent subsamples of the same size also furnish intuitive audit information through comparison of ratios, proportions, and rates among the subsamples.

If it is desired to decrease a systematic sample, it is not proper to merely disregard the last items selected. Similarly, it is not valid to discontinue the systematic selection of items until the entire universe has been covered. To do either of these would preclude the testing of the latter part of the universe. A proper procedure, after all the sample items have been selected, is to reduce the original sample by selecting a subsample from it using either a systematic or simple random sample. If the reduction is small the subsample may be used to identify excess items. If the reduction is substantial, the original selection interval could be doubled. This would reduce the sample size by one-half. For example, because the universe size is incorrectly estimated, a sample of 300 items is selected when only 200 items are desired. One method of reducing the sample size is to select every third item from the original systematic sample after a random start between one and three. Another method is to select 100 random numbers between one and 300 and identify the items in the original sample corresponding with these numbers. Those 100 items identified would be disregarded.

The techniques of reducing sample size can also be used in making interim sample appraisals. Usually it is not possible with systematic sampling to make interim appraisals, applicable to the entire universe, if every item is examined in the order of selection. The entire systematic sample must be examined to cover the universe. However, if, for example, a subsample is taken by examining only every third item or so of the larger sample, after a random start, an interim appraisal could be made after the subsample had been examined. However, the selection of three independent subsamples of the same size would be preferable. Each subsample should be large enough to stand alone in case a decision is made to stop sampling and resort to other acceptable audit techniques. In other words, each subsample should be a minimum sample for the desired precision.

Also, an "odd ball" or unfavorable condition may manifest itself more readily in a subsample than in the larger overall sample. No sampling procedure should prevent the auditor from terminating his testing at any point where a very unfavorable condition is found. This implies that the unfavorable condition is significant enough to preclude the need to draw any inference as to the probable extent of the condition in the entire universe. The condition may be obviously associated with only one operation or department. Thus, the auditor may decide to do a detailed examination of that operation or department.

IV. FOREIGN ITEMS IN THE LISTINGS

Many accounting listings include items that are not part of the audit universe, that is, "foreign" to the universe of interest. These may be items that have been identified for a 100 percent review, items of another operation not being audited, or items with zero balances such as paid-off loans.

If these items are confined to a few blocks, they do not present a problem since they may be skipped. The problem occurs when "foreign" items are intermingled with the items being sampled. One solution is to disregard a "foreign" item and proceed to the next selection point. Using this approach, the determination of the size of the selection interval must take into consideration that some items will not be useable. Otherwise, the selected sample will be smaller than desired because of skipped items. For example, a sample of Medicare patient records is to be selected for review, however, the machine listing which contains Medicare patients also contains private pay patients which are intermingled throughout the printout. The listing contains a total of 10,000 patients which are estimated to be nearly equally divided between Medicare and private pay patients. If a sample of 100 is desired, the selection interval should be 50 instead of 100. This would result in about 200 patient records being selected, of which about 100 records should be for Medicare patients.

A more feasible solution, for mechanized records, however, is to use the computer to "purify" the sampling frame by remove or stratifying the foreign items from the original listing. The computer could also select a random sample from the modified or stratified sampling frame.

V. CIRCULAR SYSTEMATIC SAMPLING

Systematic sampling has been further refined as linear systematic sampling and circular systematic sampling.

Linear systematic sampling consists of randomly selecting an initial item from among the first k items, and every kth item thereafter until the list (sampling frame) is exhausted. A better alternative is called circular systematic sampling. In using this procedure, a random start from 1 to the size of the sampling frame, N, is selected, and every kth item thereafter, where k = the largest integer less than or equal to the frame size divided by the sample size. Since the items in the frame are considered in circular arrangement, when the end of the list is reached one goes to the beginning. For example, if the selection interval, k, is 500 and the sampling frame, N, is 20,000 and the random start is 18,731, the identification of the next four randomly selected items would be 19,231 (18,731 + 500), 19,731 (19,231 + 500), 231 (19,731 + 500 = 20,231 − 20,000), 731 (231 + 500). Circular systematic sampling is preferable to linear systematic sampling since it has stronger theoretical statistical support as indicated by Sudakar [67].

To illustrate circular systematic sampling, suppose the audit approach is to select a sample of 20 stock record cards from a group of 140 to verify the residual inventory on a terminated contract.

Using two random starts, the selection interval would be 14 (140 ÷ 20 = 7 multiplied by 2 = 14). To determine the two starts, a random digit table is used. In circular systematic sampling the random starts can be any number between one and the size of the universe, 140. Suppose one random start is 049 (line 50, column 3, second digit, reading to the right, from Appendix B), and the other random start is 112 (line 83, column 10, fifth digit, reading to the right).

The next number after the first random start would be 63 (49 + 14, the interval), followed by 77, 91, 105, 119, 133, followed by 147. Since the universe is only 140, and considered circular, the next useable number would be 7 (147 − 140). In this procedure the random start should not be lapped.

The items selected for the two subsamples would be:

	Card number		Card number
1st subsample	49	2nd subsample	112
	63		126
	77		140
	91		14
	105		28
	119		42
	133		56
	7		70
	21		84
	35		98

Another variation of circular systematic sampling is to select random numbers between 1 and the size of the universe, N, and take every kth unit thereafter, going in both directions from the random starts. According to Cochran [8] this is theoretically a better approach.

VI. ADVANTAGES OF SYSTEMATIC SAMPLING

A systematic sample is easier to select and execute without a mistake than simple random sampling, especially when the frame is a tape, computer listing, punched cards, or files of ledger cards or vouchers. The items in the universe need not be numbered; only starting random numbers are required.

In many audit sampling applications, the universes not only consist of unnumbered items, but are very large. This makes the use of manual simple random sampling impractical. Also, very large unnumbered universes could make the use of manual systematic sampling increasingly more difficult and laborious to manually count from one sample item to the next as the selection interval becomes larger. Solutions for these situations, as well as other advantages of systematic sampling, are discussed next.

When documents are of somewhat uniform thickness or lines are uniformly spaced on a listing as on a computer printout, the selection of a sample may be made by measurement rather than by counting. The procedure is similar to the variations in systematic sampling previously discussed. The essential difference is that the universe size and size of the selection interval are stated in terms of measurement, such as inches, rather than by numbers of items. Using this approach, a sample selection can be accomplished easily from trays of cards or from voluminous computer listings. However, care must be exercised to assure that all items in the universe are present when selecting from trays of cards or files. When the universe records are measured, the auditor should be careful to equally compress the documents when applying the selection intervals. If files of unequal thickness are being sampled, the auditor should be aware that the thicker files have a better chance for selection.

Measurement selection or other shortcuts may result in substantial savings in time when used in manual systematic sampling where the universe is not numbered. For example, consider a computer printout that has 50 lines per page and 300 pages, totaling 15,000 unnumbered line items. If a sample of 300 line items is desired, it could be obtained by selecting one line per page. After random starts are determined and located, a strip of cardboard the length of the interval could be used to locate the subsequent sample line items. This is a very effective procedure if there are varying numbers of usable line items on the pages. For instance, if the selection interval is longer than the number of usable line items on a page, measure off page line intervals on the cardboard strip. If the cardboard strip is carefully placed, the results should be the same as from a complete counting of the universe. In this case, it is helpful to paste a strip of printout paper on the cardboard. This facilitates skipping blank and unusable lines until the desired interval is accumulated. With this scheme, stratification can be performed simultaneously with the random selection process. This stratification feature makes manual selection competitive with computer selection. Obviously, computer selection is faster, but stratification for audit purposes may be inhibited since the intervening items are not observed as in manual selection.

Measurement methods are also very useful where universes are not listed but are arranged on cards or in file cabinets. It is necessary, however, that all files be of somewhat similar thickness, otherwise a thicker file has a better chance of being selected. This situation may be avoided if the interval selection concentrates on individual documents rather than file folders, resulting in the selection of more sample items from the thicker folders.

If the selection interval is smaller than the number of line items per page, several items per page may be randomly selected from a table of random digits, and the cardboard strip used to locate the item without counting. For example, suppose that there are 50 pages of 50 lines each, and a sample of 100 items is desired, the auditor would need to randomly select two lines per page. For each page he could select two random numbers from 1 through 50, and then use a cardboard strip to locate the sample items on each page. The process goes quickly if the line spacings on the cardboard measuring strip have been prenumbered.

Systematic sampling is also simpler than simple random sampling whenever a list is extremely long or whenever a large sample is selected. One can imagine the problem of locating the 1,587th, 12,913th, and 68,125th items using simple random sampling. On the other hand, if these records are computerized, a computer program can accomplish the feat quickly. Nevertheless, many series of records to be audited are not computerized.

In addition to simplicity and time saving features, systematic sampling has the inherent feature of achieving a proportional coverage of time periods or other segments across the universe. For instance, where lists are in alphabetical order, taking every kth item is likely to give a proportionate representation (stratification) of each alphabetical grouping. For example, assume that the auditor has used systematic sampling to select a sample of 120 vouchers from a year's listing containing 12,000 vouchers. The monthly volume ranges from a low of 700 vouchers processed in February to a high of 1400 vouchers processed in December. The sample will contain approximately one percent of each month's transactions. This means that there should be twice as many vouchers randomly selected from December as from February because of the doubled volume of transactions in December.

This aspect of systematic sampling has an intuitive appeal, since it stratifies the universe into as many strata as there are sample items. Thus, the systematic sample is expected to be almost as precise as the corresponding stratified random sample with one item per stratum. The difference is that with the stratified random sample the position of the items in the strata is determined separately by randomization, whereas with the systematic sample the items occur at the same relative position in each stratum. However, the systematic sample is spread more evenly over the universe and thus may be more effective in many instances than stratified random sampling.

In stratified random sampling the universe is first divided into nonoverlapping subuniverses called strata. Then independent random samples are selected from each stratum. This is a common sampling method which is discussed in Chapter 8.

Systematic sampling is more precise than simple random sampling if the variance within the systematic samples are larger than the universe variance. In other words, if there is little variation within a systematic sample relative to that in the universe, successive items in the sample are repeating more or less similar information.

Also, systematic sampling is more efficient than simple random sampling because it eliminates autocorrelation in the sample, that is, similarity of adjacent units or items. However, if autocorrelation is significant, stratified random sampling is generally preferable. On the other hand, systematic sampling may be more efficient in forming implicit stratification which may not have been obvious initially. Another advantage, according to Kish [32] of systematic sampling is that it is better to use a fixed sampling rate or fraction rather than a fixed sample size. The sampling fraction is the proportion of the universe (sampling frame) items which are included in the sample. Since the determination of the exact size of the universe may be difficult or impractical, it is usually better to fix the sampling fraction and let the sample size vary, rather than fix the sample size and let the sampling fraction vary. If the sample size is fixed while the sampling fraction varies with the size of the universe, or stratum, needless weighting problems complicate the sampling procedure.

VII. THEORY AND SYSTEMATIC SAMPLING

There is a theoretical problem that the sampling distribution of a systematic sample consists of only k possible samples, where k is the size of the sampling interval. This theoretical problem, however, has a theoretical solution. The actual listing order of the items (sampling frame) is but one chance situation among many potential permutations of these items. To illustrate, imagine 100 5-drawer file cabinets of records along a wall. There are many ways in which those records may be accessed. One does not have to begin with the front of the top drawer of the first cabinet on the left. The cabinets may be actually or mentally rearranged. The beginning may be at the back or any other position of a drawer. Also, the initial drawer in a cabinet does not have to be the top drawer.

In fact, W. G. and L. H. Madow [39] proved that if the order of the items in a specific finite universe can be regarded as selected at random from the N! permutations, systematic sampling is on the average equivalent to simple random sampling. N! denotes N factorial which is defined as $N! = 1 \times 2 \times 3 \ldots N$, where $0!$ is assigned the value one.

VIII. PRECAUTIONS WHEN USING SYSTEMATIC SAMPLING

The major disadvantage of systematic sampling is that the precision of the estimates cannot be properly interpreted without an assumption regarding the relation between the way in which the universe is ordered and the characteristics being investigated. This assumption, however, can be avoided without increase in effort of selecting and analyzing the data if more than one random start is used. Thus, the minimum number of random starts should be two.

The auditor should be alert for two types of situations which could cause possible difficulties when systematic sampling is used. Fortunately, if either occurs in financial applications, it is easily recognized.

A monotonic trend may exist in the ordered universe list. An illustration would be home mortgages granted over a long period of time where the cost of homes and mortgages have steadily risen over the years. Thus, the mean of a systematic sample of the amount of outstanding loans could be greatly affected by the choice of random numbers, since a sample consisting of the first item in each interval (zone) could be much smaller than for a sample of the last item in each zone. In other words, a monotonic trend induces a variation among the possible sample means.

Also, a trend may be induced by the ordering of the items in the universe. For instance, if persons have been listed according to position or seniority, the result may be affected by the location of the random starts. Another example might be a payroll listing in which every 50th employee listed happened to be a supervisor. If the sampling interval also is 50 and the random start coincided with a supervisor, then all the employees selected would be supervisors.

The second type of situation to be avoided is that in which the listing has some periodic or cyclical fluctuations which correspond to the sampling fraction.

If the universe consists of a periodic trend, the effectiveness of the systematic sample depends on the value of k. The least favorable situation is one in which k is equal to the period of the trend or is an integral multiple of the period. In such cases, the sample items are similar, so that the sample is no more precise than a single random item. To avoid this pitfall, one could change the sampling fraction slightly, or better, use several random starts.

By contrast, the most favorable situation occurs when k is an odd multiple of the half-period, and thus has a mean equal to the universe mean. In such a case the sampling variance of the mean is zero. Between these two extreme situations there are various degrees of effectiveness of systematic sampling, depending on the relation between the sampling interval and the periodic trend.

Many universes exhibit periodic trends. Examples are store sales over various days of the week and traffic flow on a specific road over 24 hours. In these instances, a systematic sample at the same time on the same weekday would not be wise. However, when the periodic variations have been analyzed, a systematic sample usually can be designed to capitalize on it. Otherwise, when a periodic effect is suspected but not well known, a simple random or stratified random sample may be preferable.

It should be emphasized that only a systematic arrangement of the characteristics being tested would be of any concern. The universe may be arranged in any manner and it will not affect the sample as long as the characteristics being measured do not occur systematically and coincide with the sampling interval and starting points. For example, travel vouchers are to be tested for the allowability of costs. The vouchers could be arranged alphabetically by traveler, by travel order number, by check number or any other arrangement and it would not affect the randomness of the sample unless there were some pattern to the undocumented costs.

It is highly improbable, however, that the auditor's selection interval and the random starts would coincide exactly with any pattern in the universe being sampled. Nevertheless, the auditor should be aware of this possibility. If there is reason for concern or if it is desirous to avoid this possible difficulty, circular systematic sampling with several random starts may be used. In such cases the list is considered to be circular, so that the last item is followed by the first.

If the universe size, N, is not an integral multiple of the selection interval, k, a problem may arise. Two solutions to this problem are:

1. Use circular systematic sampling so that the last item on the listing is followed by the first. Choose random starts from 1 to N. Then apply the appropriate interval until exactly n items are chosen, going both forward and backward from the random starts, and when the end of the list is reached continue at the beginning. Any convenient interval will result in an equal probability selection of n items selected with the probability n/N. Generally, the integer closest to the ratio N/n will be most suitable. This procedure has great flexibility and can be applied to many situations.
2. Eliminate randomly, with equal probability, enough items to reduce the listing to exactly nk before selection with the interval k.

Some other alternatives for avoiding possible problems of systematic sampling are discussed next.

Randomizing the universe ordering through some shuffling procedure would permit the use of systematic sampling; however, this could destroy the beneficial aspects of implicit stratification.

It has been mentioned that there may be a remote possibility of a systematic arrangement of the characteristics being audited. Where the auditor desires to use a systematic sample and suspects that there may be a pattern or trend in the universe, there are a number of variations of the basic systematic procedure that will introduce randomization.

One method is to *randomly vary the sampling interval* instead of maintaining a fixed or systematic interval. This method requires that the interval be computed in the usual manner and the initial sample item be randomly selected as in the basic process. However, instead of using the uniform interval, a random number is selected between one and one less than double the size of the uniform interval. This random number then becomes the size of the next interval. This procedure is followed after each item is selected. The average size of the random intervals will approximate the size of a corresponding uniform interval and thus approximately the same number of sample items will be selected from the universe. For example, a sample of 100 is to be selected from a universe of 1,000 records. Using the basic systematic procedure, the uniform interval is computed at 10. A random number from one through 10 is selected for the initial sample item. In this case say the initial random number is 7. Next, rather than counting from this point and making the selection using the uniform interval, a random number between one and one less than double the size of the uniform interval, of 19, is selected. Suppose the random number selected is 14. The second item chosen in the sample would be the 14th item after the initial random item 7, or the 21st item. Then another random number from one through 19 is selected to determine the interval to the third random item. This procedure is repeated until the universe has been covered. If desired, a list of random numbers from 1 through 19 could be prepared in advance to indicate the varying interval sizes. However, the numbers should not be rearranged in numerical order nor should duplicates be eliminated.

Another method is similar to the preceding one, except that the universe is divided into equal size segments or zones. The number of items in each segment or zone is determined by the value of the uniform selection interval. Then random numbers equal to, or less than, the uniform interval are selected. These numbers indicate which item in each segment or zone is to be selected. For example, using the same information as in the preceding example, a sample of 100 is to be selected from 1,000 documents. The uniform or systematic interval is 10. A list of random numbers between 1 and 10 is selected. The list is left in its original random order and duplicates are not eliminated. For illustrative purposes, say the first three random numbers are 7, 2 and 5. In the first zone of ten items the 7th would be selected; in the second zone of ten items the 2nd, or item number 12, would be used; and in the third zone of ten items the 5th, or item number 25, would be chosen. This procedure is continued until the universe has been covered.

An adaptation of the above methods is the "multiple random start uniform interval" procedure. Under this procedure a new random start is selected each day or some other convenient time period such as each week. The frequency of changing the random starts would depend on the volume of transactions. Like the two previous methods, it has the advantages of not only avoiding potential problems associated with a pattern or trend in the universe, but of eliminating possible ramifications inherent in the discovery of any continuing fixed interval pattern of selection by the auditee.

The above procedures for randomizing systematic sampling, together with some other practical approaches, may be summarized as follows:

1. Random selections within intervals (zones).
2. Change the random starts several times to reduce the effect of any single start, yet retain most of the practical benefits.
3. Replicated selection of several different subsamples using systematic procedures for each subsample.
4. Randomly select pairs within implicit strata (zones) of size 2k items each for selection intervals of k.

If the information being sampled is on punched cards, magnetic tape or drums, it is possible to use computer programs to make the random sample selection and printout the sample listing in a prescribed order and arrangement. The computer programs to accomplish this are simple and require very little machine time. Computer random selection with a printout is a very useful and feasible approach when samples are selected during the normal processing of the transactions. These programs are also useful for stratifying by dollars and other characteristics such as accounts, activities, departments, and so on.

When using computers for sample selection, regardless of the method of selection, it is expedient, pertinent and often necessary to identify three or four times the number of sample items needed for a minimum or preliminary sample. This procedure precludes the extra time and expense required to rerun computer programs to obtain additional sample items which may be desired after an audit appraisal of the preliminary sample. This is particularly true when the sample is being selected concurrently with the normal processing of accounting data. This same technique of identifying, but not necessarily examining, a larger sample also is advisable when manually selecting a sample using the systematic sampling approach.

IX. SUMMARY

Systematic sampling is perhaps the most widely used random selection procedure among auditors, since it closely resembles the traditional fixed-percentage audit sample. The prime reason for using systematic sampling is that it is easy, simple, flexible and economical.

Another advantage of systematic sampling is that it can easily yield a proportionate sample because of its even spread over the universe. Thus, whenever stratification exists in the ordering of the sampling frame, a systematic sample will reflect it.

Systematic sampling permits the easy selection of samples from universes of unnumbered items by use of measurement procedures.

The method also provides a basis for estimating the size of the universe. For example, if a selection interval of 100 provides a sample of 366, an estimate of the universe size would be 36,600 (100 × 366).

Systematic sampling is the basic selection process used in monetary-unit sampling which is described in Chapter 12.

In most audit situations, systematic samples may be more efficient than simple random samples and also compare favorably in precision with stratified random samples. The effectiveness of systematic sampling, in relation to that of stratified random sampling or simple random sampling, is greatly dependent on the properties of the universe. There are universes for which systematic sampling is more precise and others for which it is less precise than simple random sampling. Thus, a knowledge of the universe is necessary for the most effective use of systematic sampling. This is also true for stratified random sampling.

Systematic samples can be used under the following circumstances:

1. When the universe items are essentially random or contain only mild ordering or possible cyclical arrangement.
2. Where numerous strata are used and an independent systematic sample is selected from each stratum. As an alternative, use one-half as many strata and select two systematic samples, each with an independent random start, from each stratum.
3. Where subsampling is employed.
4. Where replicated sampling is employed.

If an auditor finds evidence of periodicity in a universe, he is not advised to retreat to simple random sampling. In fact, the more that he knows about the structure of the universe, the more he is expected to benefit by taking advantage of this knowledge, choosing appropriate techniques in the systematic selection so as to avoid the periodicity.

8
Stratified Random Sampling

I. GENERAL

Simple random sampling provides a sample which assures that each item in the universe has an equal chance of being selected. One of the limitations, however, is that a large sample may be required to yield reasonable estimates.

In simple random sampling the items are randomly selected in the order that they appear in the sampling frame, without any rearrangement. The auditor's intuition and knowledge, however, suggests that some segregation or separation of sampling units is needed for more extensive testing of large dollar and sensitive items to provide extra protection against not detecting a significant deviation or error.

Stratified random sampling is one means of not only providing additional audit protection, but of improving the precision (reducing the sampling error). The procedure involves the segregation or separation of the sampling frame into subgroups or strata (layers) with similar characteristics, and then randomly selecting separate, independent samples from each strata. Stratification is also one way to use prior knowledge and experience to obtain a better estimate. Natural stratifications, such as geographical separations, may exist in the frame, or artificial stratifications may be mathematically or judgmentally induced.

Stratification is one of the more significant concepts in audit testing. It has been understood and practiced more or less intuitively by auditors for a long time, much before Neyman's 1934 paper [50]. That paper was one of the first attempts to formalize the underlying rationale of stratified sampling, and it emphasized random sampling as opposed to judgmental selection of the sample items.

If the auditor knows nothing about the universe structure except its approximate size, it is best to select a simple random or systematic sample. It is very rare, however, that an auditor knows nothing about an accounting universe. He knows from his survey that the universe consists of different kinds of transactions and values which are likely to show marked differences in characteristics. Thus, he should endeavor to use this knowledge and other information about the universe to improve the efficiency of the sample design.

Often, supplementary information about the universe permits the auditor to separate it into a number of groups or layers, and to select a random sample from each group, called a stratum. This is called stratified random sampling. This procedure usually increases the precision of the sample result. The degree of improvement, however, is dependent upon the skewness of the distribution of the universe and how skillfully the stratification is performed.

Stratification tends to isolate or separate the more extreme possibilities which may occur under simple random sampling.

Auditors have often applied the general notion of stratification without knowing its statistical advantages. When the auditor pays more attention to those transactions with high dollar values than to those with low values, he is implementing the idea of stratification.

Auditors may classify items and sort them into groups such as strata or clusters. However, these classifications merely create categories of memberships, rather than inherent relationships. These relationships which already exist among items of a universe are ignored when simple random sampling is used. These relationships are not only important for subsequent audit analysis and appraisal, but they can be used to improve the efficiency of the sampling plan.

The selection of the sample is necessarily affected by the relationships that exist among the items in the universe. These relationships occur in the records and in the accounting systems. For example, there are different systems for physicians and the health service agencies that treat some of the same patients. However, there are usually significantly different relationships among the operations and characteristics of these systems, which suggest some type of stratification of accounting records prior to audit testing.

Traditionally, auditors have separated or reconstructed items of a test area to improve testing efficiency. This is done regardless of the type of audit test or examination to be performed. When an auditor properly separates items for test purposes, his basis of separation in most cases would be equally valid for statistical stratified sampling, since knowledge and information of the universe is necessary for effective stratification under any circumstances.

If a frame is separated into at least two strata, and from each of these strata separate, independent random samples are drawn, the resulting weighted sample is a stratified random sample. Thus, a stratified random sample is basically a pooled group of random subsamples. Incidentally, those items which were previously separated from the original universe for audit stratification as described in Chapter 3, usually are kept apart, and hence are not included in the pooling or weighting process.

After the strata have been determined and each stratum sample size has been specified, the random selection is performed in exactly the same manner as in simple random sampling, except a separate and independent sample is selected from each stratum. Thus, simple random sampling is a special case of the more general concept of stratified sampling, where there is only one stratum in the universe.

All accounting systems reflect some natural stratification such as months, types of accounts, locations, activities, operations, etc. However, in considering stratified sampling, an auditor must decide whether any rearrangement of the material (frame) would be worth the extra cost and effort of the rearrangement.

Also, it is worthwhile to retain, in each stratum, the sample items in the original order in which they appeared in the frame. This will preserve the benefits of any natural stratification.

Before discussing guidance on the application of stratified sampling to auditing, it is useful to discuss briefly the principal reasons for using stratified sampling.

II. REASONS FOR USING STRATIFIED SAMPLING

The principal reasons for using stratified sampling may be summarized as follows:

A. Administrative Convenience

Frequently it is more convenient to sample from natural strata than it is to treat the entire universe as a single group. Suppose, for example, the nationwide error rate for a health service activity is to be estimated. A simple random sample from across the country could present some difficulties. It would be easier and more convenient to separate the universe into strata, having one stratum for each field office. Each office could supervise and perform the sampling for its portion of the universe.

B. Separate Information About Strata and Universe

Strata may be formed because separate estimates are desired for subuniverses or domains. A domain is a part of a universe for which separate estimates are planned in advance. For example, a nationwide audit of nursing homes may be planned such that separate estimates are published for each state. In such a case, it helps to treat the states as strata with separate and independent selection from each state. In some states the sampling rate may have to be increased to provide the desired precision for that state's separate report. The stratification process could be extended further to identify estimates for districts or counties within states. For another example, the audit objective might involve not only estimating an overall company error rate in accounts receivable, but the error rate in each company office. Also, when an auditor selects a simple random sample from the entire universe, he cannot control the sample size within each stratum. With a stratified random sample, relative representation of each stratum in the sample can be assured. This permits the auditor to also impose different precision requirements on different strata or domains. For example, he may want more precise estimates for large accounts than for small ones. To accomplish this, he would vary the strata sizes to obtain the desired strata precisions.

C. Accommodation of Different Techniques

Sampling problems may differ significantly in different portions of the universe, such as the operations and accounting systems of large companies compared with those of small firms. Substrata also may be formed to accommodate different methods and procedures within strata. Additionally, it may be desirable to employ different sampling methods or audit techniques in various portions of the universe. For example, consider the selection of a sample of health service employees in a particular state. The headquarters employees may be sampled as individuals, whereas the employees scattered throughout the state may be sampled as clusters to save travel time and cost. For another example, the audit objective may require separate information about direct employees and indirect employees. This is accomplished conveniently by separating the two types of employees into two strata.

D. Improved Sampling Precision

Stratified sampling takes advantage of the existence of different subgroups in the universe. The basic operational problem of simple random sampling is assuring proportional representation of various subgroups, especially if the sample is not large. Stratification tends to make the sampling more efficient. This is especially true with the high variability of accounting universes caused by skewed distributions. When a universe is skewed or has a high degree of variability, the sample size required to provide a reasonable degree of precision using simple random sampling may be somewhat large.

Precision is improved because each stratum has a relatively small standard deviation, and the weighted sum of the strata standard deviations is less than the standard deviation for the entire universe. Thus, if all the items within each stratum were exactly alike with respect to a characteristic, then the variance within each stratum would equal zero, and the variance between strata would account for all the variability of the characteristic. Hence, there would be no sampling error from such a stratified sample. To illustrate, consider a universe consisting of the following eight items: 1,1,1,1,1,5,5,5. The standard deviation of this universe is almost 2. However, if two strata are formed with the five one's in one stratum and the three five's in the other, the standard deviation of each stratum would be zero, and consequently the weighted sum would also be zero. However, if one knew enough about the universe in advance to stratify so perfectly, then there would be little need to take a sample.

Under most circumstances, stratified random sampling permits one to obtain the same precision as a simple random sample with a smaller sample size, or increased precision with the same sample size. In other words, stratification is used to increase the sampling efficiency and effectiveness, because it may ensure representation of all the different kinds of elements.

The reason behind this concept can be illustrated by the problem of estimating the mean pay of American servicemen. Under simple random sampling, a specified number of servicemen would be chosen from worldwide locations, and the mean pay of these servicemen would be the estimate of the mean pay for all American servicemen. However, an improved procedure would be to separate the servicemen into pay-grade groups, select separate samples and calculate means for each group, and use the weighted mean to estimate the pay. This procedure tends to be more precise than simple random sampling because it assures sample representation from all of the different pay groups.

Although there are a number of different stratification procedures, all of them may be grouped under two broad categories: proportionate and optimum allocation. Although all disproportionate sampling is not optimal, it is one of the goals of such a procedure.

In proportionate stratified sampling a uniform sampling fraction or rate is applied to each stratum. For instance, if the sampling rate is 5 percent, a separate, independent random sample of 5 percent of the items is selected from each stratum.

By contrast, in optimum allocation or disproportionate sampling, different sampling rates are applied deliberately to the various strata. When variable sampling rates are used, the precision can be improved (decreased variance) by increasing the sampling rate in the strata having higher variability for the principal characteristics being tested.

These two broad categories of stratification will now be discussed in some detail.

III. PROPORTIONATE STRATIFIED SAMPLING

A. Representative Sampling

Proportionate sampling is perhaps the most widely used testing method in auditing, even when nonstatistical testing is used. It is a common auditing practice to take a fixed-percentage sample, such as 10 percent. This type of testing has been referred to as "representative sampling," since the relative sizes of the strata samples are chosen to be equal to the relative size of the strata. This reflects the notion that the different portions (strata) of the universe are appropriately represented in the sample.

When using statistical sampling methods in auditing, the reference to a "representative sample" could be misleading. In fact, the connotation of the phrase is vague. One could consider a sample strictly representative with respect to some characteristic or variable of a universe if the distribution of the measurements of that characteristic or variable in the sample agrees precisely with the distribution of such measurements in the universe. However, if we knew whether a sample were representative in this strict sense, sampling would be unnecessary.

There is another more disconcerting aspect of a so-called "representative sample" in audit applications. Experience reveals that audit findings tend to occur in clusters of time, activities, departments, individuals, etc., instead of occurring randomly across-the-board. This is the reason that an overall error rate tends to be misleading. Individuals and operations are not equally error prone. Also, errors do not reflect similar dollar impacts. An auditor often uses random selection methods to search for cues or clues to "pockets of occurrences." For example, an audit of a health service program indicated an overall error rate of more than 10 percent, but an analysis of the occurrences revealed that less than one percent of the practitioners accounted for all of the findings.

Replication is a way to protect against atypical samples. Estimates derived from various replicates may be compared intuitively as well as statistically if the replicates are not too small. Those replicates which appear to be out-of-line can be carefully analyzed to ascertain nature, source, cause and impact of occurrences as a basis for the application of other auditing procedures.

B. Aspects of Proportionate Sampling

In proportionate sampling, each stratum sample size is made proportional to the universe size of the stratum. The sampling rate is derived by dividing the overall sample size by the universe size. In proportionate sampling, this sampling rate is applied uniformly to all strata. The variance decreases to the extent that the stratum means are different and that the items within each stratum are similar. In other words, if the means of the principal strata are nearly equal, proportionate sampling provides little benefit over unstratified sampling. Likewise, if standard deviations of principal strata are appreciably different, disproportionate sampling offers marked gains over unstratified sampling.

On the other hand, proportionate sampling seldom includes a precise proportion of the sample from all strata, because the number of items in a stratum usually is not an even multiple of the sampling rate. In practice, however, small departures from a uniform sampling rate are disregarded.

For proportionate allocation, the theoretical optimum boundary point between two strata is one-half of the sum of the means of the two strata. The actual boundary point may not be this exact value, but within a reasonable proximity based on judgment and audit objectives.

C. Advantages of Proportionate Sampling

Proportionate sampling has several desirable features. First, it is easy to administer. Not only does the use of a single sampling rate in all strata offer some convenience, but of greater importance is the ease in combining the strata sample results into estimates for the entire universe. When a uniform sampling rate is applied to each stratum, the universe estimate is merely the unweighted estimates of the subsamples. Proportionate sampling is one of a class of sample designs which leads to self-weighting samples. This is significant because the process of applying different weights to different strata is often laborious, complex and expensive. Also, in a self-weighting sample, each item in the universe has an equal chance of being selected.

Another advantage of proportionate sampling is that it minimizes the risk of getting less precise results than would be obtained from unstratified sampling. The results are more likely to be closer to the universe value than unstratified samples because larger strata are

are represented by larger portions and smaller strata by smaller portions of the total sample.

To summarize, the wide use of proportionate sampling is justified because (i) the variance cannot be greater than for an unstratified sample of the same size, (ii) it can be accomplished simply and easily, and (iii) it provides self-weighting means or other estimates. If numerous estimates have to be made, a self-weighting sample saves time and expense.

However, if the sample estimate is an attribute or percentage, the improvement over unstratified sampling is more modest. The reason for the small gain in estimating proportions (error rates) is that the variance is insensitive to moderate differences in the proportions resulting from stratification, especially when the proportions are in the range between 20 and 80 percent.

In the next section, disproportionate sampling and optimum allocation are used interchangeably. Although there is a technical difference, if an auditor relinquishes the convenience and ease of proportionate sampling for the more complex disproportionate design, it seems logical that he would be striving to attain the optimal. In a strict sense, in disproportionate sampling the sampling rate is determined by analytical considerations, whereas in optimum allocation a formula is generally used to make the strata samples proportionate to the variability within strata as well as their sizes.

IV. OPTIMUM ALLOCATION

A. Aspects of Optimum Allocation

In early applications of statistical sampling in auditing, emphasis was on simple random sampling in which each item had an equal probability of being selected. However, this emphasis began to shift when it was recognized that stratification with unequal probability procedures could produce more efficient samples.

Disproportionate sampling or optimum allocation involves the deliberate use of different sampling fractions or rates among the different strata. The aim is to assign sampling rates to the strata in such a manner as to reduce the variance. Thus, the assigned sampling rates are larger in strata where the variation among the items is larger. This is done because, for a given sample size, the variance is minimized if the sampling rates in the strata are proportional to the standard deviations of the strata.

Optimum allocation generally shows large improvements in precision when sampling from highly skewed universes, such as those usually associated with accounting systems.

The problem of determining the optimum number and type of stratifications is difficult, and tends not to provide a unique practical solution. Theoretically, a unique solution consists of having as many strata as there are dissimilar items in the universe, but such a procedure is not practical. This section, however, will provide some practical guidance in the use of more effective stratification procedures to achieve a significant improvement in sampling precision.

B. Some Optimum Allocation Guidelines

In general, as elements become larger and more complex, the variability among them increases. This suggests that activities and accounts be stratified by some measure of size, and that larger samples be drawn from those strata of larger and more variable accounts. If the variance for each stratum can be approximated from experience or from a preliminary sample, then optimum allocation would be superior to proportional stratification.

If the strata variances are not known, a measure of strata size may be based on the relative number of personnel, annual budget, case load, patient beds, occupied space, or some other size-related variable; however, the variable selected should be closely related to the audit objective. When good estimates for strata sizes are not available, weights which are inversely proportional to the known probabilities of selection within each stratum may be used.

Most of the potential improvement from disproportionate sampling is often obtained by using a few different sampling rates. In most accounting universes, the number of sampling units in the largest-valued stratum is small, but the variance is large. In such situations, it is usually best to examine all of the units in the largest stratum, which results in no sampling error for that group. Also, for skewed distributions two sampling rates may extract much of the potential improvement from stratification, such as a high rate for high dollar and sensitive item strata and a low sampling rate from the other items. The high rate may be 100 percent.

On the other hand, larger variances will result if the sampling rates among strata are significantly incorrect. However, moderate deviations of the sample allocation from the optimum do not increase the variance appreciably. Also, where the variations among strata are not large, rough estimations of the strata standard deviations may produce optimum allocation results which are worse than proportionate sampling, that is, produce a large variance.

Because of the relative instability of smaller organizations, especially in economically stressed situations, bigger relative changes may be associated with them, resulting in a large coefficient of variation (i.e., the standard deviation as a percentage of the mean). In such situations, it may be best to arbitrarily double the proportion of the sample that would have been used for such strata under optimum allocation. Also, if the optimum allocation of sample size for a small stratum is zero, it may be best to assign that stratum the smallest sampling rate that is assigned to any other stratum. Both of these procedures provide extra audit protection.

According to Hansen, Hurwitz and Madow [20], the coefficient of variation associated with optimum allocation must be larger than 33 percent to reduce the variance from proportionate sampling by 10 percent.

On the other hand, the simplicity and the self-weighting features of proportionate stratification may compensate for as much as a 20 percent loss in precision when compared with optimum allocation. Even then, proportionate stratification may be more efficient than simple random sampling.

The optimum allocation procedures discussed thus far have assumed that there is no difference in the cost to examine items in the various strata. If the costs vary among strata, optimum allocation is achieved when the strata sampling rates are directly proportional to the strata standard deviations and inversely proportional to the strata examination cost per item. In other words, if the differences among the ratios of strata standard deviations to strata unit examination costs are not substantial, disproportionate allocation may not be worthwhile. The greatest gains are for highly skewed universes where information is available for forming strata, such as dollar values of accounting records. Readers who are interested in a detailed discussion of this topic should see Hansen, Hurwitz and Madow [20].

Similar to proportionate sampling, optimum allocation is not usually efficient for estimating attributes (proportions) because the standard deviations are insensitive to moderate differences in strata proportions.

Seldom is an auditor interested only in estimates for the total universe. Often the performance of subgroups in the universe has an extremely adverse affect on estimates for the total universe. The auditor is always alert to detecting and isolating any such subgroups if they exist in a particular audit universe. However, the optimum stratification is different if the primary objective is the comparison of subgroups (domains) rather than estimates only about the entire universe.

If the primary objective is to obtain only total universe estimates, proportionate sampling should generally be used. On the other hand, if the primary objective is the comparison of subgroups, such as field offices, the optimum sample would be one where the sample size is the same for each subgroup for this minimizes the variances of the differences.

Thus, when a sample is optimum for estimating universe totals, it is not optimum for subgroup (domain) comparisons. Since auditors are often concerned with both universe totals and subgroup comparisons, a compromise solution such as the following may be used.

First, estimate universe totals using proportionate sampling, and then randomly augment the samples for the smaller subgroups (strata) to make strata sample sizes equal.

A somewhat similar allocation problem exists when more than one characteristic is to be estimated. This problem is further complicated when both totals and attributes are to be estimated from the same sample. For estimating totals, the optimum allocation should be disproportionate, but for attributes it should be proportionate.

As a compromise, first, allocate the sample in a disproportionate manner for estimating totals. Then, after analyzing how well this sample estimates significant attributes, increase the sample in those strata where disproportionate allocation provided smaller proportionality, to improve the precision of the attribute estimations.

Optimum allocation procedures make it necessary to weight the strata data to obtain estimates for the entire universe. There are computer programs available to accomplish this task for the auditor; nevertheless, a brief description of the process is useful.

The weighted universe mean is equal to the sum of the various strata means, after each is multiplied by its proper weight, where the sum of all the individual weights equals unity or one. In other words, the sample mean is estimated separately and independently for each stratum, and it is then multiplied by the stratum weight. These strata products are summed to obtain the weighted mean for the sample.

The variance of the weighted mean is obtained as follows:

1. The variance of each stratum mean is multiplied by the square of the weight for the stratum, and
2. These products are summed over the various strata.

Statistically, the minimum number of sampling units from each stratum is two. Otherwise, a stratum variance cannot be computed. In audit applications, two from each stratum would be grossly inadequate.

The strata weights frequently represent the proportions of the universe items in the strata. For computing totals, the weights may be stratum totals rather than proportions.

The weights may represent some other measures rather than the items. For example, a company's property and equipment may be sorted into strata according to applicable depreciation factors. If the objective is to estimate the company's overall depreciation rate, then it would be more appropriate to base weights on dollar values within strata rather than the number of items because of the large variation in unit dollar values within each stratum.

When good strata size estimates are not available, weights which are inversely proportional to the known probabilities of selection within strata may be used.

Computer programs are available to assist the auditor in the implementation of complex sample designs. These computer programs are not only much faster than manual operations, but they are subject to fewer nonsampling errors, especially those associated with the accuracy of mathematical computations.

V. SOME PRINCIPLES AND GUIDELINES FOR FORMING STRATA

The following principles and guidelines involve procedures for selecting valid samples as well as ways of increasing sampling efficiency:

A. Stratifying Variable

The best characteristic for forming strata is the frequency distribution of the principal variable for which information is desired. In auditing, this is usually the dollar value of the items. The next best characteristic would be the frequency distribution of some other quantity which is highly correlated with the principal variable. Computer programs are available for determining strata boundaries for both proportionate sampling and optimum allocation. The auditor generally knows enough about the audit area to make a satisfactory choice among variables available for stratification; nevertheless, the following suggestions from Kish [32] are useful:

1. Coarser divisions of several stratifying variables may be better than finer divisions of one variable.
2. Completeness and symmetry are not necessary in forming strata and substrata.
3. It is better to use unrelated stratifying variables.
4. Quantitative rather than qualitative stratifying variables are preferable.
5. Minor inaccuracies in the application of a stratifying variable are not serious; this merely decreases the efficiency of the stratification.
6. There is no need for regularity and uniformity in forming strata, but the stratifying variables should involve significant variations.
7. Objectivity is unnecessary since the use of audit judgment and knowledge of the accounting system is superior to rigid procedures in forming strata. Experience, intuition and the

judgment of the auditor are all useful in improving the sampling precision through effective stratification.

8. If no unique stratifying variable is available for all items, use some relevant variable for each stratum.
9. If the stratifying variable is unavailable, overlooked or too expensive to use for prior sorting, use post-stratification.

B. Number of Strata

A few strata yield most of the gains from a single stratifying variable. In auditing, perhaps two to six strata will suffice for any variable. In addition to natural stratification, two strata usually produce an appreciable gain in precision, especially for highly skewed distributions. Three strata may provide little improvement over two, and further stratification reflects a rapid diminishing return. In any case, the number of strata should be limited by the amount of time and effort that can be beneficially spent on stratification.

C. Collapsed Strata

If the sampling frame exhibits a high degree of variability and stratification is carried to the extent that the sample contains only one item in each stratum, then the variance cannot be computed unless the strata are at least grouped into pairs. For any two strata that form a pair, their sizes should be equal or almost equal, but the groupings should be made before the sample items are examined. This procedure is called collapsed strata.

D. Small Strata

It usually is not worthwhile to form small strata unless they are greatly different from other strata in respect to sensitive items or standard deviations. A very small stratum contributes little to the gains from stratification, since the gains are proportional to the stratum weight.

E. Equal Allocation

When an equal number of items is randomly selected from each stratum the process is called equal allocation. It is useful when strata boundaries are not fixed. For optimum allocation, the sizes of the strata can be varied to fit a desired sampling rate, while keeping the strata sample sizes equal. Also, if the strata standard deviations are somewhat uniform, and proportionate sampling is employed, then equal size samples from equal size strata are simple and efficient. Obviously, the overall estimate would be adjusted for the varying sample rate.

F. Highly Skewed Distributions

When the distribution of the sampling frame is highly skewed, it is best to separate the relatively few large valued items in the tail of the distribution, and audit this group 100 percent. Also, in auditing it is a good practice to redefine the frame to exclude the 100 percent audited group, because there is no sampling error associated with it. This action reduces the variability of the items in the delimited frame.

VI. SOME STRATIFIED SAMPLING PLANS

Which stratified sampling plan is best in a given situation depends on audit objectives, comparative costs, convenience, and the availability of computers. Formulas for computing comparative costs for various types of stratified samples are summarized and illustrated by Deming [12]. However, with the use of computer programs the auditor need not become personally involved with the formulas nor the computations.

Most of the stratified sampling techniques may be classified under one of the following plans.

PLAN A. Select a random sample from the entire sampling frame, without any rearrangement of the items. A systematic sample is an excellent procedure for accomplishing this type of stratification. For example, if July has twice as many transactions as February, then systematic sampling would provide about twice as many sample transactions from the July records as from February. Sometimes this may be the best plan of all, because of its simplicity, convenience and savings in cost of stratification. This plan permits the auditor to take advantage of any natural stratification in the sampling frame. Replication is also a means of exploiting the benefits of the natural stratification, since the replicates are separate interpenetrating portions of the frame. Each replicate is a valid sample because it randomly covers the whole frame.

PLAN B. Classify and rearrange all the items in the sampling frame into strata, and then select a proportionate stratified sample. This plan is often used because it is convenient, even where there are only trivial gains in precision. For example, in auditing an organization, the several departments may be used as strata because it is easy to do so. If, however, the strata are formed in such a manner that the homogenity within strata is about the same as random grouping, there is no gain from stratification.

PLAN C. Classify and rearrange all the items in the sampling frame into strata, and then select a sample by optimum or disproportionate sampling. An optimum allocation plan should be considered with caution, since certain advance estimations are required. Incorrect advance estimations may result in poorer precision than proportionate allocation. In using disproportionate sampling, the prime consideration is the recognition of situations where it is appropriate. Another consideration is to keep the related work of such an application to a minimum, both in the effort and in the cost of collecting, processing, analyzing and evaluating the data. However, if the examination cost per unit is the same in all strata, then optimum allocation for a specified cost reduces to optimum allocation for a specified sample size.

As a general rule, one should select in a given stratum a larger sampling rate if (i) the stratum is larger, (ii) the stratum is more variable, or (iii) the unit cost of sampling is cheaper in the stratum.

PLAN D. Select a fixed sized random sample from an unstratified sampling frame, and then group the sample items into strata by using proportionate allocation. Stratification after sampling is a significant procedure for auditors since stratification prior to sampling may be too expensive. To illustrate the procedure, suppose a simple random sample has been audited and the sample items can be classified into strata. Then the sample items should be sorted into proper strata, and strata estimates computed. Then the strata estimates are appropriately weighted and summed to obtain the overall universe estimate. A proportionate stratified sample can even be further stratified by post-stratification. For example, the original sample may have been stratified by regions. Post-stratification of each region according to workload characteristics could be introduced into the regional estimates by using auxiliary information which was obtained during the sampling operation.

PLAN E. Similar to Plan D except the sample items are grouped into strata by using optimum allocation procedures.

PLAN F. Determine strata sample sizes by proportionate allocation, as in Plan B, and then randomly select items one by one from the entire frame, sorting them into the various strata until all the desired strata sample size quotas are filled. Disregard any sampling unit which belongs to a stratum whose quota has been met.

PLAN G. Similar to Plan F except optimum allocation procedures are used.

PLAN H. Select a fairly large preliminary sample from the unstratified sampling frame, stratifying only the selected items. Then sample proportionately from the various strata of the preliminary sample to obtain a specified total sample size. This procedure is called double sampling. When the frame contains tens of thousands of items, the cost of classifying and sorting the items may far outweigh any gains from stratification. This plan, however, requires only the sorting of the sample items, which reduces the stratification cost and also speeds up the sampling operation.

PLAN I. Similar to Plan H except optimum allocation procedures are used. However, the preliminary sample must be large enough to permit the estimation of adequate weighting factors.

VII. SUMMARY

Stratified random sampling procedures are appropriate when information about strata is desired or when variances differ significantly among strata. Stratified sampling improves precision by providing a smaller sampling error for a specified sample size when compared with simple random sampling.

If the total number of items in the universe and in each stratum is known, then one of the simplest plans for stratification is proportional allocation. By dividing the total number of items required for the sample by the total number of items in the universe one derives the sampling rate or fraction, which is applied uniformly in all strata. This is called proportionate stratified sampling.

In actual practice, when proportionate sampling is used, the auditor seldom has exactly the same proportion of the sample from all strata. It is usually acceptable to disregard small departures from uniformity and to use a uniform rate equal to the overall sampling rate for the universe.

Proportionate stratified sampling is easy to administer, and the use of a single sampling rate in all strata is convenient and produces a self-weighting sample. Thus, the estimate of the sample mean is merely the simple unweighted mean of all the observations included in the sample. This is significant because the process of applying different weights to different strata is sometimes laborious and expensive.

Although in proportionate sampling the same basic procedures are used for deriving estimates as in simple random sampling, the precision is improved because the sample is allocated proportionally among the strata.

When good estimates for strata sizes are not available, weights which are inversely proportional to the known probabilities of selection within strata may be used. For this reason systematic sampling is often used to select a proportionate sample. A systematic sample will contain a relative proportion of items from each stratum, and therefore would be equivalent to a proportionate stratified sample if the items in each stratum had been mixed thoroughly before selection. However, in proportionate stratified sampling, the mixing is not necessary, since the random selections within each stratum are equivalent to a mixing or shuffling procedure.

Stratification with a uniform sampling rate almost always improves precision. Also, when variable sampling rates are used, precision will be further improved if larger sampling rates are used in the more variable and larger strata. However, oversampling the less variable strata may adversely affect the precision.

In simple random sampling each item has an equal chance of selection. In stratified random sampling one does not need to insist that each item has the same chance of being in the sample, as long as the chances for each stratum are known. The only restriction on the freedom to choose the chances of selection is that every item in the universe must have some positive chance of being selected. Thus, stratification with variable sampling rates permits one to vary the chance of selecting items from the different strata, and is called disproportionate sampling or optimum allocation, even though all disproportionate sampling is not optimal.

To obtain optimum precision of estimation, one should select strata so that their means are as different as possible, and their standard deviations are as small as possible.

In stratified sampling, the variance of the universe mean depends only on the variances of the estimates of the individual strata means. For instance, if it were possible to divide a highly variable universe into strata such that all items within each stratum had the same value, then the estimate of the universe mean would not reflect any sampling error. Also, a 100 percent review of the largest and most variable strata will result in a substantial reduction in the sample variance.

When strata variances are unknown, relevant size measures such as the number of personnel or payroll may be used as a basis for stratification.

Often estimates for strata or domains are of prime concern. In such cases, the most efficient sample is obtained by selecting the same size sample from each stratum.

There is no violation of statistical principles in the use of judgment, artistry, intuition, educated guesses, or logical decisions in the construction of strata. If the judgment is good, the sample precision will be improved.

Although good audit judgment should be exercised in the designation of strata, the actual selection of items from the strata must not be influenced by judgment; they must be selected by random methods. However, random selection and statistical evaluation do not preclude subjective audit appraisal of the individual findings for nature, source, cause, trend and impact.

```
IF YOU WANT INSTRUCTIONS, TYPE '1'; IF NOT, TYPE '0'.
1

THIS PROGRAM GENERATES UP TO 1000 DIFFERENT RANDOM NUMBERS
WITHIN A SPECIFIED RANGE AND PRINTS THEM IN ASCENDING ORDER.

IF YOU WANT THE NUMBERS PRINTED FIRST IN RANDOM ORDER AS THEY
ARE GENERATED, TYPE '2'; IF YOU WANT THEM PRINTED IN ASCENDING
ORDER ONLY, TYPE '0'.
     0

PLEASE TYPE THE QUANTITY (Q) OF RANDOM NUMBERS YOU WANT, THE
LOWEST NUMBER (L) IN YOUR RANGE, THE HIGHEST NUMBER (H), AND A
FIVE-DIGIT RANDOM NUMBER (R) FROM A TABLE OF RANDOM DIGITS.
Q,L,H,R= 20,16,155,30853


   19   22   23   26   29   31   32   34   35   39   41   43   44
   49   55   62   67   94  142  143


IF YOU WANT MORE RANDOM NUMBERS,  TYPE THE QUANTITY, THE LOW NUMBER AND
THE HIGH NUMBER IN THE RANGE, AND ANOTHER RANDOM NUMBER FROM A RANDOM
DIGIT TABLE;  IF NOT TYPE '0,0,0'.
Q,L,H,R= 20,171,4622,76102


   218    353    369    673    745    772    979   1444   1608
  1999   3149   3171   3467   3514   3588   3591   3637   3882
  3884   4443


IF YOU WANT MORE RANDOM NUMBERS,  TYPE THE QUANTITY, THE LOW NUMBER AND
THE HIGH NUMBER IN THE RANGE, AND ANOTHER RANDOM NUMBER FROM A RANDOM
DIGIT TABLE;  IF NOT TYPE '0,0,0'.
Q,L,H,R= 0,0,0


NOW AT END
```

Figure 6. Printout of computer generation of stratified random numbers.

If the examination cost per unit is the same in all strata for attribute sampling, then (i) the gain in precision from stratified random sampling over simple random sampling is small unless the proportions (error rates) vary greatly among strata, and (ii) optimum allocation for a fixed-sample size gains little over proportionate allocation if all strata proportions lie between 20 percent and 80 percent.

Stratification has many meanings and many ways of implementation. In some applications, the overall sample is randomly selected first and then the strata are formed in the sample only; however, in most audit applications, the strata are formed first. There are other applications which represent a combination of these two approaches.

In all cases, a careful balance must be maintained between the gains expected in sample precision and the additional time and resources involved in introducing a stratification scheme into the sample design.

However, computer programs can save time in applying a stratification scheme. For example, after reviewing an inventory listing, the auditor separates the items into three strata. Realizing that the sample precision can be improved if all the strata contain approximately the same dollar value, the stratification was performed as follows:

No. of items	Dollar range	Dollar value
15	$15,000.00 and over	$298,525.00
155	$ 5,000.00 - $14,999.99	290,818.25
4,452	0 - $ 4,999.99	287,341.80
4,622		$876,685.05

The auditor decided to examine all 15 items over $15,000 and to select initial samples of 20 from each of the other two strata. A computer printout of the stratified selection is shown in Figure 6. The first random selection is from the 155 items on lines 16 through 170, and the second random selection is from the 4,452 items on lines 171 through 4,622.

9

Cluster Sampling and Subsampling

I. GENERAL

When the selection of individual universe items is too laborious or expensive, the audit testing can be facilitated by selecting contiguous groups (clusters) of items. For example, from a card file equal clusters of 20 consecutive cards could be selected together. This is called cluster sampling, and it denotes methods of selection in which the primary sampling unit contains more than one universe item. That is, it involves the selection of more than one universe item at a time.

In cluster sampling, the universe is divided into groups (clusters) of items that serve as primary sampling units, and a random sample of clusters is selected from all these clusters. Each cluster becomes a sampling unit, and by defining the clusters, the auditor creates a new frame consisting of cluster sampling units. For example, if an auditor examines all the expense vouchers of a selected day or of a week, he is using a cluster sample.

The main reason for sampling clusters of items is that the cost of locating and listing the sample is usually less than if the individual items were selected one at a time. Thus, the purpose of cluster sampling is not to get the most efficient sample but to get the most precise results per unit of examination cost. To achieve this aim, selected clusters should be as heterogeneous as possible but small enough to reduce such expenses as listing costs and travel costs. Hence, cluster sampling is advantageous for large, widespread samples.

Sometimes a sample of such primary units is selected and all items of the universe within the selected units become a part of the sample. This is called one-stage cluster sampling. Other times, the selected primary units are divided into secondary units for additional stages of sampling. This procedure is known as subsampling.

Subsampling is a compromise which endeavors to decrease the sample variance by decreasing the clustering without causing a proportional increase in the cost of sampling.

II. ONE-STAGE CLUSTER SAMPLING

Although stratified sampling and cluster sampling have some resemblance since both methods divide the universe into groups, they involve somewhat opposite sampling operations. In stratified sampling, items are selected from each stratum. Thus, every stratum is represented in the sample. Since the sampling error involves variability within strata, it is advantageous to have strata as homogeneous as possible and as different as possible from each other.

By contrast, in one-stage cluster sampling, there is no sampling error within a cluster, since all the cluster items are examined. Therefore, the sampling error reflects only the variability among the clusters.

A comparison of one-stage cluster sampling extremes will be informative. Suppose, on one extreme, every cluster were heterogeneous and differences among cluster means were insignificant, then the auditor could select only one large cluster and obtain a reasonable result. However, suppose the opposite extreme existed, that is, the clusters were completely homogeneous, then one item from each cluster would suffice for a reasonable estimate.

Thus, optimal sampling precision for a one-stage cluster sample results when the items within a cluster vary as much as possible, and the clusters differ as little as possible. This principle is difficult to achieve in actual practice, since clustering tends to be based on physical contiguity which causes homogeneity by association.

A measure of homogeneity is the coefficient of interclass correlation which measures the correlation between all possible pairs of items within a cluster. If the cluster is completely homogeneous, the coefficient of intraclass correlation is equal to one, and when the homogeneity is equivalent to random sorting of items into clusters, the coefficient is zero.

It becomes apparent that the principles which make stratified sampling effective have a diametrically opposite effect on one-stage cluster sampling. In stratified sampling, the precision is improved if the stratum items are homogeneous, but in one-stage cluster sampling the improvement in precision is associated with the heterogeneity of the cluster items.

In nearly all audit situations, simple random sampling will be more efficient than one-stage cluster sampling. It may, however, cost more to locate and list simple random samples. Thus, the problem essentially is one of balancing sampling cost and sampling efficiency. Generally speaking, however, when travel costs and other costs which depend on the number of selected clusters are relatively large when compared with the costs which vary directly with the number of items examined, one-stage cluster sampling will be more economical than simple random sampling, but not as precise, because the number of independent observations is more significant in determining efficiency than the total sample size.

A major disadvantage of one-stage cluster sampling in auditing is that there tends to be a high degree of correlation between adjacent accounting transactions. For example, accounting records are usually filed in such a manner that contiguous records are likely to exhibit about the same characteristics. This may be caused by the fact that they are all processed and maintained by the same person or staff, or that they are filed chronologically, or some other reasons. However, the correlation decreases as the distance between pairs of sample items increases. Also, in sampling centrally located records, the cost advantage of cluster sampling may be insignificant, because travel cost is not a factor.

For example, suppose the auditee is a large hospital which has about 12,000 accounts receivable due from former patients. The records are on cards which are filed alphabetically in 10 drawers, each 30 inches deep. The auditor desires to select a specified number of accounts for confirmation of the balances. Based on his knowledge of the accounting system and prior experience, he decides to select a sample of 400 record cards. Since the cards are not numbered, he has a choice of using either systematic sampling or cluster sampling. Suppose he decides to use a one-stage cluster sample. Since experience has revealed that there are about 40 cards to an inch, he specifies a cluster as consisting of one-half inch of cards, that is, 20 consecutive cards per cluster. The sampling frame then consists of 600 (12,000 ÷ 20) clusters, which are the sampling units. From these units he will randomly select 20 units (clusters). Since there are 20 cards in each half-inch cluster, he will have a sample of 400 cards. However, his cluster sample would be less precise than a two-random-starts systematic sample of the same size since there is no savings in travel cost, because the records are centrally located.

III. SUBSAMPLING

After the primary cluster units are selected, the auditor can either include in the sample all of the items in the selected clusters or select a sample of smaller clusters or individual items from these clusters. When a sample of items is selected from within each cluster, this type of design is called two-stage sampling. This process could be extended to more stages in a similar manner, and is called multistage sampling. Although multistage sampling permits greater flexibility in sample design, the execution, analysis, evaluation and appraisal become more complex.

There are several reasons why subsampling is preferred over a one-stage cluster sample. Although natural clusters may exist in the universe as sampling units, they may be too large. Also, the auditor can avoid the cost of creating smaller clusters in the entire universe and limit the creation to only the selected sampling units. In other words, in a two-stage sample it is not necessary to select every item within sampled clusters.

Suppose, for example, the auditor wishes to estimate some characteristics about patient records in nursing homes. Thus, nursing homes naturally suggest themselves as an appropriate sampling unit. Since it is not feasible to examine all patient records in all nursing homes, subsampling is a practical solution. In this case, nursing homes would be the primary sampling unit (first stage), and patient records would be the secondary unit (second stage), for each selected primary unit.

In some audit applications, two-stage sampling would facilitate locating and listing the universe. Although it may be easy to list all the primary sampling units, as in the example of nursing homes, but it could be difficult and expensive to list all the secondary units, e.g., patient records in all nursing homes.

Usually, two-stage sampling is advantageous when the sampling frame exhibits the following characteristics (i) the frame can be divided into at least 20 convenient groups, and (ii) there is likely to be little variation in the characteristic of interest within any one group.

The smaller the variation within clusters, as compared to the universe variation, the greater will be the improvement in sampling precision for a specified sample size by using a two-stage sample. This means that the second stage tends to offset some of the adverse effect of the original clustering.

In a manner somewhat similar to proportionate stratified sampling, if a uniform sampling fraction or rate is applied in each cluster, a two-stage sample is self-weighting. For instance, the probability of a cluster being selected in the first stage is proportional to the total number of items in that cluster, and the probability of a particular item being selected in the second stage, provided its cluster is selected in the first stage, is inversely proportional to the total number of items in its cluster. In other words, a two-stage sample is self-weighting if selected initially with probabilities proportional to cluster sizes, and then with a fixed size sample within each selected cluster.

IV. CLUSTER SIZE

The number of items in a cluster determines its size. Clusters of equal size seldom occur naturally, but they can be artificially created in planned sampling operations. As an example, the auditor can create equal size clusters such as 20 consecutive record cards in a file. Also, equal size subsamples may be selected from unequal size clusters in a similar manner as for unequal size strata in stratified sampling.

A natural grouping of variable size clusters can be illustrated with households. Each household has one or more persons who tend to have common traits by association and relationships. Thus, the sampling precision is improved if only one person is selected from a household instead of selecting the entire household (cluster). Obviously, in some households there is only one person to select under either alternative. In this example, the primary (first stage) sampling unit is a household and the secondary unit is a person randomly selected from each selected household. However, the results must be weighted for the number of eligible household members to obtain an unbiased estimate.

V. STRATIFIED CLUSTER SAMPLING

Stratification with cluster sampling may be introduced at each sampling stage, either with proportionate sampling for all stages, or with variable sampling rates for all, or some, of the stages. The type of stratification used would be based on the auditor's judgment of the accounting system, knowledge of the characteristics of the items in the clusters, and any natural stratification. Often geographic stratification, alone, provides an improvement. Also, separate strata of unusual primary sampling units, such as larger offices, may be desirable. Stratification will improve the precision when homogeneous clusters are grouped together. Stratified multistage samples are used in auditing because of economy and flexibility. However, computer assistance is usually required in solving the related formulas.

VI. SOME CLUSTER SAMPLING GUIDELINES

Since cluster sampling is designed to facilitate the audit testing procedure by taking advantage of natural and artificial clustering of accounting records, the following principles and guidelines may be useful:

1. A sample item should belong to one and only one cluster.
2. A reasonable estimate of the size of each cluster is necessary.
3. Clusters must be small enough to provide sufficient cost savings when compared with more efficient sampling plans.
4. Clusters do not have to be identically defined. For instance, some clusters may be a city, while others may be several small towns, or even a portion of a very large city.
5. Clusters do not have to be the same size. However, when variable sample sizes are used, the sample size cannot be determined in advance. The variable sample size aspect is somewhat similar to disproportionate stratified sampling, since the results must be weighted to compensate for the unequal probabilities of selection.
6. Within a cluster, systematic sampling may be superior to simple random sampling, since adjacent groupings of items tend to be similar. Systematic sampling also reduces the likelihood of getting only transactions processed by a single individual when there is a short time frame.
7. Two-stage sampling tends to offset some of the decrease in precision caused by clustering.
8. Auditors should be especially cautious in the use of cluster sampling since the greatest homogeneity usually occurs among temporal variables and economic characteristics.

VII. RATIO ESTIMATION FOR SUBSAMPLING

Ratio estimation is useful with subsampling, because the auditor may not be interested in the average value per primary sampling unit. For example, he would seldom be interested in the average annual income per city block (primary unit), but he may be interested in the average annual personal income for the city.

The estimate of the average value per elementary unit of a characteristic may be computed as the ratio of the sample total to the number of elementary units in the sample. However, a ratio of random variables is subject to a slight bias which becomes negligible as the sample size increases. See Chapter 15 for a discussion of ratio estimation.

For more detailed technical information and appropriate formulas for use in cluster sampling and subsampling designs, the reader is referred to Cochran [8], Hansen, Hurwitz and Madow [20], Deming [11] or Kish [32]. However, one does not need the ability to solve the formulas, since computer programs are available to perform the mathematical computations.

VIII. SUMMARY

Cluster sampling takes advantage of the fact that universe items often occur in natural geographic, temporal or economic clusters.

The resemblance of cluster sampling to stratified sampling is that both procedures require a grouping of universe items; however, the selection methods are different. In stratified sampling, a separate, independent random sample is selected from each stratum; whereas, in cluster sampling, selection is from designated clusters or groups, treating each group as a single sampling unit. That is, the selection of one of the items in a group automatically requires the selection of the entire group.

In comparing a cluster sample with an individual item sample containing the same number of universe items, usually the cluster sample will (i) cost less per item examined because of the lower cost of locating and listing, (ii) the variance is higher because of the homogeneity of the items in the clusters, and (iii) the analysis, evaluation and appraisal costs are greater.

The adverse effect of clustering on sampling precision is caused by (i) the selection of groups of contiguous items which tend to be homogeneous and (ii) the generally nonrandom distribution of the universe items within those groups (clusters).

The cost savings in cluster sampling must be balanced against the loss of sampling precision due to homogeneity within clusters. However, there is no need to use cluster sampling if information is obtained by mail, phone, or at a central location, since travel cost is minimized.

When deciding whether to use a cluster sampling design or some other design, the auditor must balance cost considerations against the efficiency of the design. From a statistical point of view, whichever design provides the smallest standard error for a specified cost should be used. Since it is not necessary in a two-stage sample to select every item within sampled clusters, it may be a satisfactory compromise.

When clusters instead of individual items are selected from previously designated, strata it is called stratified cluster sampling. This procedure reduces the variability of unstratified cluster sampling.

An advantage of multistage sampling is that a sampling frame is required at each stage for only the units that have been selected at that stage.

Auditors should be extremely cautious when using the results derived from cluster sampling. The potential savings in sampling costs may be more than offset by increased effort in developing rationale to support the extension of the audit findings to those clusters which were not sampled.

In the final analysis, the audit objectives and the nature of the distribution of the universe items should be considered when determining whether the use of cluster sampling or subsampling is cost effective.

10

Detection or Exploratory Sampling

I. GENERAL

One of the explicit audit objectives mentioned in Chapter 2 is the detection objective. This objective is concerned with the location of failures to adhere to basic internal control procedures and evidence of irregularities.

In pursuing the detection objective, the auditor is not concerned initially with determining how frequently such instances occur nor the dollar impact. The primary concern is to locate at least one instance if such violations occur. The detection of the type of transaction or event that permitted the violation is the significant aspect of this objective.

The traditional sampling methods, as originally applied to auditing, concentrated on estimating the frequency or dollar amount of occurrences or errors. Detection sampling, however, is concerned with determining the necessary sample size to have a specified probability of finding at least one example of some kind of critical event when it occurs with some minimum frequency in a universe. The detection sampling approach has also been called exploratory sampling [73] and discovery sampling [2].

It is possible to calculate the probability that at least one example of a specified kind of event will be detected by an indicated sample size if the event occurs with a specified frequency in the universe. The table in Appendix C was designed for use in detection, or exploratory, sampling. This table is extracted from more extensive tables which were prepared by the Air Force Auditor General [78].

Other detection, or discovery, sampling tables are available. Some are based on the rate of occurrence or events in the universe, while others, such as Appendix C, are based on number of occurrences. Experience reveals that auditors are usually more concerned with the absolute number of critical events regardless of the universe size. If the auditor makes the decision that he wants a high probability of detecting at least one of the critical events if as few as 20 such events are in the universe, then this decision applies whether the universe is 500, 5,000, or 20,000. In other words, critical events which an auditor seeks are not ordinarily a function of the universe size. Also, occurrence rates can be estimated by dividing the number of sample events (errors) by the sample size.

The detection sampling objective can be integrated with other objectives provided the sample size is large enough to satisfy all objectives simultaneously. However, there might be occasions when the sole objective of the sample is detection. For example, after dismissing an employee, a company might perform a special audit to determine whether irregularities occurred in the employee' work. If the employee approved disbursements up to $5,000 from a special bank account, one might select a sample large enough so that if one percent or more of the approved disbursements contained irregularities, one would have a 99 percent probability of finding an example. Detecting one example of fraudulent activity would certainly cause the investigation to be expanded.

Although the purpose of detection is to reveal an example, one may evaluate the results using other statistical estimation procedures. Thus, suppose that the sample fails to disclose any irregular activities. In this case one can estimate an upper precision limit based on a zero rate. This procedure is discussed in more detail in Chapter 13.

II. REQUIRED DECISIONS

Use of detection sampling requires the auditor to make the following determinations:

1. Definition of critical event (error).
2. Critical number or rate of occurrence of the event which would cause concern.
3. Desired probability of finding at least one example of the event.
4. Estimate sample size based on considerations in 2 and 3 above, and using a table such as Appendix C.

A critical event might be defined as an irregularity representing an unauthorized transaction or manipulation. This definition could be expanded to include accounting errors, such as incorrect postings.

In order to estimate the sample size appropriate for the detection sampling objective, one must specify the critical number or rate of occurrence of the event and the desired probability. The critical number or rate of occurrence is that number or rate in the universe which must occur, or be exceeded, to achieve the desired probability of detecting at least one example of the event.

III. DETECTION SAMPLING TABLE

The table in Appendix C presents the probabilities that an indicated sample size will contain one or more units with a designated characteristic if the sample is drawn at random from a universe which contains a specified number of units with that characteristic. For convenience, the items which possess the designated characteristic are referred to in the table as "errors."

The table presents selected convenient universe sizes from 200 to 200,000. One universe size is presented on each page with up to 40 selected sample sizes ranging from 5 to 700. For each universe and sample size combination, the number of "total errors" range from 1 to 2,000, as appropriate. The probabilities in the body of the table are presented as percentages.

The first row of numbers across the top of each page indicates the total number of "errors" in the universe. The first column on the left of each page lists the various sample sizes. For example, page 2 of the table is for a universe size of 500, as indicated in the page heading. Thus, if a sample of 50 units is drawn from a universe of 500 units with 10 units in error, then the probability of the sample containing one or more units in error is .655. Since the probabilities are presented as percentages, the number appears in the table as 65.5. Table entries are not shown when either the sample size or number of units in error exceed the universe size; dashes are printed in those cells. It should be noted in the table that the computerized computations rounded the probability percentages to one decimal, thereby producing some 100 percent, or absolute certainty, probabilities. However, one never has absolute certainty when using a sample, even if all transactions except one have been examined. That remaining one may be the culprit.

The detection sampling table in Appendix C is used as follows:

1. Select the proper page for the universe size. Always be conservative; that is, if the universe is 8,000, use the table for 10,000.

2. Select the column on the page for the number of critical events in the universe (total errors).
3. Going down the column, find the first number which is greater than or equal to the desired probability (assurance).
4. The corresponding value in this row of the first column (Sample Size) in the maximum sample size for the stipulated specifications in 2 and 3 above.

Since the table in Appendix C is based on the hypergometric probability function, the random samples should be selected without replacement, that is, no duplicate sample items.

A close examination of the table reveals the futility of endeavoring to detect rare occurrences in a large universe using small sample sizes. For example, if as many as 30 critical events were in a universe of 5,000, a sample of 50 has only a 26 percent probability of detecting one of the events (see page 8, Appendix C). Under the same circumstances, a sample of 100 has less than a 46 percent probability of revealing such an event. Hence, very rare characteristics or events are feasible to sample only if locating them is relatively inexpensive.

IV. SEARCHING FOR RARE EVENTS

Rare characteristics or events that are expensive to locate create sampling problems. The following approaches are illustrative of some available procedures for searching for rare characteristics or events. One must, however, consider the rarity of the event, available resources, and knowledge of concentrations of the event in determining the optimal procedure.

1. Multipurpose samples may detect several rare characteristics or events in a single sample and divide the costs among them. This procedure may also provide information about relationships among the characteristics. In market research, for example, samples often cover several products, each of which may be relatively rare.
2. In statistical sampling large clusters are usually avoided because they adversely affect the sample precision. However, large clusters can decrease the cost of locating characteristics, but even large clusters may reveal only a few of a rare characteristic. If the rare characteristic or event is concentrated in small areas, these areas may be recognized from prior audits and thereby stratified for more intensive examination.
3. It is efficient to use disproportionate stratified sampling if one can locate small strata that contain large proportions

of the rare characteristic or event. In other words, if most of the rare characteristics or events can be located in small clusters, these clusters should be sampled more heavily. Significant gains in precision accrue if most of the rare characteristics or events are founded in a relatively small area. For a cluster of errors to have a material impact, the cluster must consist either of a few large amounts or a significant number of small amounts.

4. Disproportionate stratified sampling may be combined with two-phase, or double, sampling. The first phase is a screening to identify strata in which the proportions of the rare characteristic are significantly different. Then disproportionate sampling is used in the second phase. In phase sampling, the sampling unit retains the same designation throughout the operation.

V. ILLUSTRATIONS OF DETECTION SAMPLING

Example 1

Assume that an auditor is examining a universe of 1,000 transactions for any irregularity. Naturally, if there were only one irregularity, the auditor would like to find it. However, the result of a 100 percent examination is seldom commensurate with the additional cost when compared with a well designed sample. He decides that more than 10 transactions with manipulations would indicate a serious problem. Also, he desires to have a 90 percent probability of finding at least one irregularity if there are 10 or more in the universe. Looking down the column for 10 "total errors" on the page for a universe of 1,000, he locates 89.4 and 92.3. To be conservative, he chooses 92.3 for his probability. The corresponding sample size in the first column is 225. The disclosure of an irregularity may occur on any transaction during the testing procedure. At that point, the auditor may wish to stop testing and to investigate the circumstances. Afterwards, he may modify his sampling plan or resort to other auditing procedures. If, however, he examines the entire sample of 225 transactions without detecting any irregularities, he has a 92 percent assurance (probability) that there are less than 10 transactions in the 1,000 containing any irregularity.

Example 2

Detection sampling tables may be used to determine the assurance (probability) assigned to a statement that no more than a specified number of erroneous transactions are in a universe when an error-free sample of a specified size is examined. For instance, if a sample of 300 transactions detects no errors when selected at random from a universe of 3,000 transactions, the table may be used to interpret the results. Looking at the row for a sample of 300 on the page for a universe of 3,000, one observes various probability percentages that no more than a certain number of erroneous transactions exists in the universe. Hence, there is a 65 percent probability that no more than 10 erroneous transactions exist in this universe. This same row reveals that there is a 10 percent probability that no more than one erroneous transaction exists, and almost certainty (99.5 percent probability) that the universe does not contain 50 erroneous transactions.

Example 3

A bank desires to ascertain whether any account checks were signed without authorization. It was decided that the bank would be alarmed if there were more than 25 checks in the universe with unauthorized signatures. Although the primary objective was to search for fraudulent signatures, an unauthorized signature is not necessarily an indication of fraud. The universe consisted of 5,000 checks and a 90 percent assurance was specified.

By referring to Appendix C for a universe of 5,000 with an assumed total of 25 errors (which is a 0.5 percent error rate), a maximum sample of 450 is indicated. This sample size was obtained by moving down the column headed 25 (on page 8 of the table) until the specified assurance of at least 90 percent was reached. In this case, the closest figure to it, but not less, is 90.6 percent. The corresponding sample size, therefore, is 450.

Since the sample was to be taken prior to any rearrangement of the checks, a systematic sample was selected.

The signature on each sample check was compared with the signature on file in the bank. On the 223rd sample check an unauthorized signature was detected; a son had signed for his mother. However, an investigation revealed that the son had signed the check with his mother's approval. Ordinarily, when one of the critical examples is found, detection sampling ceases. It is interesting to note from the table that if a sample of 225 reveals an error, then there is only a 68.5 percent assurance that there are less than 25 errors in a universe of 5,000. If the bank adheres to the 90 percent assurance, then the number of errors could be 50. This is determined by moving from a sample size of 225 across that row until a figure of at least 90 percent appears. In this case, 90.1 appears under the column headed by 50.

Since the one finding was not a fraudulent signature, the sampling continued for the 450 checks and no additional unauthorized signatures were detected. Thus, it was concluded that the search for fraudulent signatures had been satisfied.

The reader is referred to Kaufman [28] for a detailed discussion on the use of detection (discovery) sampling for the confirmation of commercial loans and savings accounts at branches of a large bank.

VI. SUMMARY

The objective of detection sampling is to provide a specified probability or assurance of finding at least one example of a characteristic if its occurrence in the universe is at or above a specified number or rate. Detection sampling is appropriate when the occurrence frequency or rate is low, yet one desires a specified probability that the sample is large enough so one can expect to detect at least one example.

In traditional survey sampling, the sample size is related to the variability of the characteristic of interest, rather than the universe size. In detection sampling, the sample size is related to the universe size as well as the rate of occurrence.

Aside from the need for large sample sizes, detection sampling requires a reasonable estimate of the frequency or rate of occurrence of the critical characteristic. This is similar to searching for "a needle in the haystack," provided one knows how many needles are in the haystack. Flexible sampling, described in Chapter 13, avoids this problem and uses smaller sample sizes. The procedure is based on the same probability function (hypergeometric) as detection sampling, but it involves a no error assumption. The sample sizes are reduced substantially through stratification, replication, error analysis and audit appraisal.

11
Work or Random Time Sampling

I. GENERAL

Work sampling is a method of random observations to obtain information about an activity or operation. It enables one to obtain information without continuously watching or reporting on everybody and everything.

The use of random observations to estimate the proportion of time spent on activities of interest was originally described by Tippett [75], and he called it the "snap-reading" method. This random observation technique was also called the ratio-delay method because the purpose was to ascertain the ratio of various kinds of delays in machine operations to total available time. When the technique was initially extended to human activities, the studies were also called ratio-delay because the procedures were borrowed from those used in machine downtime studies. However, the purpose was to find causes as well as ratios of delays.

When work sampling was first introduced, the observation was an objective attribute which consisted of observing whether a machine was operating or idle. With the expansion of activity categorization and the introduction of variable elements of cost, the binary aspect of sampling for attributes was no longer appropriate.

Later work sampling techniques involved the classification of the types of productive effort, the classification of nonproductive time into categories of avoidable and nonavoidable delays, and even the reporting of causes for reduced productivity due to fatigue or other influences. This procedure was developed to highlight the extent of nonproductiveness and to reveal the major causes.

Due to this broadening concept of the method in the early 1950's, the term "work sampling" was suggested as being more appropriate, and it was first used in a 1952 article by Brisley [7]. Subsequently, many other terms have been used including random time sampling, random moment sampling and activity sampling.

Richardson [56] defines work sampling as a measurement technique for the quantitative analysis of nonrepetitive or irregular activities, where no frequency description information is available. He also indicates that it is useful as an overall preliminary survey of office activities.

Thus, work sampling may be useful for an initial analysis of an activity that is not standardized or to estimate unavailable operational data. Since the ultimate objective is to develop work standards, many work sampling applications are not repeated. They are followed by more detailed work measurement techniques. For example, work sampling has been used with queuing theory and regression analysis to refine indirect labor standards and demand-labor situations. Therefore, it is neither economical nor practical to conduct repetitive work sampling applications in the same area. Other techniques, such as multiple-regression analysis, are more economical, effective and efficient for repetitive studies.

Industrial engineers have used work sampling to estimate clerical, service and technical costs because much of this work is noncyclic. They contend that work sampling is an economical technique for estimating and controlling indirect costs, especially where methods have not been standardized or the activities involve a high level of individual craftsmanship or expertise. Obviously, this assertion is predicated on the assumption of a validly designed sampling plan which is properly and carefully executed by trained, knowledgeable observers. Unfortunately, many random time (work) sampling applications do not satisfy this assumption.

The growth of work sampling in industry is attributable to the need for very gross estimates of maintenance and service activities, especially in appraising effort and funds extended in productive and nonproductive activities.

Industry has been interested in work sampling as a technique for revealing opportunities for cost-cutting improvements as a means of boosting productivity. This objective may be viewed as akin to operational auditing where one of the primary purposes is cost avoidance.

In industry, work sampling methods have been used to supplement or replace other more costly and time-consuming work measurement techniques. In auditing, work sampling studies may be used to supplement other procedures such as regression analysis, perambulations and floor checks. It could provide an adjunct to established procedures used in the review and evaluation of direct and indirect personnel utilization practices. On the other hand, where adequate written records are available, other methods usually are better than work sampling, which provides a means of gathering data where there are no written records.

Proponents of work sampling methods cite the following as some of the problems associated with a time-and-attendance reporting system:

1. Time sheets require a surprising amount of effort to fill out daily and summarize monthly. However, the use of mark-sense cards with activity and other relevant codings will reduce the amount of time involved.
2. There may be considerable guessing about certain times and certain activities. This may apply also to work sampling observers if they are not knowledgeable of the activities and are poorly trained or disinterested.
3. Reports may be filled out hurriedly at the end of the day or even later when a faulty memory causes errors.
4. Errors may arise when more than one job is being performed simultaneously. However, this would also be a problem with work sampling.

II. CONSIDERATIONS IN DESIGNING WORK SAMPLES

In statistical sampling, units are selected from sampling frames, which often are readily available written records. These units may be objects, areas, time, or activities. Objects include persons, records, articles, automobiles, and so on. Also, sampling units are either individual units or clusters of units. Individual units include persons, vouchers, patient records, and so on. A cluster sample unit is illustrated by employees in an office, students in a university, patients in a hospital, employees in a factory, and so on.

However, there are problems if the frame is not easily defined or does not consist of such a written record, work activities, and personnel behavior. Under such circumstances, a time frame may be used. That is, randomly selected slices of time may be used to estimate work characteristics and related costs. The technique may be applied to workers, machines or equipment. These may be referred to as subjects.

Random time in work sampling may be applied to either objects or areas, or both.

A time frame has to be specified. However, when the frame consists of a continuous set of records such as invoices, time is automatically considered by sampling the records for the desired fiscal period. When no written records exist, time must be explicitly sampled, such as minutes, half-minutes, hours, days, and so on.

All working minutes in a fiscal period is the time frame often used. Therefore, the fiscal period is determined by the appropriate number of days in a work week, such as a five-day work week or a six-day work week. Tested characteristics of cost estimates which are influenced by seasonal or monthly variations require a minimum time frame of 12 months to balance these influences. For example, budget planning (and estimates) require annual frames consisting of all subjects. In other ways, the time frame and its scope must be consistent with the purpose of the study.

The minute is a convenient sampling unit for use in a time frame because it is appropriate as an instant or as a short duration. As an instant, it can express an activity in proportions. Since most work activities have a duration longer than one minute, the minute sampling unit may be used to estimate aggregates, means and ratios.

For example, annual time frames may be expressed in minutes as follows:

1. One year is equivalent to 365 days or 525,600 minutes (365 × 24 × 60). In cases of 366 days, there are 527,040 minutes (366 × 24 × 60).
2. A work year of 250 days is 120,000 minutes (250 × 8 × 60).
3. A work year of 250 days with 30 minutes for lunch is 127,500 minutes.

The totality of these minute sampling units represents the frame. In using such a frame it is assumed that the observed activity will continue for the duration of the minute. Where this is not the case, a shorter time unit should be used as the sampling unit, say a second or a fraction of a second.

According to Rosander [58], the four types of employee frames found most often in practice are the following:

1. Each employee works at an assigned desk or table. The employee may or may not work with machines or equipment, and mobiiity is somewhat restricted. In these cases employess can be identified by name, location and organizational unit. Thus, a feasible procedure is to identify the employee frame by organizational unit or by location numbers, and all minutes on all employees in the unit, or at the location, at random

times. This is considered better than calling random times on a sample of individual employees because when everyone in a unit is covered at the same random time, it is easier to manage and control the procedure and the results. Either anonymous self-reporting or observers may be used; however, anonymous self-reporting is cheaper and may be better because the employee knows more about what he is doing than an observer.

2. Employees work in small groups at the work site, as in construction. Under such circumstances, the employee works in a somewhat physical area during a day. With such an employee frame, an observer records the required data for all employees in the group for every randomly selected minute. This procedure assumes that the number of employees is small enough to permit the observer to quickly observe and record each employee's activity. However, if the number of employees is more than ten, then the next method may be better.
3. Employees work individually or in small groups but not in a small physical area, such as a hospital. Under these circumstances, the work area is divided into a number of smaller mutually exclusive and exhaustive subareas. If the identity of these subareas is unknown, a study of the work area is necessary to determine feasible subareas. Then randomly selected minutes are assigned to randomly selected subareas.
4. Some employees, such as repair and field workers, move from one work area to another. For these kinds of employees it may be necessary to use 100 percent reporting.

Where employees are working with machines or equipment, the utilization of the machines or equipment can be built into the sampling plan.

The area frame consists of mutually exclusive and exhaustive subareas such as offices, factories, city blocks, counties, hospitals, construction sites, and so on. The number, size and shape of subareas depend on the concentration of employees in various parts of the work area during the work day. The following guidance is useful in setting up subareas:

1. Use natural boundaries such as four posts or pillars.
2. Use subareas sizes which prevent the concentration of too many employees.
3. Avoid subareas which contain no employees most of the time.
4. Use subareas which can easily be identified and observed.
5. Since subareas are not strata, they should be as heterogeneous as possible for activities and wage rates.

6. If heterogeneous subareas cannot be formed, consider the use of disproportionate stratified sampling so that employees contributing most to certain activities or wages are in separate strata and sampled at higher fractions.
7. Select subareas so that boundaries are clear and obvious.

For more detailed discussions and guidance on setting up frames for use in work sampling, the reader should refer to Rosander [58].

III. STEPS IN CONDUCTING A WORK SAMPLING STUDY

The following steps are recommended for work sampling applications:

1. Determine the work sampling objective. Establish a need for a study and then determine whether work sampling will provide the necessary information at a reasonable cost.
2. Decide on the output unit to be estimated. Prior to commencing the study, decide on the desired output unit to be measured for a specified time period.
3. Train personnel. Observers must be knowledgeable of the activities being measured and all personnel must be aware of the purpose of the study and the appropriate procedures to be followed. Observers should be instructed to base their observations upon the first instant they see an employee. This minimizes observation bias as well as employee's reaction. Aside from objectivity, the observers' manner in making the observations is important. They should use randomly selected alternate routes to the study area and be as inconspicuous as possible. For example, in a factory the observers should be dressed in factory type attire.
4. Stratify activities into observable categories. Activity categories must be observable and consistant with accounting and personnel classifications to facilitate audit checks and reviews. Some categories are considered productive activities, while others are classified nonproductive. Only the direct work is related to the accomplishment of the operation, but some amount of the other activities is inevitable. One of the aims of management is to seek an optimization of the proportion of direct work. Also, to assure consistent results, it is important that the activity categories be carefully predefined in familiar operational terms. Both self-reporters and observers need a complete and precise set of category decision rules.
5. Design reporting forms. Forms should be designed for the particular study with adequate descriptions of activity categories. Consider the use of punched cards and anonymous

self-reporting. The observation sheets (or self-reporting cards) and summary sheets should be designed to facilitate daily recapping. This procedure provides daily and cumulative proportions and relationships.

6. Estimate the number of random observations. The sample size is based on the desired precision. The time length of the sampling should correspond to the purpose of the study. For example, if there are quarterly allocations or cycles, then the sample should extend over the entire quarter. It is significant that an adequate sample be selected, since it cannot be increased after-the-fact.
7. Randomly select observations. For the results to be statistically valid, the observations must be random and independent of each other, so that the selection of each has no relationship to the selection of any other observation. However, some work sampling plans are so poorly designed that after the initial random observation, the times of the subsequent observations are obvious, and, therefore not random nor independent. Additionally, observations are biased when supervisors or others alert the staff just prior to the observations. Since there is a gradual reduction in human productivity during the course of a work day, observations should be made both in the morning and in the afternoon. Where anonymous self-reporting is used, a random telephone dialing device is useful.
8. Perform a test sample. This validates the activity categories and clarifies vague instructions.
9. Use simple procedures. Punched cards and anonymous self-reporting are examples of simplified procedures.
10. Use a control chart. To control nonsampling errors in reporting, a specified percent confidence limit control chart may be useful. The chart provides for management by exception as well as indicating progress and providing a basis for making comparisons.
11. Analyze, evaluate and appraise results. The analysis provides information on consistency, relationships and bases for improving future samples. Replication is recommended for estimating the proportion and the aggregate dollars for each specific activity category, and their standard errors. For example, four random minutes, being independent, constitute four replicates. More replicates may be used if desired. For a detailed explanation of the use of replication in work sampling, with examples and illustrative computations, the reader should refer to Rosander [58].
12. Develop a follow-up procedure. This provides a means of checking progress and noting trends and potential problem areas.

IV. IMPACT OF NONSAMPLING ERRORS

The subjectivity of activity observations is probably the major source of nonsampling errors. In some other test procedures the observations are objective, such as testing light bulbs; they either light or they do not. This degree of objectivity is not present in the activity categorization of work sampling.

Observations which are not essentially binary are relatively subjective, and therefore, require considerable observer job knowledge and judgment. Also, experience reveals considerable variation among observers, as well as by the same observer over time. Thus, it is imperative that the variaticns among observers be measured, analyzed, and adjustments made. Otherwise, the work sampling results are much more imprecise than the computed evaluation indicates.

For this reason, Parish [51] contends that the traditional method of computing confidence limits for attributes is not appropriate for work sampling. He recommends that the percent a given category is observed by a single observer, or in a single tour, or on a particular day, be regarded as a single data point, and that the confidence limits be based on the standard deviation between these data points and the "t" Distribution. The calculation should be made for each source of variability that can be observed. Also, the confidence limits used for a work sampling study should be the widest calculated interval of a specified confidence level. Using this method on many of the work sampling studies would produce confidence limits so wide as to render them useless. Yet, one must endeavor to compensate for the wide variations associated with the subjective nature of most work sampling observations. Otherwise, the results will be misleading.

Furthermore, the behavioral pattern of workers being observed can distort the results. One would think that the distortion would always favor more direct work, but in highly unionized shops there is a tendency to stop or retard working when an observer appears. Also, groups of employees may temporarily cluster together to prevent definitive observations of their activities. Behavioral problems may be minimized by using anonymous self-reporting.

There is a tendency in work sampling applications to ignore the control and analysis of nonsampling errors, such as erroneous activity categories, observer bias, careless recording, missing data and incorrect calculations. These factors, however, may have a far more adverse impact on the results than the computed sampling error (precision).

In addition, data gathered from a single, isolated work sample may be of limited value. The auditor and evaluator should compare the results of the given study with previous or similar studies, with industry ratios and norms, or with other bases of comparison. Furthermore, many single, isolated work sampling studies are either not statistically evaluated or the appraisals are inappropriate because significant non-sampling errors are not reflected in the decision. However, one should use caution in comparing work sampling studies where there were different observers, different definitions of categories, different training methods, and so on. There should even be concern about comparing results with similar studies at the same location, with the same observers, because observers tend to change their viewpoints over time.

V. SOME WORK SAMPLING APPLICATIONS

Rosander [58] includes work sampling applications among his collection of case studies. Summaries of several of these cases follow.

Work Sampling in a Large Work Area

The large work area was a railroad freight platform where boxcars were being loaded and unloaded. The work day extended from 8 a.m. to midnight.

The platform was divided by fixed objects into twelve easily identifiable areas, none containing more than ten workers. Sample areas were balanced over two eight-hour work periods: 8 a.m. to 4 p.m. and 4 p.m. to midnight, so that each of the twelve areas was selected once and only once in each four-day period. In accomplishing this, each eight hour period was divided into four two-hour periods, and a random minute was selected from each.

Since the observers knew the workers and the work activities, it was easy for them to observe the activities of the cluster of workers in each area when the random minute occurred.

A ratio estimator was used because the total number of workers observed as well as those engaged in a specific activity varied among work areas, and in the same work area from one random observation to another.

Work Sampling to an IRS Division

Rosander, Guterman and McKeon [59] describe in detail a random time sampling application to the activities of the Statistics Division of the Internal Revenue Service. Rosander [58] has summarized the report as a simplified case study, including the appropriate computations of activity proportions and aggregate dollars, as well as their standard errors.

When the primary purpose of a work sampling study is to allocate costs to activities or to estimate cost reimbursements, the procedures used in this case study are appropriate.

In this study, the employees were located in contiguous offices where telephones were conveniently located. Management needed information on employee activities and utilization for planning, scheduling and budgeting. Since budgeting was involved, every work day of at least one calendar year had to be included. Also, every minute of the work day had to be included in the time frame. A work day consisted of 8.5 hours, or 510 minutes, including a 30-minute lunch period.

The final sampling plan ran continuously for 26 months. It involved anonymous self-reporting with random times (minutes) telephoned from a central location. Each employee was furnished an instruction and activity coding book together with a pad of checksheet-type data cards, so that the employee could post the proper code to the data card whenever a random time was telephoned. Every minute in the work day was sampled, including the lunch period. Personnel were stratified between operating and planning. A random minute was used as an indicator of an instant when estimating proportions of time, and as a duration or interval when estimating salary costs.

Since anonymous self-reporting has been mentioned several times, the experience from this study with the procedure should be of interest. It was concluded that anonymous self-reporting works well if employees are reassured that it will not be used for performance rating or similar purposes. Its low cost is far less than the use of observers. Independent validation checks can be made of the information given on the data cards, such as proportions of time spent at lunch and on leave. Also, the immediate review of the data cards by supervisors or secretaries reduces misreporting, inconsistencies and omissions. Usually a secretary prepares data cards for employees on leave. Thus, the entire office is accounted for through a control check.

Work Sampling of Clerical and Machine Operations

This application involved the preparation and mailing of bills to customers of a large telephone company.

Prior to the introduction of work sampling, monthly time sheets were kept by each employee. These sheets identified various activities and the amount of daily time spent on each activity. The monthly results were used to estimate the relative time spent on various activities, and for comparison with past experience and with the performance of other companies in the Bell Telephone System.

A test was conducted using both the time-sheet system and work sampling simultaneously on the same operations and clerks. Both methods tended to produce similar results for clerical production hours. However, where many different operations were performed on a variety of machines, it was decided, based on the test, that random time sampling provided superior data for the kind of information required. Consequently, the decision was made to use random time sampling in the miscellaneous machine units of the telephone company.

This case study is summarized from a paper by McMurdo [43]. Readers who desire more details and standard error computations, should refer to that paper or to Rosander [58].

According to Kennedy [30], the Bell Telephone System has used work sampling methods in its functional accounting program to allocate expense or to estimate the average time per transaction for cost allocation. Under this program, the average time per type of transaction is estimated by determining the proportion of time spent on a particular type of transaction and dividing that estimated time by the known volume of such transactions processed during the study period. The average times are used to determine the costs of processing various types of transactions.

This work sampling method is used to estimate the mean and variance of the results. The transactions may be processed by either employees or machines. In brief, a sample of a specified number of observations is selected from all employees and machines included in the study. The observations are classified as transactions being processed or not being processed. The total number of transactions completed during the study period are usually available from records.

Kennedy describes two ways of estimating the average time per transaction from a fixed time period work sample.

One method is replication, which is described in Chapter 13. It involves forming equal size subsamples of time periods and computing the estimates of the average transaction time for each subsample and for the overall average of all subsamples.

The confidence limits of the overall average estimate resulting from replication reflect the effects of the number of transactions as well as differences among time periods.

The subsamples could be groups of employees; however, each subsample must have its own total count of transactions, and this count may not be readily available for each group. By contrast, it is more likely that the total number of transactions is available for subsamples of time periods, such as a day or a week.

Another method of estimating the average time per transaction involves identifying the transaction when it is observed in process and tallying the number of times that it is observed. It may, however, be difficult to identify a specific transaction and determine the number of times it has been observed.

Mandel [41] describes a continual application of work sampling by the Post Office Department to estimate and allocate the costs of handling each class of mail as a basis for recommending postal rate changes. A stratified, multistage sample was used. The primary sampling unit consisted of post offices which were stratified by annual revenue. The second stage was composed of employees within the selected post offices, selected on the basis of the last two digits of their social security numbers. Finally, random instants of time were selected for observing employees. A selected employee was observed four times during the work day he was in the sample; however, only the first observation was randomly selected within the first two hours of the work day. The other three observations were at a two-hour fixed interval.

The proportion of time devoted to a specific class of mail was estimated by dividing the number of observations of that activity by the total number of sample observations. The number of hours spent on a specific class of mail was estimated by multiplying the proportion of all observations made on that activity by the total number of work hours spent on all activities. Finally, the corresponding cost for a specific class of mail was estimated by multiplying the estimated number of work hours by the average cost per hour for that activity.

There was an additional problem in estimating the in-office costs of mail carriers, since in-office salary costs were not available for carriers. In these cases, estimates were derived for the proportion of time carriers spent on each class of mail in the office and on total work time out of office. These proportions were multiplied by total salaries of carriers to estimate in-office costs by class of mail to total (not by class of mail) out-of-office costs.

The standard error for this stratified three-stage sample study was computed by multiplying three times the simple random sampling error. However, replication would seem to provide more appropriate estimators and standard errors.

VI. ADVANTAGES AND LIMITATIONS OF WORK SAMPLING

A. Advantages of Work Sampling

1. Ease of application.
2. It is more economical than a time study.
3. Cost is minimized and reporting knowledge is optional with anonymous self-reporting.
4. It can be used to estimate nonproductive time.
5. Does not require the use of a stopwatch.
6. If properly designed, the sample size and the length of the time period will reveal unusual conditions.
7. It can be used to estimate gross activity in irregular demand-type labor activities.
8. It can be used to estimate overall performance.
9. If the sample design is valid and properly executed, the reliability (precision) of the estimates can be calculated.
10. With proper controls, it can be used to measure progress over time.

11. It can be used over long-cycle operations.
12. Less interference with operation, and consequently, more favorable employee reaction when compared with other work measurement methods.

B. Limitations of Work Sampling

1. Work sampling is not a direct measure of productivity. It only measures what employees appear to be doing. It does not measure the amount of work accomplished, the quality of the work, the relative merits of the methods used, nor whether the activity is essential.
2. It is a poor substitute for time studies in direct labor measurement.
3. It will not replace conventional methods for job analysis.
4. It does not reduce the need for planning, programming, and scheduling.
5. A very large number of observations over a long period of time is required for reasonable precision.
6. When dollar estimations are required, the attribute sampling aspects of work sampling are inappropriate.
7. Random observations must be distributed over all work hours for the study period.
8. Time intervals between observations must be sufficient to provide independent readings.
9. It will not identify problem areas not associated with the utilization of personnel, machines or equipment.
10. Results may be inconclusive or misleading due to invalid sampling design or execution.
11. It is difficult to control significant nonsampling errors due to the subjective aspects of the observations.
12. The sampling results may be difficult to properly evaluate and explain.
13. There is generally a lack of sufficient documentation and detailed information for audit purposes.
14. If a study period of a relatively short duration is judgmentally selected for making random observations, the observations may not be a random sample. To avoid this problem, the sample should be selected from the entire fiscal period, especially when used for budget purposes.
15. Applying work sampling in a large work area may pose a hardship for observers. For instance, Parish [51] revealed that a work sampling tour route at a power plant construction site was about five miles, and that each observer walked two tours a day for nine consecutive work days. Incidentally, there were data for only eight days because a rain storm

stopped the study one day. On the bright side, it was easy to determine which craft a worker belonged to by the markings on his hardhat.

VII. SUMMARY

Work sampling involves the use of random observations to estimate the proportion of time spent on various work activities. It can be used to estimate the average time per unit of effort when the total units of effort are known.

Work sampling does not measure the time required to perform some sequence of activities. It, however, can be used to describe possible work activities in terms of general categories, and then estimate the proportion of the total average work day spent in each category. Some early work sampling studies used only two general categories: productive activities or nonproductive activities. In subsequent studies the two general categories of activities have been divided into subcategories which are pertinent to the particular study situation.

During random observations the state or condition of the employee, machine and equipment is noted, and this state is classified into one of the predefined subcategories. From the proportion of observations in each category or subcategory, inferences are made concerning the total activity or operation.

It is important to realize that the universe to be sampled is not the work staff. The sampling unit is what each employee is doing at any particular instant of time.

Work sampling, if properly designed and executed, may be considered a compromise between the extremes of subjective opinion and 100 percent time-and-attendance reporting and analysis. However, an adequate and properly executed work sampling study is expensive. Aside from the activities of observers, there is the cost of training, quality control, supervision, data reduction, and analysis and evaluation of results. On the other hand, where there are no written records, work sampling provides a means of gathering management information which may be impractical to obtain by any other method.

In work sampling applications, one should be careful not to introduce a bias which could result from (i) observations taken at the whim of the observer rather than in accord with a random selection scheme, (ii) sampling from only a portion of the universe, and (iii) not designing the sample to provide for all variable work activities and wage elements.

Many activity categorization work sampling studies have either failed to include statistical evaluations or have tended to overstate the usefulness of the results by ignoring the significant impact of the nonsampling errors of subjective observations, erroneous or incomplete activity categories, careless recordings, missing data and incorrect calculations.

12

Monetary-Unit Sampling

I. GENERAL

In monetary-unit sampling any monetary unit such as a dollar, pound, mark, franc, yen, etc., may be used as the sampling unit. In this chapter, dollars will be the unit used.

Basically, in dollar-unit sampling the auditor randomly selects individual dollar units from the universe and examines the audit units (vouchers, invoices, line items, etc.) containing the selected dollar units to determine the status of the dollars. Thus, an audit unit's probability of selection is related to its size (dollar value). For example, a transaction with a book value of $5,000 has a chance of selection ten times greater than a transaction of $500.

In sampling for attributes, the results of an audit are expressed as error rates. That is, one incorrect voucher in a sample of 100 would be a one percent sample error rate. The sample error rate is considered as the error rate (point estimate) of the universe. Also, if a random sample is selected, the auditor can compute a probabilistic upper error limit which the actual universe error rate is not expected to exceed.

When sampling for variables, one is dealing with variation in some measurement possessed by every number of the universe. Examples of sampling for variables are: the average value of a group of invoices or the total amount of undocumented costs in a set of records. Similarly, as for attributes, upper dollar error limits can be computed for random samples.

An unstratified sample of dollar units requires the random selection of individual dollars in the universe, rather than the selection of audit units. With this modification, the unstratified selection of dollar units is equivalent to the unstratified selection of audit units. Thus, the universe size is now the total number of dollars recorded in the universe. Since the sampling unit is an individual dollar, the maximum allowable cost (overstatement) for any unit is $1.

In a sample to estimate the total dollar error in some amount of an audit universe, the conventional audit-unit methods used to compute confidence limits are likely to be unsatisfactory unless the sample size is very large or there is extensive audit stratification. Thus, if the estimate is computed by expanding the error amounts in the sample or by employing the ratio of these amounts to the related book amounts, the validity of the usual procedure for computing confidence limits is likely to be impaired, because often all, or nearly all, the error amounts in the sample are equal to zero and the universe is highly skewed.

Auditing experience indicates that the error in a particular book value included in some total amount on a financial record practically never exceeds that book value. Taking advantage of this assumption, dollar-unit sampling is a procedure for selecting the sample and computing an upper dollar confidence limit.

The original method of dollar-unit sampling was strictly an attribute approach with dollar implications. The modification involved defining the sampling unit as an individual dollar instead of an audit unit, such as a voucher. For example, suppose the auditor were sampling from $1 million of vouchers. Instead of considering the universe as 10,000 individual vouchers, he regards it as $1 million. If he selects a simple random sample of 100 of these individual dollars and finds no errors, then the cumulative upper error limit attribute factor in Table 12.1 indicates that there is a 95 percent assurance that there are no more than 3 percent errors among the 1 million individual dollars. If there were 3 percent of undocumented dollars, the total error for the $1 million would be $30,000. Thus, based on no error in a random sample of 100, the auditor has 95 percent confidence that the total undocumented cost in the vouchers is not more than $30,000.

Using the same information in the above illustration, in systematic sampling for attributes, every 100th (10,000/100) voucher would be selected, after a random start. In dollar-unit sampling, by analogy, every 10,000th (1,000,000/100) dollar (that is, the voucher on which such a dollar appears) is selected for audit examination. Instead of the sampling interval being every 100th voucher as in audit-unit sampling, it is every 10,000th dollar. If there is a large number of transactions, a computer may be required to add progressively through the universe of dollars.

To illustrate dollar-unit sampling, suppose a balance sheet shows \$2 million as the amount of accounts receivable. The auditor desires to verify no more than 100 of these accounts and then use the sample data to compute an upper confidence limit for the amount of dollar error in the \$2 million so that he can be at least 95 percent confident that the actual amount of dollar errors will not exceed this limit.

The first step is to select a sample of 100 dollar units; by considering each dollar among the \$2 million as a sampling unit and selecting 100 of these units randomly. The account receivable corresponding to each of the 100 dollar units selected is then verified. Next, the amount of any error in the book value shown for that amount is determined. Thus, the number of accounts verified may be as large as 100, but will be less if more than one of the selected dollar units belong to the same account receivable.

In the example of \$2 million, if no error is detected, the auditor has 95 percent confidence that at worst 3 percent (or \$60,000) undocumented or erroneous dollars exist in the universe. Each one of these potentially erroneous dollar units is assumed to be 100 percent erroneous. Although no errors are found, it is conservative to compute the worst possible errors, assuming no voucher could be overstated by more than 100 percent of the reported book value.

In practice, however, the auditor may find errors in the sample. In the example, if one 100 percent error is detected in the sample of 100, the auditor would have a 95 percent confidence that the universe error rate does not exceed (4.75 ÷ 100) × (\$2 million) or \$95,000.*

For a particular transaction, the ratio of the dollar error to the book value, expressed as percentage, is referred to as the tainting percentage. This percentage is attached to each dollar unit making up the book value for a particular transaction and is assumed to be no greater than 100 percent for any dollar unit in the universe sample. The objective is to compute an upper confidence limit for the mean of the tainting percentages in that universe.

For example, suppose the one error resulted from a \$500 reported book value that should have been \$400. The transaction has been overstated by 20 percent of its reported book value. Thus, the 500 dollar units on that transaction are overstated by a 20 percent error, not a 100 percent error. One must, however, realize that 100 percent errors are still possible in the universe even though the only error found was a 20 percent monetary error.

In this example, all of the 100 tainting percentages in the sample are zero except for one which is equal to 20 percent. Using these data

---

*The 4.75 is the one-error factor shown in Table 12.1.

Table 12.1. Poisson Cumulative Upper Error Limit Rates

Number of errors	Confidence level	
	80 percent	95 percent
0	1.61	3.00
1	3.00	4.75
2	4.28	6.30
3	5.52	7.76
4	6.73	9.16
5	7.91	10.52
6	9.08	11.85
7	10.24	13.15
8	11.38	14.44
9	12.52	15.71
10	13.66	16.97
11	14.78	18.21
12	15.90	19.45
13	17.02	20.67
14	18.13	21.89
15	19.24	23.10

Note: The above factors are the products of sample sizes and error rates for the particular confidence level. For example, if one desires the sample size for 95 percent confidence that the error rate does exceed 3 percent, then:

$$\text{Sample size} \times \text{error rate} = 3 \text{ (from above table)}$$
$$\text{Sample size} = 3 \div \text{error rate}$$
$$\text{Sample size} = 3 \div 0.03$$
$$\text{Sample size} = 100$$

with the most prevalent method of computing the upper confidence limit, a 95 percent upper error limit is computed in Figure 7 using the appropriate factors from Table 12.1.

The upper confidence limits for the error rates are based on the Poisson distribution. For an extensive table of the Poisson distributions,

Number of Errors in Sample (a)	Tainting Percentage of Error (b)	UEL* Incremental Factor for 95% Confidence (c)	Projected Effect of Error (b) x (c)
0	100	3.00	3.00
1	20	1.75	.35
	Total Upper Error Limit (UEL) Factor		3.35

* The incremental factor for one error (1.75) is obtained by subtracting applicable adjacent cumulative factors in Table 12.1. In this example, 4.75 - 3.00 = 1.75.

The following formula is useful in computing the UEL:

$$\text{UEL} = \frac{1}{\text{sample size}}\,(\text{zero error factor + tainting ratio x incremental factor})$$

$$\text{Thus, UEL} = \frac{1}{100}\,[(3.00 + 0.20\,(4.75 - 3.00)]$$

$$\text{UEL} = \frac{1}{100}\,(3.00 + 0.20 \times 1.75)$$

$$\text{UEL} = \frac{1}{100}\,(3.35)$$

UEL = .0335 or 3.35 percent

Figure 7. 95 percent upper confidence limit for one 20 percent monetary error.

showing upper and lower limits for more error rates and several confidence levels, see Leslie, Teitlebaum and Anderson [37].

Hence, the 95 percent upper confidence limit for the total dollar error is 3.35 percent of $2 million, or $67,500, for a sample of 100.

II. EVOLUTION OF DOLLAR-UNIT SAMPLING

Prior to 1960, most applications of statistical sampling to auditing involved classical sampling methods based on large-sample, normal-distribution assumptions. Stringer [66] and others began to advocate other sampling methods as being more appropriate for the low error rate, highly skewed universes usually encountered in audit activities. Although Stringer never published his techniques, his public accounting firm, Haskins & Sells, implemented them. The essential feature of this sampling strategy, as described by Meikle [44], is based on proportionate sampling.

It appears, however, that van Heerden [81] of The Netherlands was the pioneer in the field of monetary-unit sampling. He developed the guilder-unit method in the 1950's. It was strictly an attribute approach since each sampled guilder was considered either totally correct or totally incorrect. Also, all errors were treated as maximum errors. For example, suppose the universe consists of 10,000 accounts with a book value of $2 million where the largest account balance is $1,000. Also, suppose that a sample reveals a two percent overstatement error. Then, two percent of the 10,000 accounts, or 200, would be priced at the value of the largest account balance of $1,000, resulting in an upper limit, or bound, of $200,000.

In the adaption of proportionate sampling (described in Chapter 8) to auditing, the probability of selection is made proportional to the book value of a transaction. This is achieved by designing the sample such that each stratum is relatively narrow, thus having all book values in a stratum close to the largest stratum book value.

Except for the Haskins & Sell's programs, apparently little research effort was devoted to this area until 1973 when Anderson and Teitlebaum [1] suggested the use of an individual dollar as the sampling unit. The auditor would still examine the entire transaction to which the sample dollar belongs, but he would allocate discovered errors uniformly to all dollars in the transaction. This is called the tainted dollar approach.

Dollar-unit sampling may be viewed as an extension of proportionate sampling. For example, when the universe is stratified by dollar value, one method is to allocate the sample in proportion to the dollar value of the transactions in each stratum. If this approach is carried to the point where each transaction constitutes a stratum, the result would be dollar-unit sampling. Consequently, dollar-unit sampling optimizes the efficiency obtained through dollar stratification. Furthermore, it eliminates the problems associated with determining strata boundaries and in allocating the sample among the strata.

Since 1973, there has been a proliferation in the number of articles and presentations on dollar-unit sampling and related subjects as shown among the references culminating in the publication of the first book on the subject in 1979 by Leslie, Teitlebaum and Anderson [37].

III. AUDITOR DECISIONS IN APPLYING DOLLAR-UNIT SAMPLING

While dollar stratification is unnecessary in dollar-unit sampling since monetary stratification is automatically achieved by the process, audit stratification or separation of sensitive, problem and zero-valued transactions is still required. The amount of the monetary stratification threshold, which the auditor specifies, should not be larger than the average sampling interval or cell width. Thus, all transactions with values equal to or greater than this threshold will be automatically stratified.

In many audit situations, it may be desirable to also accumulate negative amounts or credits in a separate stratum. This procedure will obviously facilitate the evaluation of the sample results when there are both overstatements and understatements.

Additionally, the audit objective may require the separate stratification of transactions associated with special and sensitive operations or activities where the monetary amounts vary significantly.

Generally, dollar-unit sampling will involve only one or two explicit strata. It could, however, involve more if the universe were stratified by accounts or activities. Even then, the number of strata may be kept small by separating the accounts into broad categories according to expected quality, such as above average, average and below average.

Although the mechanical aspects of dollar-unit sampling are easy and simple, several judgments and decisions are necessary prior to its effective implementation. For example, when dollar-unit sampling is used for substantive tests, the strategy involves the following actions:

1. Determination of an appropriate materiality limit, which is a prime consideration, since one of the objectives is to provide a reasonable assurance that the financial statements are not materially in error. Therefore, the monetary upper error limit used should not exceed the minimum total amount of errors which would be considered material in the particular audit. However, the auditing concept of materiality is a judgment determination which is independent of the sampling procedure. Leslie, Teitlebaum and Anderson [37] suggest a limit of 5 to 10 percent of pretax net income. For governmental agencies and nonprofit organizations some other criterion is necessary.
2. Based on experience and the appraisal of other sources of reliance, estimate the total expected monetary error.
3. Based on the above information, a sample size is estimated for the desired confidence level that the total expected monetary error will not be exceeded.

4. The average sampling interval, or the width of the cell, is then determined by dividing the total book value by the sample size.
5. The sample result is evaluated in terms of a probabilistic upper monetary error limit or bound.

IV. SELECTING DOLLAR-UNIT SAMPLES

A. Some Selection Methods

There are several methods of randomly selecting dollar-unit samples. The most obvious is the fixed-interval, or systematic, selection with a single random start. Some auditors, however, avoid this selection procedure for fear of pattern or trend problems. Multi-starts with systematic selection minimize these problems.

Other methods are simple random selection, cell selection and varying-interval selection. The four methods of randomly selecting dollar-unit samples will be discussed and illustrated in this section. Of the four methods discussed, the cell selection method is generally preferred because the sample items are somewhat evenly distributed throughout the universe while avoiding the potential problems associated with systematic sampling.

In the illustrations in Figure 8, which is adopted from Leslie, Teitlebaum and Anderson [37], each rectangle represents a universe of $50,000 from which a sample of four dollar units is selected by the indicated methods.

For the systematic, or fixed-interval, sample, a random start within the $12,500 ($50,000 ÷ 4) interval is selected; in this example, say, dollar number 3,098 is the first dollar unit. This is followed by every 12,500th dollar. Thus, the sample would consist of dollar units 3,098, 15,598, 28,098 and 40,598. The sampling process involves counting dollars throughout the universe to pick every 12,500th dollar after the random start. Hence, to select a dollar-unit sample it is necessary to add through the universe, using an adding machine or a computer.

In simple random selection, random points (dollar units) are randomly selected separately and independently throughout the universe. In Figure 8b, the first randomly selected dollar unit is 30,405; the second is 8,362; the third is 21,885; and the fourth is 38,534.

As in the case of audit-unit sampling, simple random selection gives each dollar unit an equal chance of selection; however, the units need an identity to permit this type of selection. If they are not identified, it is necessary to count through the universe to find the randomly selected points. Thus, the use of simple random selection for dollar-unit sampling is not usually as economical and convenient as varying interval selection.

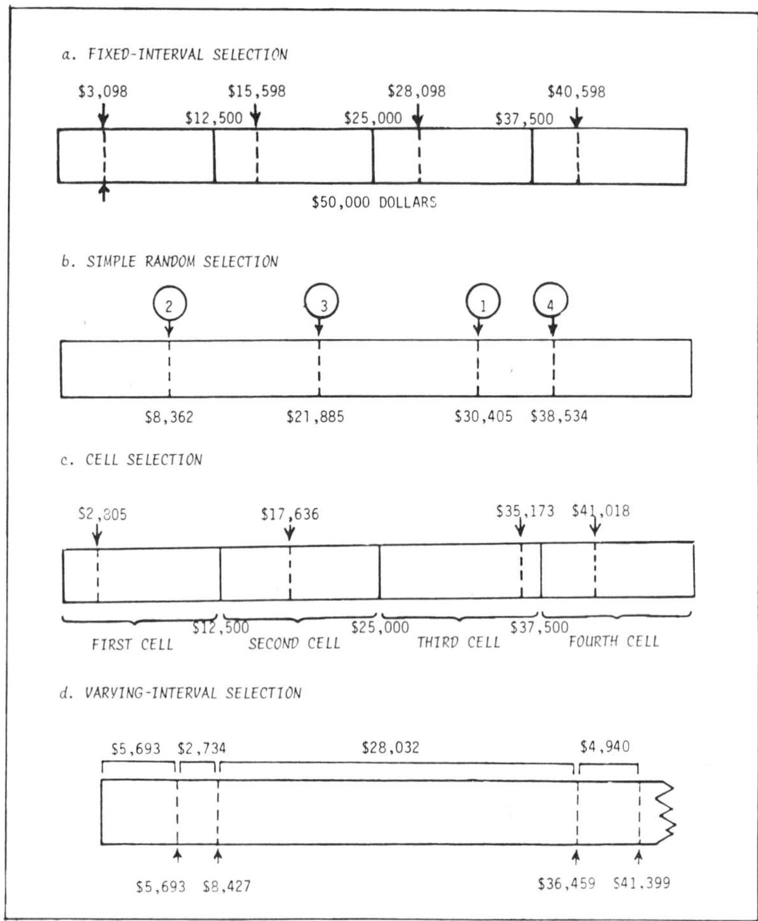

Figure 8. Methods for selecting dollar-unit samples.

For cell selection, the universe is divided into equal size cells starting at the beginning of the universe. Then random points (dollar units) are selected within each cell. To locate each of the selection points (units), it is necessary to determine the interval between adjacent points. As shown in Figure 8c, the first interval is the distance from the beginning of the universe to the first random unit, the second is the distance from the first to the second random unit, and so on to the end of the universe. The determination of intervals for cell selection is significant, because the objective is to achieve the proportional coverage inherent in systematic sampling while avoiding the potential problem of monotonic or periodic trends in the data coinciding with the selection interval.

To illustrate cell selection, suppose the universe consists of \$2,500, see Figure 9, and one desires to select a sample of 5 units. Then the width of a cell, also referred to as the average sampling interval, is \$500 (\$2,500 ÷ 5). To obtain intervals for the cell selection method, the auditor needs to use a random digit table, such as Appendix B. He selects random numbers between 1 and the cell width, in this case between 1 and 500, until the sample size is reached.

Using the illustration in Figure 9, the first randomly selected number is 127; therefore, the first interval is \$127 into the first cell. Thus, 127 is both the cumulative selection point and interval value. The second randomly selected number is 432, and it is \$432 into the second cell. Since the 432 is in the second cell, it is one cell width (500) plus the 432 into the second cell, resulting in a cumulative

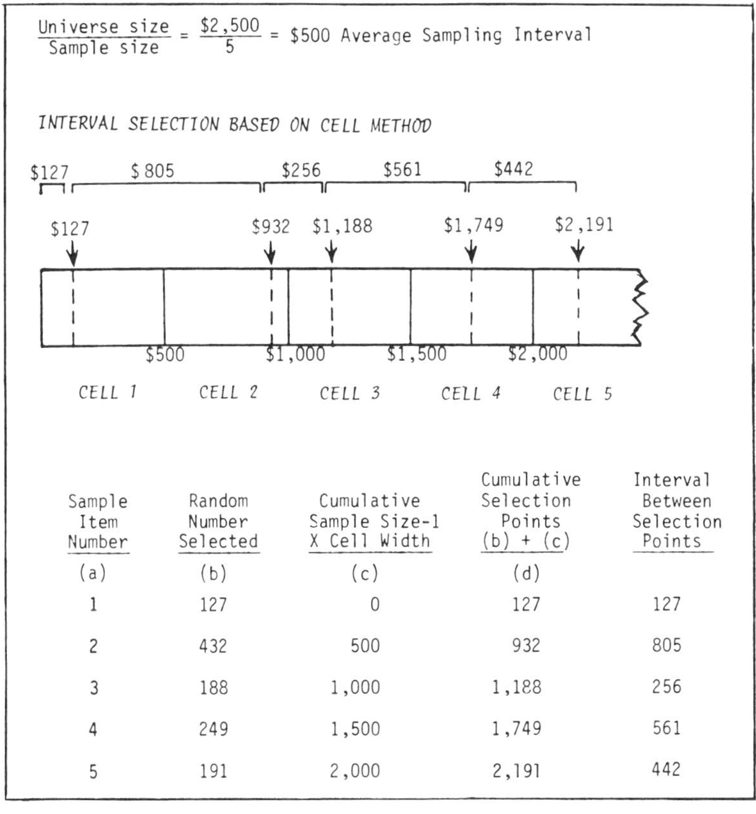

Sample Item Number	Random Number Selected	Cumulative Sample Size-1 X Cell Width	Cumulative Selection Points (b) + (c)	Interval Between Selection Points
(a)	(b)	(c)	(d)	
1	127	0	127	127
2	432	500	932	805
3	188	1,000	1,188	256
4	249	1,500	1,749	561
5	191	2,000	2,191	442

Figure 9. Determination of interval for cell selection.

selection point of 932. Therefore, the interval between the first and second selection points is $932 minus $127 or $805. Each of the other cumulative selection points and intervals is computed in a similar manner. Obviously, these cumulative and interval determinations could be easily accomplished by use of a computer.

As mentioned before, cell selection is equally applicable for audit-unit sampling as for dollar-unit sampling. In dollar-unit sampling the auditor accumulates dollars to locate a particular sample unit while in audit-unit sampling he counts or measures through the transactions to select the units.

Varying-interval selection is similar to cell selection except there is no constraint on the distance between adjacent selection points. Some varying-interval procedures simulate simple random selection. For instance, one could use a random digit table to select a sample of random points within the universe size, arrange them in ascending order, and compute the values of the intervals. This would be one method of varying-interval selection. In Figure 8d, these intervals are 5693, 2734, 28,032 and 4940.

In summary, cell selection is generally preferred over the other selection methods. In the first place, the sample is somewhat evenly distributed throughout the universe while avoiding the potential problems of systematic selection. Neither simple random selection nor varying interval selection provides this assurance. Both of these methods permit very large intervals between some selection points, especially if the sample is not very large. Also, cell selection permits no more than two multiple selection points in any audit unit, and this can only occur when the audit unit is in two adjacent cells.

Several short-cut methods for selecting dollar-unit samples are described and illustrated by Leslie, Teitlebaum and Anderson [37]. These include the use of cumulative totals, frequent subtotals, page totals and a two-stage audit-unit/dollar-unit method. These methods save time in adding through the universe, especially when there are many relatively small value items in the universe.

B. Illustration of a Dollar-Unit Sample Selection

Although fixed-interval selection is not the preferred procedure for selecting a dollar-unit sample, it is easy and simple, and therefore, will be used to illustrate dollar-unit selection.

Suppose in a given sampling frame the auditor wants to examine all transactions over $1,000 or show a credit balance, and give all other transactions a probability of selection equal to the book value divided by $1,000. One of the first steps would be to select a random start between 1 and 1,000, say 230. He would then accumulate the values of transactions in the sampling frame, selecting those

Amount of Transaction	Random Start & Cumulative Interval	Cumulative Debits in Sampling Frame	100% Selection Debits Over $1,000	100% Selection All Credits	Sample Transaction Selected
$ 263	$ 230	$ 263	$		$263
522		785			
368		1,153			
1,833			1,833		
498	1,230	1,651			498
(353)				353	
120		1,771			
832	2,230	2,603			832
2,859			2,859		
45		2,648			
382		3,030			
942	3,230	3,972			942
102		4,074			
(467)				467	
96		4,170			
58		4,228			
142	4,230	4,370			142
783		5,153			
etc.	etc.	etc.	etc.	etc.	etc.

Figure 10. Illustration of fixed-interval dollar-unit selection.

transactions containing the 230th dollar, and so on. It is expeditious, however, to provide one stratum for values greater than the sampling interval ($1,000) and another stratum for credits. The auditor can decide later whether to examine these strata in detail or select replicate samples from them. An illustration of a fixed-interval selection of a dollar-unit sample is shown in Figure 10. Although separate strata are provided for material amounts and credits, the procedure does not provide adequate assurance for the selection of sensitive or problem transactions of varying dollar amounts, such as those for ineligible participants in a program or operation.

One of the advantages of dollar-unit sampling is that it can eliminate the need for two-stage sampling. For example, if the transactions selected in the sample in Figure 10 were vouchers on which a number of different invoices were paid, the examination could be limited to the invoice containing the 230th, 1230th, 2230th, and so on, dollar unit in the sampling frame. That is, the 230th dollar in the $263 voucher, the 77th (1230 − 1153) dollar in the $498 voucher, the 459th (2230 − 1771) dollar in the $832 voucher, and so on.

It should be emphasized that if the auditor could limit his examination in this example in Figure 10 to merely the 230th, 1230th 2230th and so on, dollars in the frame, he would be examining a randomly selected sample of dollar units. Hence, if it were possible to examine 100 randomly selected individual dollar units and no errors were detected, the auditor would have 95 percent confidence that the number of erroneous dollars in the frame did not exceed 3 percent of the book value. In practice, the examination would normally cover all of the dollars in each transaction selected. Thus, the auditor is sampling clusters of dollar units, but the evaluation procedure does not consider the fact that these additional dollar units are examined. In other words, the auditor has additional evidence, and therefore, additional assurance, from the sample which is not reflected in the evaluation of the results. To take advantage of this additional information, however, would recreate the distributional problem which dollar-unit sampling endeavors to avoid. The point to be made is that the dollar-unit sampling evaluation procedures currently used tend to be conservative.

V. SAMPLE SIZE DETERMINATION FOR DOLLAR–UNIT SAMPLES

There is no specific sample size for dollar-unit sampling since it depends upon the auditor's estimate of the underlying error rate as well as materiality and desired confidence. Also, the statistical evaluation is related directly to the errors found in the sample. By contrast, in audit-unit sampling for variables, the sample size and the sample evaluation depend upon the variability of the items and less on errors detected. However, the auditor should investigate for audit significance and trends any errors detected by any procedure.

There are several interrelated ways of estimating the sample size for a dollar-unit sample. The simplest way is to divide the universe dollar total by the average sampling interval. Thus, if the universe consists of $3 million and the desired sampling interval is $20,000, then:

$$\text{SAMPLE SIZE} = \frac{\$3{,}000{,}000}{20{,}000} = 150.$$

Another way to compute the sample size is to divide the Poisson distribution confidence factor by the tolerable error rate (upper error limit.) The tolerable upper error limit is the ratio of dollar materiality to total universe dollars. To illustrate, suppose the auditor has a materiality of $60,000 for a universe of $4 million, then:

$$\text{TOLERABLE UPPER ERROR LIMIT} = \frac{60{,}000}{4{,}000{,}000} = 0.015.$$

The Poisson factor for a 95 percent confidence level, assuming no errors, is 3, as shown in Table 12.1. Dividing this factor by the tolerable upper error limit of 0.015, results in the following:

$$\text{SAMPLE SIZE} = \frac{3}{0.015} = 200.$$

If the upper error limit formula is combined with the sample size formula in the preceding paragraph, one obtains another approach to sample size determination; that is, multiply the universe dollars by the appropriate Poisson factor and then divide the product by the materiality amount. For example, using the same information in the last illustration, and assuming no errors, the result is:

$$\text{SAMPLE SIZE} = \frac{4{,}000{,}000 \times 3}{60{,}000} = 200.$$

The above calculations are based on the assumption that no errors are found in the sample. If the auditor could tolerate one error in his sample, then the Poisson factor would be 4.75, as shown in Table 12.1, and the revised sample size would be:

$$\text{REVISED SAMPLE SIZE} = \frac{4.75}{0.015} = 317.$$

Also,

$$\frac{4{,}000{,}000 \times 4.75}{60{,}000} = 317.$$

The actual dollar-unit sample selected may be less than the sample size used to compute the sampling interval. From Appendix D, a sample of 150 would be required to obtain 95 percent confidence that the error rate does not exceed 2 percent, assuming no error is detected in the sample. Dividing this 150 into an account balance of, for example, $15 million would yield an interval of $100,000. If all transactions were $100,000 or less, 150 transactions would be selected. However, if 10 transactions for more than $100,000 accounted for $10 million, only 60 transactions would be examined: the 10 over $100,000 and a sample of 50 under $100,000 ($5,000,000 ÷ $100,000 = 50).

If the upper error limit obtained from a dollar-unit sample is deemed too high, the examination could be extended by reducing the sampling interval. Suppose, for example, that a $1,000 interval has been used to select the sample, and the upper error limit is considered to be too much. The interval could be reduced to $800, which would exclude all items over $800 from the sample. If this results in a satisfactory sampling interval, the audit objective could be achieved by extending the examination to all items over $800. Since these items had at least an 80 percent probability of selection initially, this modification should not result in a significant increase in audit effort. An alternative is to purposefully use a smaller selection interval which would provide a larger sample.

VI. EVALUATION OF DOLLAR-UNIT SAMPLES

A. Precision Limits

In classical applications of statistical sampling, precision is usually expressed as an upper and a lower limit above and below a point estimate of some characteristic in the universe from which the sample is selected. In the use of dollar-unit sampling, however, the upper limit is the only precision limit recommended because of additional assumptions necessary for estimating monetary understatement error limits.

In computing an overstatement error on a transaction, it is assumed that the monetary error limit is the amount of the transaction. By contrast, there is no monetary limit on the amount by which a universe may be understated by missing transactions or understatements on available transactions. Also, the lower error limit factors are valid when errors are only possible in one direction. That is, all possible errors in the universe are either all overstatements or all understatements.

B. Probability Factors for Computing Upper Error Limits (UEL)

Since cumulative probability factors are essential in the computation of upper error limits, a brief review of the essential features of three probability distributions will provide insight for preference for a particular set of factors.

The binomial probability distribution, discussed in Chapter 14, provides the approximate probability of finding x errors in a sample of size n drawn from a universe containing a specific occurrence rate p. However, it is based on the assumption that the occurrences are independent; that is, the probabilities remain constant from one observation to another. If samples are selected from an infinite (or extremely large) universe or from a finite universe with replacement, the occurrences are independent and the binomial may be used.

In auditing, however, sample items are generally not replaced before other sample items are drawn. This sampling procedure is called sampling without replacement. To adjust for the slightly changed universe after each sample item is selected, the hypergeometric probability distribution should be used instead of the binomial. For this reason the probability tables in Appendix C and Appendix D are based on the hypergeometric distribution.

In both the binomial and hypergeometric distributions, the terms add to unity, but each probability will differ depending on how large the universe is compared with the sample size, except for extremely large universes. The essential distinction between the mathematics of the two distributions is that in the binomial, the p and q terms are repeated, where p is the universe error rate and q is 1-p. While in the hypergeometric, the factors vary continually as one sample item after another is drawn, reflecting the gradually decreasing universe.

On the other hand, if the sample size, n, is large, and the probability of an occurrence is close to zero, then the occurrence is said to be a rare occurrence. In practice, if $n \geqslant 50$ and $np < 5$, where p is the error rate, the occurrence is considered rare.

For rare occurrences, the Poisson and hypergeometric distributions yield good approximations of the normal curve. The Poisson distribution is more useful, however, when the product np is known or can be estimated but neither n nor p is known. It provides the approximate probability of finding x errors when the expected mean is equal to np (sample size times error rate).

Although the formula for the hypergeometric produces the precise probability of finding x errors in a sample of n, it requires three variables in addition to the number of errors in the sample. They are the universe size, the sample size and the number of errors in the universe. By contrast, the Poisson distribution formula requires only one other variable, the product np. This permits the construction of a single-page Poisson table compared with a range of universe size tables for the hypergeometric, such as shown in Appendix D. This is the reason that most proponents of dollar-unit sampling use the Poisson factors. The reader, however, has a choice between the hypergeometric factors in Appendix D or the Poisson factors in Leslie, Teitlebaum and Anderson [37] or in a statistics textbook. However, if the concern is for 80 percent or 95 percent upper confidence limits, Table 12.1 will suffice.

It should be noted that Appendix D provides precise upper limits of error rates, such as less than 3 percent; whereas, the Poisson table used in dollar-unit sampling provides precise confidence, or probability, levels, such as 95 percent confidence.

C. Components of Upper Error Limit (UEL) Factor

Since there are likely to be areas which will be examined judgmentally or on a 100 percent basis, it is necessary to specify the amount of the overall materiality limit to be expected from the statistical sample. This requirement, however, is not unique to dollar-unit sampling, for it relates to the expected sampling precision. In audit-item sampling for dollar amounts, precision is related to the sampling variance. A somewhat similar concept in dollar-unit sampling is called the precision gap widening component, and it represents the proportion or amount by which the total precision gap or interval increases each time an error is found. In computing upper error bounds, this relates to the definition of precision as the interval between the point estimate and the upper error limit. The precision gap widening and other components of the upper error limit will be discussed next.

As in all sample estimations, the point estimate is simply a projection of the sample result to the universe. Where rates are being estimated, it is assumed that the sample rate is also the universe rate. However, the sample error rate is modified by the monetary tainting percentage of errors in computing a point estimate for a dollar-unit sample.

As mentioned, the precision gap widening represents the proportion or amount by which the total precision gap increases as a result of finding errors. The incremental factor in the UEL formula shown in Figure 7 automatically reflects the precision gap widening interval as well as the error rate and tainting percentage. It is noted in that figure that the incremental factor for one error at the 95 percent confidence level is 1.75, resulting from subtracting 3.00 from 4.75. The 1.75 consists of two components: error rate and precision gap widening. In this case, one error was detected in a sample of 100, which reflects an error rate of one percent. Subtracting this rate from the 1.75 incremental factor gives a result of .75, which is the precision gap widening component of the upper error limit factor.

To illustrate the three upper error limit components, consider a sample result where there is one 100 percent tainting error, from a sample of 100, and where the average sampling interval is $1,000. For the 95 percent confidence level, one obtains the following:

	Upper error limit component	Average sampling interval	Dollar projection
Basic precision (for zero error)	3.00	$1,000	$3,000
Sample error rate	1.00	1,000	1,000
Precision gap widening	.75	1,000	750
Upper error limit	4.75		$4,750

D. Stringer Method of Computing Upper Confidence Limit

Prior to the Stringer method, the computation of upper confidence limits for monetary-unit sampling used only attribute information; that is, whether or not a sampling unit did or did not contain an overstatement error. Also, it had been assumed that all overstatement errors were maximum ones. However, when a sample reveals errors it also provides information about the relative magnitude of the errors. The Stringer method uses the sample information to reduce the upper confidence limit according to the excess of the maximum possible error over the actual error.

To illustrate, suppose a sample of 100 dollar units is selected from a universe of \$2 million and one overstatement error is found. Also suppose the error ratio—error amount divided by the corresponding book value—is 0.5, such as a \$250 error in a \$500 book value. Thus, the upper monetary 95 percent confidence limit using the Stringer method is:

$$\$UEL = \frac{1}{\text{sample size}} - \begin{array}{l}\text{(zero error factor + tainting ratio} \times \text{incre-}\\ \text{mental factor) universe value, or}\end{array}$$

$$= \frac{1}{100}[3.00 + (0.5)\ (1.75)]\ \$2{,}000{,}000 = \frac{1}{100}\ (3.875)\ \$2{,}000{,}000$$

$$= \$77{,}500$$

It should be noted that except for the universe value and the \$ sign the above formula is the same one used in Figure 7, only the tainting is different. Thus, the upper monetary error confidence limit is the product of the appropriate total upper error limit factor and the universe value.

Since the first term in the formula reflects no errors, such a term is necessary even for an error-free sample. In fact, this is an advantage over classical procedures which would give zero for an error-free sample.

The following steps summarize the extension of the procedure to evaluate more errors with different tainting ratios:

1. Price basic precision (no error factor) at 100 percent.
2. Rank errors in descending order of tainting ratios to be conservative, since the upper error limit increments occur in descending order.
3. Multiply each tainting ratio by the paired upper error limit incremental factor.
4. Add the products from Step 3 to obtain a total upper error limit factor.
5. Divide the total upper error limit, from Step 4, by the sample size.
6. Multiply the result from Step 5 by the total value of the universe.

7. Compute separate evaluations for overstatements and for understatements.

To illustrate, suppose a sample of 200 from a \$500,000 universe yields three overstatement errors with tainting of 0.80, 0.60, and 0.85, and a 95 percent confidence is desired for the projected upper dollar error limit.

Using the factors in Table 12.1 for a 95 percent confidence level and following the guidance in the above steps, the upper dollar error limit is computed as follows:

$$\$UEL = \frac{1}{200}\,[3.00 + 0.85\,(4.75 - 3.00) + 0.80\,(6.30 - 4.75) + 0.60\,(7.76 - 6.30)]\ \$500{,}000 =$$

$$= \frac{1}{200}\,[3.00 + 0.85\,(1.75 + 0.80\,(1.55) + 0.60\,(1.46)]\ \$500{,}000$$

$$= \frac{1}{200}\,(3.00 + 1.485 + 1.24 + .8760)\ \$500{,}000$$

$$= \frac{1}{200}\,(6.601)\ \$500{,}000 = .033 \times \$500{,}000$$

$$= \$16{,}500.$$

Thus, the upper error monetary limit at the 95 percent confidence level is \$16,500, based on the sample size, number and tainting of errors, and the total universe value.

Research has focused on endeavors to obtain less conservative computations of upper confidence bounds. Notable among these efforts are multinomial bounds by Feinberg, Neter and Leith [16], the cell method by Leslie, Teitlebaum and Anderson [37] and combined attribute variable (CAV) bounds by Goodfellow, Loebbecke and Neter [19].

Other evaluation methods have been suggested but they are either more conservative than the Stringer approach or they have not been proven sufficiently valid for use with accounting universes.

E. Combined Evaluation of Results from Different Samples

In combining the results from different dollar-unit samples in the same audit, the following procedure has been used:

1. Compute the point estimate and upper error limit for gross overstatements, ignoring understatements.
2. Compute the point estimate and upper error limit for gross understatements, ignoring overstatements.
3. Net overstatement and understatement point estimates to obtain the universe point estimate.
4. Reduce each upper error limit by the point estimate in the opposite direction to obtain the net upper error limits.

The following example illustrates the above approach for combining overstatement and understatement projections and evaluations.

	Point estimate	Gross error limit	Net over-statement error limit	Net under-statement error limit
Over-statements	\$9,000	\$30,000	\$30,000 −5,000	
Under-statements	−5,000	−19,000		−\$19,000 +9,000
Net estimate	\$4,000		\$25,000	\$−10,000

In this example, the point estimate nets to \$4,000 and the net overstatement limit to \$25,000 and the net understatement limit to \$−10,000.

However, there is no unanimity among practitioners on how to combine overstatement and understatement evaluations. Another method involves offsetting the upper error limit of overstatements with the lower error limit for understatements.

VII. ADVANTAGES AND LIMITATIONS OF DOLLAR-UNIT SAMPLING

Although some advantages of dollar-unit sampling have already been mentioned, those and other advantages are summarized below:

1. The units sampled and the value of any errors found are related directly to the monetary value of the universe, thus

avoiding the difficulties of converting attribute frequencies into dollars.

2. It does not rely on any assumption about the distribution properties of the universe dollars which is unlike traditional estimation sampling plans.
3. Since it provides direct projections of point estimates and error accumulation, interim consolidations, appraisals and evaluations are possible.
4. Since the average sampling interval is in dollars rather than number of audit units, the sample selection may be concurrent and continual.
5. No dollar stratification is required since it is accomplished automatically, thus avoiding problems involved in determining dollar strata boundaries and in allocating the sample size among strata.
6. Optimum sample sizes are generally permitted because the procedure achieves infinite dollar stratification automatically.
7. The extent of sampling is related to the relative accuracy of the transactions rather than the variability among them.
8. It solves the problem of detecting the large but infrequent error, since audit units with large dollar values will be selected.
9. It permits an estimation of the probable magnitude of total dollar errors even when no error is found in the sample.
10. The sampling mechanics are relatively simple compared with other sampling plans.
11. If the universe is small, the selection may be performed manually; however, computers may be used in the selection and evaluation.
12. It can eliminate the need for two-stage sampling.

The limitations of dollar-unit sampling may be summarized as:

1. The upper dollar error limit is conservative; it always exceeds zero even when no errors are found.
2. Since it is necessary to accumulate the universe dollars progressively in selecting the sample, a computer may be required for a large universe. However, this would not be a disadvantage if the records are already computerized.
3. There is the problem that missing and zero book value transactions escape selection, since the chance of selection of any transaction is roughly proportional to its value, but missing items is always a problem.
4. A large ratio tainting in a small transaction can significantly increase the computed upper error limit. When this is the

case, it is recommended that small value transactions be grouped together to assure that the sampled intervals or cells contain a specified minimum amount.

5. If there are a lot of monetary errors, dollar-unit sampling may not be appropriate.
6. It does not permit the selection of dollars with unequal probabilities, as might be desirable if tests of compliance indicate a difference in the relative quality of work in various offices or activities.
7. Dollar-unit sampling may not be effective in compliance testing. If an estimated rate of compliance errors exceeds a specified limit in audit-unit sampling, no reliance would be placed on a control. Stratifying by dollars is appropriate in compliance testing if controls are intended to differ by various values of transactions; however, if dollar-unit sampling is used in compliance testing, the auditee can be expected to adjust to the monitoring procedure. In other words, under dollar-unit sampling small value transactions will not be examined to the same extent as large value transactions. Thus, the auditee would circumvent a control by concealing erroneous statements on small dollar transactions on the assumption that there is little chance of detection by dollar-unit sampling.
8. If large or small understatements or overstatements are equally likely on each audit unit, then dollar-unit sampling introduces some bias if used to estimate understatements in a universe. In such a case, audit-unit sampling is recommended in estimating understatements.
9. Although lower error limits may be computed for dollar-unit samples, care should be exercised in such computations since the lower error limit (LEL) factors are valid when errors are *only possible* in one direction, such as either all overstatements or all understatements. If the lower error amount is crucial to an audit decision, audit-unit sampling is recommended. Even then, there are problems which are inherent in the nature of accounting universes and auditing techniques.
10. Dollar-unit sampling assumes that all dollars in the audit unit will have the same characteristics, and hence, it does not provide data for error rate analysis. The procedure makes it easy for the auditor to combine and then estimate an error rate for completely unrelated dollar unit values—say those of apples, oranges and rocks for an average error rate for fruit-rock salad—and it would make it easy for the auditor to overlook the complete lack of significance of such an error rate. In other words the assumption that all

dollars in the audit unit will have the same characteristics may not be valid. The most common example is in the examination of vouchers. In most cases, the audit unit will be a voucher. Each voucher will often contain several invoices and each invoice, several line items. The selected dollar unit would be associated with one invoice line item; but this line item could not be regarded as an audit unit, unless the other line items in the voucher are also considered separate audit units. Neither can the auditor assume that all other line items on the voucher invoices will have the same characteristics. They will often be completely different and unrelated. This difference is not related to differences in the identity of items; it is related to differences in error rate.

11. The selected dollar sampling unit may be associated with a cluster of many other dollar units, thus expanding the sampling effort. The auditor cannot confine his examination to only one part or portion of a voucher. If he did and thereby overlooked a significant error, his failure to examine the voucher and to take the error into consideration in his audit findings, if questionable, would be indefensible. Furthermore, having once examined the entire audit unit, the auditor must evaluate and take into account as part of his findings all data, including all errors, composing the unit. Thus, the selection of a dollar sampling unit may bring into audit consideration a cluster of many other dollar units, some of which may have materially different characteristics. These clusters of dollar units might be treated as a single unit containing any number or types of errors; or, possibly, the individual units composing each cluster might be treated as separate sampling units with their corresponding errors. On the other hand, dollar-unit sampling theory is associated with individual dollar units, when in fact the auditor is getting a cluster of dollars, thus he has additional evidence from the sample which is not reflected in the statistical evaluation.
12. The sample design should permit the auditor to take advantage of known information; however, dollar-unit sampling assumes that the problems are primarily related to high dollar transactions when they could be associated with middle dollar or low dollar transactions.

VIII. SUMMARY

Since the early 1960's, researchers have indicated that classical sampling methods using difference and ratio estimators based on large sample, normal distribution assumption may be inappropriate for auditing. Accounting universes are usually highly skewed with low error rates. For example, when no error is found in a sample, the difference estimator gives point estimates and confidence limits of zero.

Attribute sampling theory is one approach which avoids the classical distribution assumption in deriving confidence limits. Flexible sampling, described in Chapter 13, applies this theory to audit units, such as individual invoices, vouchers, line items, etc. Monetary-unit sampling applies the theory to individual monetary units.

Those readers who are interested in a detailed discussion of different sampling methods for estimating monetary confidence bounds or limits based on attributes and variables should see Fienberg, Neter and Leitch [16].

Monetary (dollar)-unit sampling is a sampling strategy formulated to satisfy the auditor's desire for a reasonable assurance that the total monetary errors in a given accounting universe does not exceed a specified threshold of monetary materiality. The sampling unit is defined as an individual dollar on a transaction rather than an individual transaction. Detected errors are expressed in terms of error ratios, called tainting percentages, of the selected dollar-units.

The dollar-unit sample size may be computed in several ways depending upon the available information. If the average sampling interval has been determined on the basis of experience and other sources of reliance, the sample size is computed by dividing the total universe dollars by the sampling interval, which is related to materiality. Since flexible sampling may be considered the attribute counterpart to dollar-unit sampling, the flexible sampling tables in Appendix D also may be used in planning sample size and in the evaluation of the results.

The method gives each transaction in the universe a probability of selection proportionate to its book value. To achieve this, there are several interval procedures for selecting dollar-unit samples. However, a less conservative (more precise) method of sample evaluation results when the universe is divided into equal size cells, and a random point (dollar unit) is selected within each cell. If there is the possibility that the initial sample may be too small, it is advisable to divide the cells in half and select a separate random sample from each half.

The sample result is evaluated by determining an upper monetary error limit for the ratios of tainted dollars in the sample transactions examined. The upper monetary error limit is a one-sided confidence interval for a desired confidence level.

As for all statistical sampling plans, the desired confidence level represents a judgment. Likewise, the determination of how many errors an auditor will tolerate in a sample and still consider the results acceptable is a judgment or management decision. However, the auditor must appraise each occurrence for audit significance, nature and possible trend before he can make a decision on the acceptability of even a single error.

Although overstatement and understatement errors are evaluated separately, there is no agreement on the best way to combine overstatement and understatement projections.

Caution should be exercised in computing lower error limits since the probability factors are valid for such computations only if all possible errors are in one direction, that is, either all overstatements or all understatements.

13
Flexible Sampling

I. EVOLUTION OF AUDIT SAMPLING

Flexible sampling is an outgrowth of the evolution of modern sampling theories and concepts. Many advances in statistical sampling applications have resulted from the successive removal of constraints that had been imposed on audit sample selection and estimation. In the early days of testing, it was customary for auditors to examine, as thoroughly as time permitted, the records they had chosen to be the most "representative" sample. Later, the emphasis was shifted to randomly selecting samples.

In the initial audit applications of random sampling, emphasis was on unstratified simple random selection which stresses that every item in the universe should have an equal chance of being included in the sample. This emphasis began to wane when it was recognized that stratification and variable probabilities for both individual items and strata could result in more precise estimates while focusing on material amounts and problem areas.

Later, the strict requirement that all bias should be avoided was modified with the recognition that some bias is acceptable if it is more than offset by an improvement in sample precision. It is for this reason that many audit objectives favor the use of the ratio estimation procedure, which is discussed in the next chapter.

Also, the initial notion that a statistical sample should have a fixed, predetermined size was relaxed in the face of the more sophisticated theory of sequential sampling first advocated by Wald [82]. Incidentally, many nonstatisticians using statistical sampling for the first time have the erroneous notion that a predetermined sample size automatically produces the desired sampling precision.

Sequential sampling theory, contrary to classical notions, reveals that it is not unethical to peek at the sample while it is being selected and to modify the sampling strategy based on interim observations. In fact, transactions can be examined only one at a time.

Numerous statistical sampling methods have been proposed as a means of assisting the auditor in estimating the monetary error in a stated book value. Most of these procedures use adaptations of classical statistical methods.

As first noted by Stringer [66] in 1963, and subsequently by other authors such as Anderson and Teitlebaum [1], Kaplan [27] and Neter and Loebbecke [47], these methods may be unreliable when applied in an auditing environment, especially when developing a statistical inference about the total monetary error. The unreliability is associated with the usually low error rate, highly skewed universes and relatively small sample sizes experienced in auditing. Also, Kaplan indicates that the basic problem is that the classical methods are designed for a homogeneous universe; whereas, the auditing universe is a mixture of two quite different universes: one consisting of correct items and the other, much smaller one, of erroneous items. He further states that any techniques which do not explicitly recognize these two fundamentally different universes seem inadequate for audit purposes.

The audit universe distributional and low error rate problems have compelled the reintroduction of more control and greater use of available information and judgment in the audit sampling procedures. Statisticians associated with this movement are called Bayesians, after an eighteenth-century Englishman, Thomas Bayes, who derived a formula for handling what are known as prior probabilities.

Traditionally, probability has been associated with relative frequency, but to Bayesians, probability is primarily a matter of personal conviction. For example, if prior knowledge and opinion about a condition are strong and a preliminary sample supports this opinion, a Bayesian may cease further testing and concentrate on the investigation of the nature and cause of any misstatements, seeking indications of patterns, clusters or trends for the purpose of sustaining or modifying the prior opinion. In other words, to a Bayesian, estimates of probabilities of events are based on his subjective beliefs as modified by empirical data. By contrast, probability estimates in classical statistics are based solely on objective data. Thus, the Bayesian approach appeals to decision makers, since the point of sufficient information to make a decision is reached quicker. Also, the approach considers the cost of obtaining additional data.

The apparent reason for the reintroduction of more control in the sample selection procedures is to accommodate, at a more sophisticated level, the objectives of purposive or judgmental testing. This is especially significant for areas of large dollars and known problems. Audit stratification is a means of satisfying these objectives and retaining the fundamental theoretical basis of statistical sampling for the other areas or strata of the universe or frame.

The trend towards using more of the auditor's knowledge of the structure and nature of the universe in the designing and execution of the sample is the conceptual basis of flexible sampling.

It appears that statistical sampling audit application problems result from an attempt to use statistical techniques, per se, to formulate direct audit conclusions. However, if statistical sampling is to be used successfully, it must be employed in conjunction with other acceptable auditing procedures and the auditor's judgment and knowledge of the particular audit environment. In other words, a proper blending of statistical techniques with the auditor's judgment and knowledge will produce high quality, reasonable audit conclusions and decisions, even if the statistical techniques are not as precise as desired. This means that the nature, source and impact of each misstatement detected must be individually investigated, analyzed and appraised. If the monetary error amount is immaterial, and this is consistent with other sources of reliance, then the reliability of the statistical confidence intervals is secondary.

Thus, it would seem that the auditor's primary concern is to have an auditing strategy which will protect him against accepting a materially erroneous set of records. This is the strategy of flexible sampling. It permits the integration of appropriate statistical procedures with the auditor's judgment, experience and knowledge of the particular audit environment, and with all other acceptable auditing procedures, including judgment testing through audit stratification as described in Chapter 3.

Briefly, audit stratification involves setting aside (stratifying) for separate examination those transactions or groups of transactions which judgment deems material, sensitive or otherwise significant, and thus necessitates more detailed examination and analysis in the formulation of an audit opinion.

Audit stratification can frequently be determined from experience, an audit survey, or by a casual review of the records (universe). In those instances where these procedures are not practical, the auditor can select a preliminary sample as a guide for determining audit stratification. This preliminary sample may be considered the first replicate in a replicated sample as described in a subsequent section.

In some cases prior stratification of the universe may not be practical but stratified sampling is desirable. In these cases, a sample may be selected from the entire frame without stratification and the sample results stratified. This is called post-stratification. Kish [32] describes in detail practical ways of performing post-stratification without imperiling the integrity of the sample results.

II. THE FLEXIBLE SAMPLING STRATEGY

The flexible sampling strategy embodies the use of all sources of reliance, prior information and knowledge for planning and for effective stratification. Operations and transactions to be selected subjectively or examined 100 percent are separated from the original frame by audit stratification. The delimited frame may then be statistically stratified by techniques explained in Chapter 8.

The sample is selected by replication, a technique of drawing the total sample by selecting a number of independent samples of the same size, as discussed in the next section. The sample size may be either equal components or multiples of a minimum sample, assuming no error is found. The notion of a minimum sample and its use are explained in a subsequent section.

It is rare that an auditor approaches an auditable area without some prior knowledge of the reliability of its content. He may know from prior experience that the administrative controls established to guard against errors are well designed and rigidly followed. Previous audits may have revealed random errors of no consequence. If there have been no changes in the system, controls, or in operating conditions, then the auditor may feel justified in accepting a greater degree of risk in his current examination. Flexible sampling provides a means of measuring that risk and taking advantage of what is considered an acceptable degree of risk without hindering the ability to diminish the risk by additional replications.

The auditor by examining a sample selection in accordance with the flexible sampling strategy can determine with a minimum effort and with a specified degree of precision whether the individual charges to a specific cost are reasonable accurate and acceptable; whether a specific step in an accounting, estimating, or a related procedure is operating in an acceptable manner; and whether the product of such a procedural step is reasonable and acceptable.

This sampling strategy may also be used in consonance with the initial phase of a dollar-unit sampling plan, especially in the determination of the sample size. A dollar-unit sample is designed to estimate from the examination of a sample the maximum expected dollar values of misstatements. A detailed discussion of dollar-unit sampling is presented in Chapter 12.

Flexible sampling is a strategy which permits the auditor to subjectively appraise and statistically evaluate the results of individual sample replicates and stop, or continue the testing or redirect the audit effort as the appraisal and evaluation of the replicated results may dictate. In practice, emphasis is to stop sampling as soon as indicators reveal that further testing is unnecessary, and to resort to other acceptable audit techniques. The strategy uses the concept of the upper error limit or worse condition—similar to dollar limits in dollar-unit sampling—that the auditor may reasonably expect to find in the universe.

Conversely, the approach estimates the risk that the auditor is taking that the conditions existing in the universe are worse than estimated from the sample. Thus, if the minimum sample reveals no finding, the test-checking phase may be terminated with a known degree of risk associated with the decision. If there is no finding, the sample size is the smallest practical statistical sample for the audit objectives. On the other hand, each and every finding must be thoroughly investigated for audit significance, trends, patterns, and clues for systematic accounting or procedural deviations or errors as a basis for redirecting audit effort. It should be noted that the minimum sample is only part of the total audit testing. The other testing is accomplished through audit stratification of material and sensitive areas or strata.

By combining flexible sampling with traditional audit analysis techniques, the sampling procedures may be used to guide the auditor to areas of clustered questionable costs, or to unacceptable accounting or management practices. It can also be used to supplement other audit techniques by providing a prescribed degree of assurance that material amounts of questionable costs or unacceptable practices are not being overlooked.

Flexible sampling permits a minimum of advanced planning and assumptions. In most cases, the amount of work involved will be no greater, and frequently considerably less, than that required by other sampling plans including the intuitive sampling approaches. In fact, the flexible sampling strategy is similar in its operation to the intuitive sampling plans which auditors are accustomed. The primary difference is that random selection procedures are substituted for the intuitive procedures used for the selection of the sample items. However, intuition and scanning may also be employed effectively with flexible sampling.

Various auditing steps are performed by intuition. For example, suppose there is a large number of transactions to review in various accounts, and the auditor wishes to test some frame (listing) to assure its adequacy. In attempting to establish a suitable source document he may perform a rapid review and inspection of the documents. He may then make a subjective choice of a source document to test it.

Finally, he scans and reviews the frame to establish its approximate size and to obtain other relevant information.

During this preliminary review, the auditor is alert for audit leads and clues. As he scans the material presented for his consideration, he gains knowledge about the frame and some of its characteristics which will enable him to more effectively stratify. Also, if he makes a manual systematic selection, he can again scan transactions to improve his knowledge of the frame while making the selection. Furthermore, if during his examination of sample items, the auditor identifies one or more sampling units as more properly requiring separate audit consideration, he may still set them aside (audit stratification). His identification of these units results from the recognition of the fact that his initial description of the audit operation did not contemplate the inclusion of such units.

After the selected items are examined, the detected misstatements are classified and grouped for error analysis and audit appraisal by some device similar to an evaluation matrix, illustrated in Chapter 2. Error analysis and audit appraisal, significant aspects of auditing, are usually not addressed by classical statistical sampling procedures, where the emphasis is on estimating and evaluating errors rather than ascertaining their nature and cause. For example, the audit handling of a material monetary error consisting of a large number of small errors affecting many accounts is different from a material monetary error composed of large discrepancies in only one or a few accounts. A detailed discussion of error analysis and audit appraisal are in the latter part of this chapter.

Often an audit conclusion or decision, based on audit appraisal and rationale, is possible after an analysis and investigation of the nature, source and impact of disclosed misstatements. In cases where a statistical evaluation of the sample results is required or desirable, it would follow the error analysis and audit appraisal phases in flexible sampling. Some common statistical estimation and evaluation techniques are described and illustrated in Chapter 15.

In summarizing, the flexible sampling strategy integrates judgment, knowledge (prior information) and statistical procedures in focusing audit effort on material and sensitive activities and transactions; it employs replication to the delimited frame, using minimum samples; it involves error analysis and appraisal of sample results to ascertain nature, cause, trend and impact of each error; and when necessary, it provides a statistical evaluation of the sample results.

III. REPLICATED SAMPLING

A. Concepts

It is possible to replicate any sample design by halving the total sample and repeating the same sampling procedure to obtain two independent replicates. Thus, replication involves the selection from the entire universe of two or more independent random samples of the same, or about the same, size. Replicates should be separately identified so that they can be evaluated and appraised separately, as well as combined.

Replicated sampling is facilitated if the universe is divided into zones of equal, or approximate equal, size. A new set of random numbers may be selected in each zone or random starts may be selected in the initial zone. Then add the zone width to these random starts to obtain the sampling units in the other zones.

It is important to understand that since the replicates are separate interpenetrating portions of the entire universe, each is a valid sample. Each replicate is randomly selected from the whole universe.

B. Steps in Replications

The essential steps in replication may be summarized as follows:

1. Decide on the sampling unit.
2. Construct the frame—the list of sampling units—and assign an identification number to every sampling unit, or derive a rule which will provide an identification code to any sampling unit when required.
3. Compute the zoning interval Z—which denotes the number of sampling units in a zone—by dividing the product of the universe size and the number of replicates by the desired sample size.
4. Within each zone select as many independent sampling units as there are replications, such that each replication consists of one unit from each zone. The replication of the sampling design in every zone creates the simplicity in the theory, preparation and computations.
5. Compute the desired statistics for each replicate.

6. Use either formulas or a computer program to estimate and to evaluate the results.

C. Basic Factors in Replication

Prior to presenting an illustration of replication, it is useful to define the following factors.

$$\text{Zone width} = Z = \frac{kN}{n} = k\,\frac{1}{f} = kw$$

where

k = number of replicates

N = total number of sampling units in frame

n = total sample size

$w = \frac{N}{n} = \frac{1}{f}$ = weight or multiplier

$f = \frac{n}{N}$ = sampling rate or fraction

The sources of values of selected factors are as follows:

The value of N is known or is estimated.
The value of Z is calculated from known or assumed values of k, n, N.
The value of n is derived from formulas, from tables, from experience, or from all these sources.
The value of k is determined on the basis of the objective, experience, how many replicates are to be used, or for other reasons.

The selection of sampling units from each zone may be independently within zones—as in the case of cell selection for dollar-unit sampling—or from k random starts in the first zone, plus successive additions of the zone width Z until the frame is exhausted, as in systematic selection. If the number of zones m is an integral multiple of the sample size, then $n = km$.

D. Illustration of Replication

Suppose the auditor wishes to select 200 records using five random starts to produce five independent samples, or replicates from a universe of 20,000. The zone width is determined by dividing the product of the universe size (20,000) and the number of replicates (5) by the desired sample size (200). Thus, the zone width = Z = (20,000 × 5) ÷ 200 = 500. The first line of the table reveals the five random starts to which a zone width of 500 is added successively to obtain the following replicates:

		Five samples or replicates				
Zone	Zone width	1	2	3	4	5
1	1–500	209	306	8	141	219
2	501–1000	709	806	508	641	719
3	1001–1500	1209	1306	1008	1141	1219
4	1601–2000	1709	1806	1508	1641	1719
5	2001–2500	2209	2306	2008	2141	2219
6	2501–3000	2709	2806	2508	2641	2719
.	.	.	.	.	.	.
.	.	.	.	.	.	.
.	.	.	.	.	.	.
40	19,501–20,000	19,709	19,806	19,508	19,641	19,719

If the records were computerized, the computer could count to 8 and read out the eighth record which is coded 3 for the third replicate. The process continues until the 141st record is reached, read out, and coded 4 for the fourth replicate; then to the 209th record which is coded 1, then to the 219th record which is coded 5, then to the 306th record which is coded 2, then to the 508th record which is coded 3, and so on. A computer can be programmed to give counts of the total universe, the total sample size, the replicate code number of each item selected, and the number of items in each replicate, as well as print out all the individual numbers. If the universe is not an even multiple of the zone width, the last line will not have items for all the replicates, but this causes no problem since the number of sample items in each replicate will not differ by more than one. Also, if the zones within a stratum contain an equal number of sampling units, then every sampling unit in the stratum will have the same probability of selection; consequently, so would all of the employees, all of the activities, and so on, unless one deliberately changes the probabilities. Thus, one of the major advantages of replication is that it eliminates the series of weights required for the varying probabilities of selection involved in multistage sampling. The sampling procedure automatically assigns probability in proportion to the number of sampling units in each primary unit. To raise the probability of any primary unit, one decreases the size of the sampling units, so that there are more of them. For a practical illustration of assigning sampling units to primary units, the reader should see Deming [12, p. 91].

The computations of estimates and evaluations from replicates are discussed in Chapter 15. Deming [12] and Rosander [58] discuss and illustrate several accounting applications and case studies involving replicated sampling.

E. Flexibility of Replicated Sampling

The great flexibility of replicated sampling makes it an integral part of the flexible sampling strategy. The flexibility of the selection method requires only that the replications be similar in design and be independent within each zone. Also, it provides a simple and direct way of calculating various kinds of estimates and their standard errors, which contributes to the flexibility and usefulness of the method. The estimates may be means, totals, frequencies, proportions, differences, ratios, regression coefficients, and the like. The use of a computer greatly facilitates the audit application of replicated sampling.

The flexible nature of the procedure permits many variations. Both audit stratification and statistical stratification may be related to zones, since zones are implicit strata. As mentioned, replication may be used to eliminate the complexities associated with multistage sampling. For example, one replicate may be used for the total universe or frame, while another replicate may be used for a subclass or domain. Furthermore, based on the auditor's judgment and knowledge, one or more replicates may be used to test for or detect an abnormal situation or a material error where an unusual value occurs in a replicate. Thus, replicated sampling is a powerful technique, permitting the auditor to use his judgment and knowledge more efficiently, effectively and economically.

According to Deming [12, p. 91] disproportionate sampling fractions may be used if needed. This is accomplished by varying the zone sizes; thus, the smaller the sampling fraction, the larger the zone size. Deming calls this "thinning of the zone."

If the number of replications is small, there are several additional advantages for auditors. For example, some variable nonsampling errors can be automatically included with the variance computations by randomizing them among the replications. For instance, the variability among results for different auditors can be included in the variance by randomizing auditors over the replications; for details, see Deming [12, p. 91]. Additionally, the visual evidence shown in the consistent results of the replications provides the auditor with intuitive support for his decision, which is an advantage when the auditee does not comprehend the statistical aspects of the sample evaluation. Also, Deming [12, p. 425] has used replication in the evaluation of the bias in an estimating method.

F. Number of Replicates

Those who advocate the use of few replicates, such as Mahalanobis [40] who often used four, cite the following advantages: (i) the selection is simplified; (ii) when zones are used as strata, better stratification is possible; (iii) the visual display of replicated evidence is easy; (iv) the computation of standard errors is facilitated; (v) the randomization of nonsampling variables, such as the performance of different auditors, is possible; and (vi) the replicated estimates, based on larger samples, for each estimate, more closely approach normality.

Deming [12] and others often use ten replicates for the following reasons: (i) the standard error has greater precision than few replicates; (ii) the dangers of periodic variations in systematic selection are reduced; and (iii) the normality of the sample mean can be based on more replications.

In the use of flexible sampling for auditing purposes, with the usual constraints on sample sizes, larger sample sizes for fewer replicates would seem to be preferable to much smaller sample sizes for a fairly large number of replicates. In other words, four replicates of 50 sampling units each would seem to provide more audit information than 25 replicates of eight sampling units.

G. Size of Replicates

The sample sizes used in flexible sampling are based on the notion of a minimum sample. As used in this book, the minimum sample is the smallest sample which satisfies the audit objective, assuming no misstatements are detected during the examination.

Since at least two replications are required, it is often efficient to use half-minimums as the size of replicates. Hence, a third replicate would not double the minimum; it would increase the minimum sample by one-half. Similarly, if the minimum sample is large, each replicate may be one-quarter or some other component of the minimum.

However, it is recommended that no replicate be less than 50 when dealing with accounting data.

The determination of size for flexible sampling is somewhat similar to dollar-unit sampling with the primary difference being the type of sampling unit. In flexible sampling, the unit is an audit item in lieu of an individual dollar. In fact, if one assumes no errors in the sample, the sample size is the same for both sampling strategies. Also, the two strategies may be used to augment each other to obtain both error and monetary limits derived from replicates.

The difficulty of specifying the sample size in advance becomes more acute when sampling a new area or new material. One solution is to use two or more replicates. In designing the sample, specify a size that is large enough to provide the desired precision. Then randomly divide the sample into at least two replicates. Next, examine and evaluate the first replicate. If the desired precision is achieved from the first replicate, stop the sampling; otherwise, proceed with the second replicate. However, when the estimate of the sampling precision is critical, Hansen, Hurwitz and Madow [20] recommend a coefficient of variation no greater than fifteen percent in the estimate of the standard error from the initial replicate before relinquishing the second replicate.

H. Determination of Minimum Samples

In determining sample size, the auditor must reconcile two competing factors. He desires relatively high sampling precision, but from a relatively small sample. To reconcile these factors he must first determine the tolerance or confidence limit and the probability which best expresses the minimum sampling reliability he considers acceptable. From these guidelines, assuming no error will be found, he can determine the minimum sample using the table in Appendix D. For example, from a universe of over 1,000 items and a probability of 95 percent that the expected number of occurrences of a characteristic in the universe will be less than three percent (the specified tolerance limit), the auditor will need a minimum sample of 100, assuming the sample reveals no error. However, if an error is detected, this minimum sample is not sufficient. The auditor then should investigate each error for nature, source and impact to determine audit significance.

Because of limited resources, the auditor must frequently limit the sample size. When using intuitive selection procedures, he has no way of knowing the additional risk that he is taking that his conclusions from a limited testing operation might be inappropriate or unreasonable. By using flexible sampling, he can statistically estimate the probability and the complementary risk. For example, suppose a universe consists of over 1,000 items, and time and resources might not permit the selection and examination of more than 100 sample items. Assuming that the examination of a simple random sample of this size disclosed three random (no pattern or source clusters) findings, the probability of less than a five percent occurrence rate in the universe is 74 percent (from Appendix D).

A rule of thumb, based on auditing experience, indicates that the absolute minimum sample for attributes (rates or proportions) should be at least 50, and 100 for variables or dollar estimates. (A similar approach to flexible sampling may be used in dollar-unit sampling where the sampling unit is a dollar.) Smaller samples usually are not sufficient to allow the characteristics to manifest themselves, patterns or clusters to form, and occurrence rates and ratios to stabilize. This rule assumes audit stratification and no sample finding.

Also, auditing experience reveals that a feasible maximum overall size of a flexible sample as eight replicates, each consitituting a one-half minimum sample size. This sampling strategy permits the termination of the sampling operation at the end of any replication that the findings become determinate. The strategy also permits the auditor to modify his audit objectives based on the nature, scope, impact and trend of the findings. Since the replicates are of equal size, the results may be intuitively appraised as well as statistically evaluated. From the nature of the findings in the sample, the auditor can decide with a known degree of confidence whether he should reschedule his audit work to permit a more extensive examination of the universe, probably with some further stratification, or because of reliance on other sources of evidence, accept the results of the limited examination. If the occurrences are of a random nature, occasional or chance mistakes from various sources with no pattern, and if they do not materially affect a cost, the auditor might accept the lower sampling reliability possibly with the recommendation that internal controls be tightened to reduce the probability of future misstatements. On the other hand, if the findings are from a single source, department, person, operation, or form a pattern or trend, then the auditor should stratify the frame in such a manner that the problem area (source of most findings) may be examined and reported on separately as not being typical.

The above discussion of sample sizes is related to the table in Appendix D. To use the table, the auditor must decide upon the maximum tolerable error rate he will accept even if the sample reveals no error. Using this rate, the table shows the minimum sample size for a specified probability that the universe error rate is less than the maximum tolerable error rate, provided the sample reveals no error. Thus, one purpose of the table in Appendix D is to minimize the size of the sample necessary to obtain sufficient evidence that a system or a procedure is working reasonably well. This objective relates to selecting the smallest sample from which one can determine the maximum tolerable error rate in a universe or frame at a specified probability level. Flexible sampling emphasizes the minimum necessary for a statistical evaluation concerning the upper tolerance limit, and the need to investigate and analyze the sample results to determine the nature, cause and impact of errors revealed by the sample.

There are different pages in Appendix D for selected universe sizes, as indicated in the headings. The table is related to four ranges of universe sizes: 100 to 199, from 200 to 399, from 400 to 1,000, and all universes over 1,000. Different sizes are used because the corresponding probabilities are slightly larger for smaller universes. Thus, it would be conservative to overestimate an unknown universe size.

To illustrate, suppose the maximum tolerable error rate is five percent for a universe size over 2,000. Using the appropriate page for the universe size, one looks for the column headed "5%." Next, he considers the rows of the table which indicate number of errors found in a sample for the specified sample sizes. By specifying a sample size and the number of errors, he identifies a particular row in the table. For instance, finding the intersection of the column corresponding to 5% and the row for a sample of size 50 with no error, one observes the number to be 92.31.* This number is the probability, in percentage, that the universe error rate is less than five percent. Hence, if the auditor requires a probability of at least 95 percent, a sample of 50 items would not be sufficient. He would need a minimum sample of 70 to achieve at least a 95 percent (97.24) probability, while detecting no error in the sample.

It is important to note that this table does not require the auditor to specify an expected universe error rate as some other sampling strategies.

For a sample of size 50, with no error, the auditor is only 78.19 percent confident that the universe error rate is less than three percent. Also, if a sample of 100 items contains one error, he cannot be as confident (69.62%) that the universe error rate is less than three percent. Also, the minimum sample objective cannot be accomplished. The auditor may then strive towards an optimum sample size, which is the smallest sample which satisfies the audit objective. Under such circumstances, he would select additional replicates. On the other hand, he may stop testing and resort to other auditing techniques.

IV. ERROR ANALYSIS AND AUDIT APPRAISAL

A. Concepts

Error analysis and audit appraisal involve separately investigating, analyzing and appraising the nature, cause and audit significance of

---

*On page 23 of Appendix D.

each individual sample finding. An evaluation matrix, illustrated in Chapter 2, provides a simple device for tabulating and classifying audit findings, which often occur in discernible clusters or patterns.

Whether the auditor uses an evaluation matrix or not, it is imperative that he thoroughly investigates and analyzes each misstatement (error) revealed by the sample. This procedure is referred to as an appraisal because it involves a value judgment.

It is important to realize that in auditing an overall error rate is generally of little value, since it does not represent a typical condition of the entire universe. Compliance deviations and monetary misstatements must be stratified and associated with sources, causes and impact if they are to make a worthwhile contribution to the auditor's opinion and the supporting rationale. The process for accomplishing this in reviewing sample disclosures is called error analysis and audit appraisal.

A compliance deviation is a deviation from the internal control procedure which is essential to sustain the effectiveness of the internal controls. In analyzing both compliance and monetary deviations, judgment is required in the appraisal of the qualitative aspects of the deviations. Some of the considerations involve the nature, cause and impact, such as whether the errors are in accounting principles or in application, are deliberate or unintentional, are repetitive or nonrepetitive, are due to misunderstanding or carelessness, are found in clusters or exhibit no pattern, are consistent with other audit evidence, or caused a monetary error of what magnitude. It is significant if a compliance deviation results in a monetary misstatement, since only monetary errors affect financial statements. Although compliance deviations increase the risk of material misstatements in accounting records, such deviations do not necessarily result in material misstatements. According to SAS No. 39, *Audit Sampling*, ". . . deviations from pertinent control procedures . . . would result in errors in the accounting records only if the deviations and the errors occurred on the same transactions." [3]

If the auditor makes a preliminary judgment that internal controls can be relied on to prevent material errors, he can lower the confidence level for the related substantive tests. If there is an audit trail, he tests for compliance, and if he finds any deviations, each should be analyzed and appraised for audit significance. If there is no audit trail, he should resort to corroborative inquiries and floor checks of personnel and routines.

Substantive testing involves (i) tests of transactions and balances and (ii) the analytical review of significant ratios and trends and the investigation of unusual fluctuations, patterns and clusters of questionable items or errors. Individual transactions and balances are tested to ascertain if they are valid, reasonable, and are processed according to acceptable accounting procedures.

Analytical review is related to audit appraisal since it is usually subjective even though statistical techniques, such as regression analysis and cluster analysis, may be used in the process.

B. Some Categories of Errors

1. Repetitive and Nonrepetitive Errors

The auditor's primary concern involves two types of basic errors: (i) compliance deviations (attributes) and (ii) monetary differences (variables); however, for analytical review, accounting errors should be classified as repetitive or nonrepetitive. For instance, a misunderstanding of office instructions in programming a computer to journalize some kind of transaction will cause repetitive errors affecting a number of transactions in one or more accounts until the program is corrected. On the other hand, a fingering error in inputting the amount of some transaction into the computer is likely to affect only one or two transactions, resulting in a nonrepetitive error.

When statistical sampling is used, the risk of not detecting a particular nonrepetitive error is usually large. For example, if the testing of inventory records consists of selecting a five-percent simple random sample and then comparing the quantities shown in the sample with the actual quantities, the risk of not detecting a nonrepetitive error on a particular item will be .95. However, if a repetitive error affects each of 50 items in these records, the risk of not detecting the error on at least one of those 50 items will be somewhat less than $(.95)^{50}$, or 7.7 percent. And if the cause and source of a detected repetitive error are determined through audit appraisal, it is usually possible to correct all the items affected with little effort, and to prevent further repetition of that particular kind of error.

Misstatements due to deceptions may be repetitive, since more than one item is generally sufficient to disclose the deception. For this reason, each apparent error must be thoroughly investigated and analyzed. However, unless there is strong supportive rationale, the auditor should not project nonrepetitive sample errors to the universe, even though random selection methods were employed in drawing the sample.

2. Intentional and Unintentional Errors

In audit analysis and appraisal, unintentional noncompliance would probably be considered less serious than deliberate noncompliance. Also, errors resulting from a misunderstanding would have a different impact and would be considered separately from those caused by slipshod compliance.

Unintentional monetary errors and intentional irregularities can make financial statements misleading. In audit appraisal, the difference between unintentional and intentional monetary errors is very significant. It may be possible to determine the cause of unintentional errors and to estimate the probable total amount involved. By contrast, it would be more difficult to estimate the amount of cleverly concealed intentional misstatements. Nevertheless, the detection of one deliberate error should alert the auditor to the possibility that the universe being audited may contain many intentional errors. Thus, the detection of a single intentional error should alert the auditor to the immediate need to either modify or supplement his audit plan, since the risk of not following up on such an error is too great. Obviously, the analysis and appraisal would be intensified.

3. *Recoverable and Nonrecoverable Monetary Errors**

Since monetary misstatements directly affect financial statements, it is important that monetary errors be detected if they exist, and their potential impact be carefully analyzed and appraised before any statistical evaluation is performed. In performing audit analysis and appraisal, it is useful to classify monetary errors into (i) permanently concealed errors and (ii) temporarily concealed errors.

A monetary error is regarded as permanently concealed if the system has not caught it, and, unless an examination is performed, it will remain undetected. Some permanently concealed errors are recoverable while others are not. In any case, they are normally associated with the income statement. For example, if a product has been erroneously underpriced, the total revenue will be less than it should have been, and will not be recoverable on the transaction tested nor any similar transactions. On the other hand, certain types of expense overpayments are recoverable.

Although an estimate of unrecoverable dollars resulting from errors may be desirable to emphasize the cost impact of the breakdown in internal control, a statistical evaluation of the lost revenue or excess expenses is vacuous. Also, management may not seek legally recoverable dollars because of possible adverse public relations.

In the case of recoverable permanently concealed monetary errors, a dollar estimate and a statistical evaluation is more rational. However, if the cost of recovery is greater than the recoverable dollars, there would be little motivation for management to take action. It should be noted that a small number of errors may not provide a reliable evaluation, unless the amount of each error is about the same and is considered by the auditor to be typical under the circumstances.

---

*For a more detailed discussion, see Taylor [68].

Temporarily concealed monetary errors are normally associated with the balance sheet. For instance, if the inventory value is overstated, the overstatement increases the net worth on the balance sheet. However, the error is not permanently concealed since it is subject to be detected every time the inventory is examined. Another example is an overstatement of accounts receivable because of a lag in recording returned sales. The auditor, through audit stratification, can isolate and localize this category of error by examining, separately and in detail, accounts where returned sales are posted. This permits him to estimate the amount of overstatement caused by the delay in recording returned sales. Since audit stratification and a detailed review are involved in isolating and handling this error situation, a statistical evaluation is neither appropriate nor necessary.

4. *Errors Affecting One or More Than One Financial Statement*

There are monetary errors which affect either one or more than one financial statement. In making decisions about monetary precision limits it would seem that auditors would be more concerned with errors which affect more than one financial statement. As an illustration, if an individual sale were recorded twice, both revenue and accounts receivable would be overstated, thereby affecting both the balance sheet and the income statement. Thus, in the audit appraisal, the errors affecting one financial statement should be separated from those affecting more than one financial statement because the audit significance may be different.

5. *Systematic and Random Errors*

Systematic errors—those caused for an apparent reason—should be analyzed and appraised separately from those which occur randomly without any discernible reason or pattern. Systematic errors would result from using an obsolete price list. The cause of a systematic error permits the auditor to stratify for analysis and appraisal.

Inventory counts could represent random errors; some counts would be too high, while others would be too low for no apparent reason. However, if all the counts are in one direction, such as too high, then the errors should not be considered random even if there does not appear to be a discernible reason. Unlike systematic errors, random errors are difficult to analyze and appraise unless a discernible pattern emerges which could lead to stratification. For instance, suppose 20 inventory errors are detected, each less than $500, and ten are overstatements and ten are understatements. The auditor may conclude that the errors are offsetting. However, his conclusion would be different if an evaluation matrix of the errors reveals that practically all the errors are from the same operating activity.

V. SUMMARY

The auditor is primarily concerned with an organization's accounting and administrative controls for safeguarding assets and the reliability of the financial records. The safeguarding of assets involves protection against losses from intentional and unintentional errors, repetitive and nonrepetitive errors, and systematic and random errors.

To accomplish this goal in an effective, efficient and economical manner, flexible sampling is recommended. This strategy involves the integration of judgment, experience, knowledge and statistics in the formulation of an audit opinion. The basic techniques employed are audit stratification, replication of components of minimum samples, and error analysis and audit appraisal. When necessary, statistical evaluation, discussed in the last chapter, is a phase of the strategy. However, it is not rational for disclosed errors to be statistically evaluated if the result is obviously beyond acceptable monetary precision limits. Even when the result is within acceptable limits, the auditor would seem to be obligated to analyze and appraise each error because of the possible legal consequences if he shuns this responsibility.

Flexible sampling provides the auditor with a strategy which improves his effectiveness by (i) a saving of time and effort in reaching a conclusion of acceptability of records and procedures, and (ii) assisting in directing his effort into audit areas most likely to provide the highest return for the time expended.

The flexibility of this sampling strategy in permitting more effective utilization of audit effort is probably more significant than the saving of time represented by a reduction of effort needed to support a decision. This sampling strategy provides the auditor with a degree of flexibility which is lacking in rigidly prescribed sampling plans. Under flexible sampling, the auditor who encounters a faulty document will pause to consider the source and nature of the error. If his judgment dictates, he may initiate, in the area of the faulty document, a subsidiary purposive or statistical sampling procedure or do a 100 percent review to support or negate the hypothesis established upon detection of the initial error. In other words, the auditor may, at any time, temporarily conclude his sampling of the entire frame to pursue a more likely clue, knowing that he can return to or modify his original plan if he concludes that the clue provides no additional evidence. This approach permits the auditor to optimize the use of his time in a manner that assures the greatest possible audit impact from his efforts.

Accounting practices usually vary among operations and even more markedly from location to location. Thus, a single sample, however well designed and performed, provides information from only one of nearly an infinite number of similar samples. Consequently, to assure that audit conclusions, based at least partly on sample information, are reasonable and practical, several replications of the frame are recommended as an additional protection against using atypical sample data.

In performing an analysis and appraisal, the concept of an overall error rate or amount is confusing to many auditors who have a responsibility to appraise the financial impact of individual categories and sources of errors. Under these circumstances, the statistical concept of "on the average" is of little value to them. Also, they realize that their estimates are suspect if they include nonrepetitive occurrences.

Furthermore, it should be clearly understood that sample testing of account balances is no substitute for the audit of accounting practices, and in no way decreases the significance of this part of the audit program. In other words, the auditor should rely on his judgment, experience and knowledge, as well as all other available sources of reliance, in searching for possible weaknesses in an accounting system.

14
Some Basic Statistical Concepts

I. INTRODUCTION

With the assistance of computers, auditors can identify a randomly selected sample and obtain an estimate of the precision of the results of their audit of the sample items. However, this "cookbook" approach fails to provide an appreciation or an understanding of the basic underlying concepts and theories, thereby permitting abuses of the statistical procedures.

The statistical evaluation of sample results, which is discussed in the next chapter, involves the use of both descriptive and inferential statistical concepts and principles. Thus, it is appropriate to provide a brief review of relevant statistical concepts and principles prior to discussing the evaluation of sample results. This chapter, therefore, is designed for readers with little or no statistical training and for those who desire a brief review of basic statistical concepts. The discussions are brief introductions to statistical concepts as they apply to sampling procedures. Hence, this chapter is intended to be optional for those readers who do not need a review of basic statistics.

The discussion in this chapter, as in the rest of the book, are not intended to be mathematically rigorous, but are simple explanations and illustrations of the principles and theories involved. Those readers who desire more detailed discussions should refer to Coxton, Cowden and Bolch [10], Neter and Wasserman [46] or Snedecor and Cochran [62].

The appropriate uses of statistical sampling methods as they apply to auditing and accounting involve, at least, acquiring an understanding and an appreciation of statistical concepts discussed in this chapter. As a prerequisite for acquiring this knowledge, an auditor needs only a minimal level of mathematical skills. Although many simple manual computations will be used to illustrate principles and procedures, in actual practice laborious and complex sampling operations can be accomplished with the aid of a computer.

Statistical methods may be classified as either descriptive statistics or inferential, or analytical, statistics. Descriptive techniques provide data in some summarized format such as tables, charts, graphs, indices of relationships, and the like, for the purpose of describing a situation. Statistical inference methods, as applied to auditing, are primarily used to evaluate conclusions about a universe from which a sample has been drawn. Thus, statistical inference techniques permit the auditor to characterize a universe from sample results, and to state how well the sample estimated, that is, the precision of the sample. Although statistical inference methods are useful in hypothesis testing as well as in estimation, experience reveals that hypothesis testing procedures generally provide insufficient information for most audit decisions, because they do not reveal whether a rejection is with a whisper or a shout. Estimation procedures provide values which can be assessed in the formulation of an audit opinion.

The two major statistical classifications are related to each other but are distinguished by method and point of view. Estimates from samples may be considered as both descriptive statistics in characterizing the sample results, and as inferential statistics in describing the items in the universe which have not been audited.

Those readers who are interested in the application of significance testing to accounting data should refer to Elliott and Rogers [13] and Kinney [31].

II. DESCRIPTIVE STATISTICS

The methods of descriptive statistics may be grouped into two general categories: graphical and mathematical. Although some graphic forms (tables, charts and graphs) will be used, this chapter emphasizes computed quantities.

Raw data appear as unorganized masses of figures without form or structure. Even when data have been arranged in a general table, the format may not satisfy the audit requirements. Data must be put in a relevant format and given coherent structure before analysis and inference are feasible.

The simplest organization of data is an array, which arranges values in order of magnitude, indicating the range covered and the general distribution within the range. Although the array presents data in a more suitable format for analysis than a haphazard distribution, more organization and structure are required before the significance of the data becomes apparent.

A. Frequency Distribution

By grouping the data within class limits, a simplified and more compact presentation—known as a frequency distribution—is obtained. The purpose of a frequency distribution is to show, in a condensed form, the nature of the distribution through the range covered. For illustrative purposes, consider the values of the 50 accounts receivable shown in Figure 11.

Using an arbitrary class interval of $3,000, the same data may be arranged in a frequency distribution table as shown in Figure 12. This figure shows class intervals, class boundaries with tallies, interval medians, or midpoints, and class frequencies for the same 50 accounts receivable.

Frequency distributions are important in presenting a compact summary of data and in preparing the data for further analysis. Such distributions are not limited to a tabular format, but may be presented graphically, in which the data characteristics become more apparent. Two common ways of graphically presenting frequency distributions are the frequency histogram and polygon. Thus, using the data from Figure 12 the histogram and polygon in Figure 13 are obtained. The sides of the columns represent the upper and lower class boundaries,

$4,600	5,000	4,400	4,700	5,700	4,800	3,300	5,300	5,300	5,300
6,300	6,000	6,900	5,400	3,600	3,700	4,500	5,000	3,100	4,900
2,900	4,000	5,000	5,000	5,500	5,900	3,800	4,300	5,300	4,700
5,300	6,300	6,400	4,700	6,600	7,100	6,700	5,200	4,000	3,400
5,600	5,000	4,700	5,700	4,300	6,000	4,000	5,000	4,700	5,100

Figure 11. Values of 50 accounts receivable.

Class Intervals	Class Boundaries	Tally	Interval Median (Midpoint) ($ 000)	Frequency
2,800-3,000	2,750-3,050	/	29	1
3,100-3,300	3,050-3,350	//	32	2
3,400-3,600	3,350-3,650	//	35	2
3,700-3,900	3,650-3,950	//	38	2
4,000-4,200	3,950-4,250	///	41	3
4,300-4,500	4,250-4,550	////	44	4
4,600-4,800	4,550-4,850	~~////~~ //	47	7
4,900-5,100	4,850-5,150	~~////~~ ///	50	8
5,200-5,400	5,150-5,450	~~////~~ //	53	7
5,500-5,700	5,450-5,750	~~////~~	56	4
5,800-6,000	5,750-6,050	///	59	3
6,100-6,300	6,050-6,350	//	62	2
6,400-6,600	6,350-6,650	//	65	2
6,700-6,900	6,650-6,950	//	68	2
7,000-7,200	6,950-7,250	/	71	1

Figure 12. Frequency distribution table of values of 50 accounts receivable.

and their heights (and areas) are proportional to the frequencies within the class intervals. The associated frequency polygon consists of a series of straight lines plotted from the midpoints of the tops of the columns of the frequency histogram.

By extending the information in the frequency distribution table, Figure 12, relative and cumulative frequencies may be obtained as shown in the table in Figure 14.

If the vertical frequency scale in Figure 13 were replaced with the corresponding relative frequencies in Figure 14, the same graphs would become a relative frequency histogram and polygon. As will be discussed in a later section, one way of defining probability is in terms of relative frequencies.

B. Indices of Central Tendency

In working with frequency distributions, it is apparent that most series of data tend to group, or cluster, about some interior or central

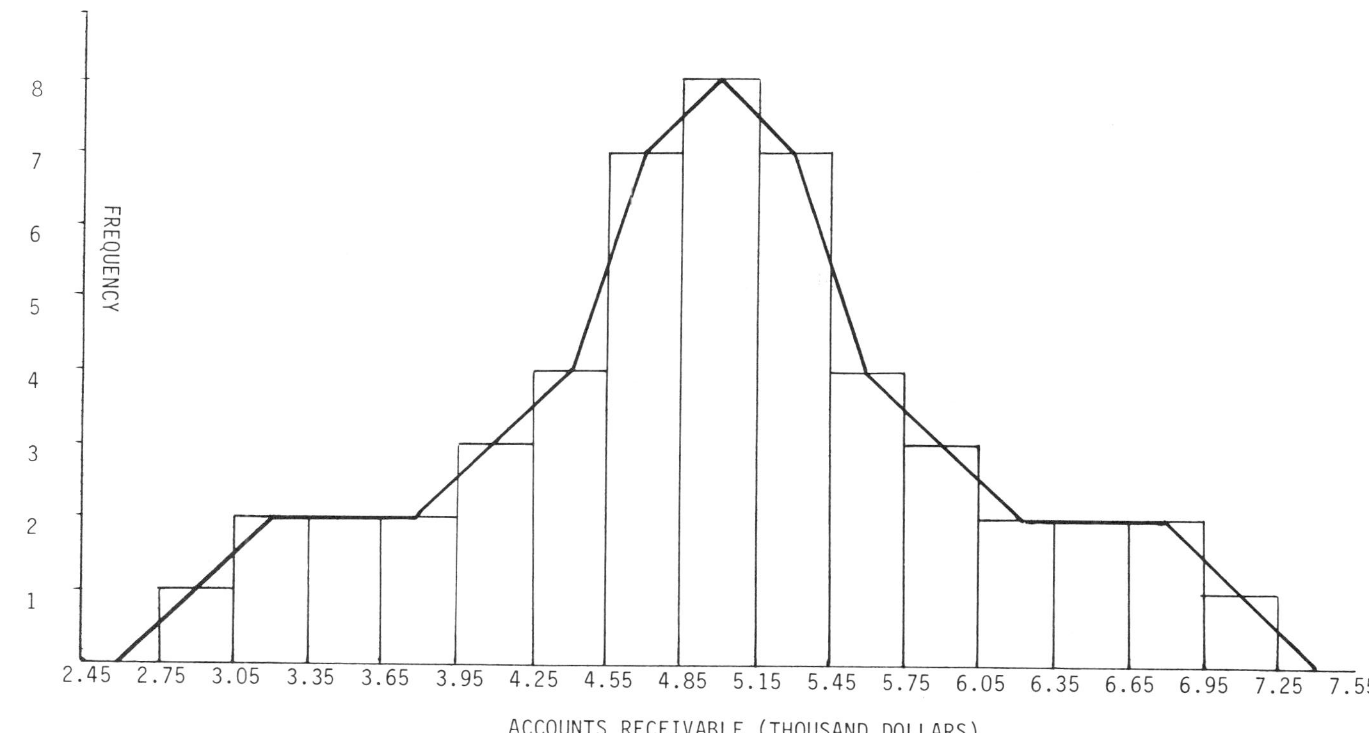

Figure 13. Frequency histograph and polygon of values of 50 accounts receivable.

Class Boundaries	Frequency	Cumulative Frequency	Relative Frequency (Percent)	Cumulative Relative Frequency
2,750-3,050	1	1	2	2
3,050-3,350	2	3	4	6
3,350,3,650	2	5	4	10
3,650-3,950	2	7	4	14
3,950-4,250	3	10	6	20
4,250-4,550	4	14	8	28
4,550-4,850	7	21	14	42
4,850-5,150	8	29	16	58
5,150-5,450	7	36	14	72
5,450-5,750	4	40	8	80
5,750-6,050	3	43	6	86
6,050-6,350	2	45	4	90
6,350-6,650	2	47	4	94
6,650-6,950	2	49	4	98
6,950-7,250	1	50	2	100

Figure 14. Cumulative and relative frequency of values of 50 accounts receivable.

value. This characteristic, referred to as central tendency, may be used to describe the data.

Mathematical descriptive statistics are computed quantities which summarize masses of data. The most common summarization takes the form of providing an index of central tendency and an index of dispersion of the data. The most common indices of central tendency and dispersion, or variability, will be discussed briefly and illustrated.

Before discussing these indices, it is useful to differentiate between a point estimate and an interval estimate. A point estimate is a single value such as a total, mean, median, ratio, variance, standard deviation, and the like. An interval estimate is a range of values, which indicates, with an associated probability, the limits within which an actual value is expected to fall if repeated samples of the same size and same procedures were selected from a universe.

The arithmetic mean, which henceforth will be referred to as the mean, the median and the mode are the three most common indices of central tendency. The advantages and limitations of each of these indices will be discussed.

1. Mean

For ungrouped data, the mean is obtained by adding all observations and dividing by the number of observations. For example, the mean of 2, 4, 6, 12 and 16 is

$$\text{mean} = \frac{\text{sum of items}}{\text{number of items}} = \frac{2 + 4 + 6 + 12 + 16}{5} = \frac{40}{5} = 8$$

Thus, the mean is a single value, like a center of gravity, which is used to characterize a set of data. It is the most commonly used index of central tendency—value around which other values tend to cluster.

The desirable properties of the mean may be summarized as follows:

a. It is the most commonly used and the most easily understood index of central tendency.
b. The mean is relatively easy to compute.
c. The mean can be computed from ungrouped data.
d. Only total values and total number of items are needed for its computation.
e. The sum of the deviations of the individual items from the mean equals zero.
f. The weighted mean of two or more combined distributions can be calculated from their individual means.

Example:

Number students	Mean income
3	\$5,000
12	\$4,000

$$\text{The mean income of the 15 students} = \frac{(3 \times 5000) + (12 \times 4000)}{15} = \$4{,}200$$

g. The sum of the squared deviations from the mean is a minimum, i.e., less than the squared deviations from any other number in a distribution.
h. The means of the sums of two or more values is the sum of their means.

Example:

Salary income	Interest income	Total income
12,000	700	12,700
15,000	800	15,800
18,000	900	18,900
45,000	2,400	47,400

$$\text{Mean salary income} = \frac{45,000}{3} = 15,000$$

$$\text{Mean interest income} = \frac{2,400}{3} = 800$$

$$\text{Mean total income} = \frac{47,400}{3} = 15,800$$

Thus, salary mean (15,000) + interest mean (800) = 15,800, the total income mean.

i. If a constant, or fixed amount, is added to all items in a distribution, the mean will be increased by that amount.

Example:	Original data	After adding 5 to each item
	16	21
	14	19
	15	20
	18	23
	17	22
	80	105

$$\text{Mean} = \frac{80}{5} = 16 \qquad \text{Mean} = \frac{105}{5} = 21$$

Thus, adding 5 to each item increases the mean from 16 to 21.

j. Since the mean is a computed value, it may be manipulated algebraically.

The major limitations of the mean are:

a) The mean cannot be computed for an open-end distribution because one is unable to determine the midpoint of the open-end class.
b) The mean is affected by extreme values.
c) The mean may not be typical, since it may not be one of the actual numbers in a series.
d) Being a computed value, it cannot be computed graphically.

2. Median

Other indices of central tendency may be used. One of these is the median which is the value of the middle item in an array. If there is no unique middle number, as whenever the total number of items is even, the mean of the middle two values is taken as the median. Thus, the median may be considered the balance point with 50 percent of the items on each side. In other words, for ungrouped data, the median location formula is:

$$\text{Median location} = \frac{\text{number of items} + 1}{2}$$

Example (i): Given series: 2, 9, 4, 7, 2, 1, 8, 3, 6
Arrange in an ascending array: 1, 2, 2, 3, 4, 6, 7, 8, 9

$$\text{Median location} = \frac{9 + 1}{2} = \text{5th item, i.e., 4.}$$

Example (ii): Given array: 1, 2, 3, 5, 9, 10, 11, 12

$\text{Median location} = \frac{8 + 1}{2} = 4\frac{1}{2}$, where 5 is the fourth item and 9 the fifth item.

Thus $\frac{5 + 9}{2} = \frac{14}{2} = 7$, the median, which is midway between 5 and 9.

For grouped data, the median is the number that corresponds to the point on the horizontal scale through which a vertical line divides the total area of a histogram into two equal parts. A formula similar to the following one may be used to compute the median from grouped data:

$$\text{Median} = \begin{bmatrix}\text{lower limit of} \\ \text{median class} \\ \text{interval}\end{bmatrix} + \frac{\left|\frac{\text{Total items}}{2}\right| - \left|\begin{matrix}\text{Total items in pre-} \\ \text{ceding intervals}\end{matrix}\right|}{\left|\begin{matrix}\text{Number of items in interval} \\ \text{containing median}\end{matrix}\right|} \times \begin{bmatrix}\text{Width of} \\ \text{interval} \\ \text{containing} \\ \text{median}\end{bmatrix}$$

The desirable properties of the median may be summarized as follows:

a. The median of a set of data always exists.
b. The median is not easily affected by extreme values.

Example: In both of the following series the median is 6:

(i) 1, 2, 4, 6, 10, 12, 15
(ii) 1, 2, 4, 6, 10, 25000, 50000

c. The median may be found from any distribution, even if the distribution is open ended provided the total number of values is known.
d. The median can be used to describe qualitative observations which are not expressed numerically.

Example: Individuals may be asked to rank five samples of laundry detergents according to cleaning power. The detergent ranking third is the median.

e. Being a positional value, it can be estimated graphically from an ogive.*

The major limitations of the median are:

a. The median cannot be determined from unorganized data.
b. The median may be unreliable if there are only a relatively few items in a series.
c. The median cannot be manipulated algebraically. For instance, medians of subgroups cannot be averaged to find the overall median. To find an overall median, the data for all the subgroups must be combined into a single array.

Example: Series 1: 1, 3, 5, 9, 13; median = 5
Series 2: 3, 7, 10, 14, 17, 19; median = 12
Series 1 and 2 combined: 1, 3, 3, 5, 7, 9, 10, 14, 17, 19; median = 9
Whereas, averaging 5 and 12 gives $8\frac{1}{2}$.

d. The value of the median tends to vary more from series to series than the mean.

Example:

Series	Median	Mean
2, 6, 4	4	4
1, 9, 2	2	4
1, 4, 4	4	3
1, 6, 2	2	3
6, 1, 5	5	4

e. When there is an even number of items in a distribution, the median may not be one of the actual values, and thus, it may not be typical.

---

*An ogive is a graph of a cumulative frequeny distribution.

3. *Mode*

Another index of central tendency is the mode which is the most frequent, or popular, value in a sequence. It can be determined from ungrouped or grouped data. For grouped data, the mode is the midvalue of the class with the highest frequency.

The desirable properties of the mode may be summarized as follows:

a. If a unique mode exists, it represents the most typical value of a distribution.
b. For ungrouped data, the mode can be determined without any calculations.
c. The mode is entirely independent of extreme values.
d. The mode can be determined graphically.
e. Similar to the median, the mode can be used to describe qualitative data.

The major limitations of the mode are:

a. The mode cannot be determined from unorganized data.
b. The mode may not exist.

   Example: The series 1, 2, 3, 7, 10, 14, 90 has no mode.

c. There may not be a unique mode.

   Example: The series 1, 2, 2, 3, 5, 14, 14, 21, 30 is bimodal, i.e., 2 and 14.

d. The mode may be hidden by grouping in a frequency distribution.
e. As in the case of the median, the mode cannot be manipulated algebraically. For example, modes from two or more subgroups cannot be averaged to find the overall mode.
f. The mode may be unreliable unless there is a large number of items in the distribution.

4. *Comparison of Indices of Central Tendency*

There are several ways of indicating the central tendency of a distribution. One way is to construct a distribution table, and by inspection, observe the interval or point where the highest frequency occurs. This point, or midpoint of this interval, would be considered the mode. However, in some distributions there may be more than one peak or mode. Such distributions are called bimodal or multimodal, making it impossible, or at least misleading, to use the mode as an

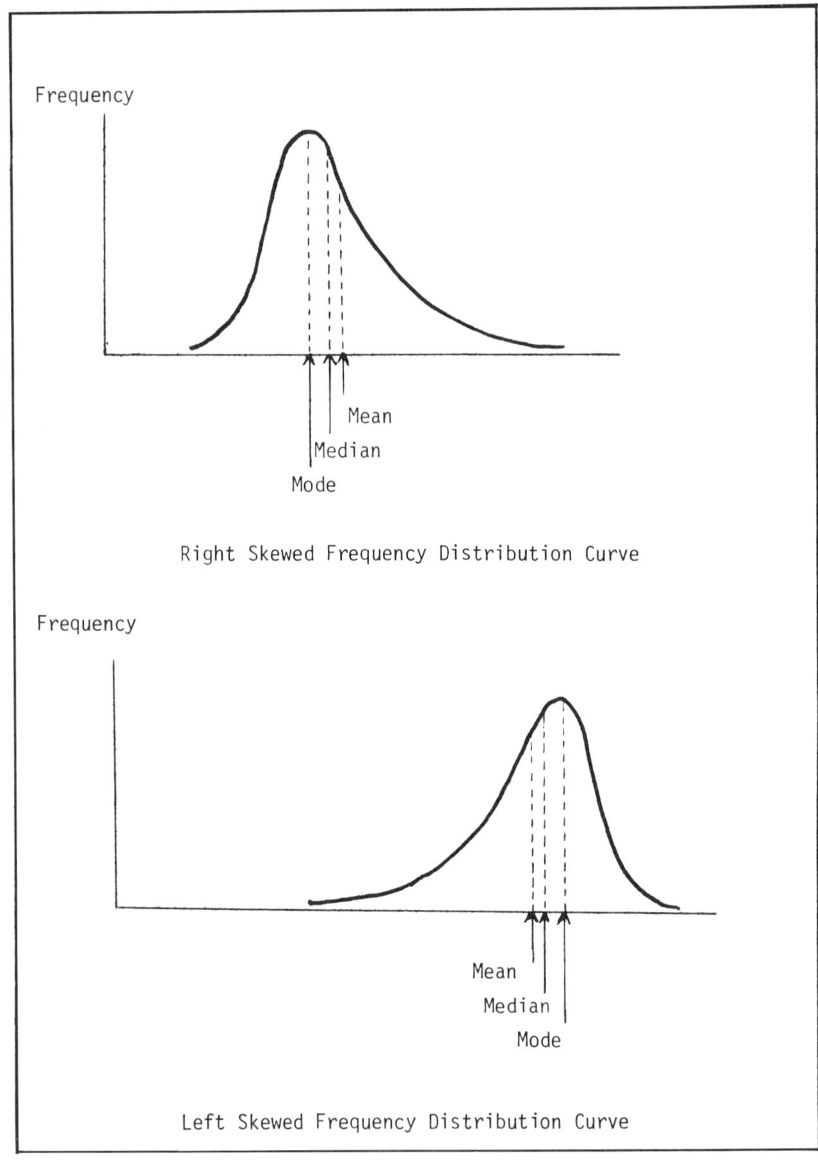

Figure 15. Relationship of mean, median and mode in skewed frequency distributions.

index of central tendency. Also, there are some cases when the highest frequency may occur at the beginning, or at the end of the distribution. Here, too, the use of the mode would produce a distortion. On the other hand, the median more closely approximates the true central tendency since it divides the data into halves of equal numbers. Often, however, the number of items is too large to justify the effort of arranging them in an array and counting to find the median. Also the median does not give proper weight to variations which occur within a distribution, since individual values are not reflected in the results. Thus, for reasons indicated above, and the fact that it can be used in computational analyses, the mean is the most widely used index of central tendency. Nevertheless, for a mere representation of the central tendency of a distribution, the median is often used because it is less affected by extreme values. The most important characteristic of the mean is its precise relationship to the total value and to the number of items involved.

In a right (positive) skewed distribution, which often occurs in accounting data, the mean is larger than the median, which in turn, is larger than the mode. If there is moderate skewness, the distance between the median and the mode is about twice that between the median and the mean. This relationship which permits the estimation of any one of these three indices from the other two, shown graphically in Figure 15, may be summarized as follows:

a. The mean and median always move in the direction of the skewness, or tail, of the frequency distribution curve, with the mean being closest to the tail.
b. In a moderately skewed frequency distribution, the approximate distance between the mean and the median is 1/3 of the distance between the mean and the mode.
c. The mode is the value of the highest point on the distribution curve.

C. Indices of Dispersion

Indices of central tendency help to characterize a distribution of data but more information is needed. For instance, two sets of data may have the same mean, yet be very dissimilar in characteristics. In other words, sets of data often differ from each other in their extent of dispersion or scatter about the mean of the distribution. Thus, there is a need for a suitable index of dispersion, or variability, in addition to the mean.

1. Range

The simplest index of dispersion is the range or limits within which values of a distribution fall. It is usually stated as the difference between the two extreme values in a distribution. In the case of grouped data, the difference would be between the lower limit (boundary) of the first class and the upper limit of the last class. The advantages of the range are that it (i) requires little computation, (ii) provides a quick visual measure of the spread of the data, and (iii) is used in quality control operations. The range, however, is a poor index of dispersion because (i) it is affected by extreme values and (ii) it does not reveal how the data are distributed between the two extreme values.

2. Average or Mean Deviation

The average, or mean, deviation is the mean of the absolute deviation of each value in a series from either the mean or the median of the entire series. That is, it is the sum of the absolute deviations of individual items in a series divided by the number of items in the series. Absolute deviations consider all signs to be positive. As mentioned before, if the negative signs are not neglected, the sum of the deviations of the items from the mean would be zero. While the average deviation of a sample shows how the items in the sample are dispersed, it does not provide a satisfactory estimate of how widely the items in the universe from which the sample is selected are likely to be dispersed. Since most statistical sampling is performed for the purpose of making inferences, the average, or mean deviation is seldom used as an index of dispersion.

3. Standard Deviation

For purposes of generalization from inferential statistics, the most useful and frequently used index of dispersion is the standard deviation of the mean, which is also known as the root mean squared deviation since it is defined as the square root of the mean of the squared deviation of individual items from their mean.

In words and symbols, the standard deviation of a series of values is computed as follows:

a. Obtain the individual deviations by substracting the value of the mean of the series, from the value of each individual item, X:

$$X - \mu$$

b. Then square each individual difference or deviation:

$$(X - \mu)^2$$

c. Sum all of the squared differences:

$$\Sigma(X - \mu)^2$$

d. Divide the sum of the squared differences by the total number of items, N, in the series to obtain the variance of the items:

$$\text{Variance} = \sigma^2 = \frac{\Sigma(X - \mu)^2}{N}$$

e. The square root of the variance is the standard deviation of the items:

$$\text{Standard deviation} = \sigma = \sqrt{\frac{\Sigma(X - \mu)^2}{N}}$$

It should be noted that when the standard deviation is computed from sample data, the N in the above steps is replaced by N-1. The purpose of this substitution is to provide a better estimate of universe standard deviation. However, when the sample is composed of many items, there is practically no difference in using N or N-1.

The above computational steps can be best illustrated by the following simple example:

Compute the standard deviation of 11, 5, 8, 7, 9
First the mean is calculated to be:

$$\text{Mean} = \mu = \frac{11 + 5 + 8 + 7 + 9}{5} = \frac{40}{5} = 8$$

Calculations for the first three steps in a tabular format such as the following are helpful:

Item (X)	Item minus mean $(X - \mu)$	Item minus mean squared $(X - \mu)^2$
11	11 - 8 = 3	$(3)^2 = 9$
5	5 - 8 = -3	$(-3)^2 = 9$
8	8 - 8 = 0	$(0)^2 = 0$
7	7 - 8 = -1	$(-1)^2 = 1$
9	9 - 8 = 1	$(1)^2 = 1$
40	0	20

The next step is to divide the total of the last column in the above table by the number of items to obtain the variance.

$$\text{Variance} = \sigma^2 = \frac{20}{5} = 4.$$

Finally, the square root of the variance, $\sqrt{4} = 2$, is the standard deviation of 11, 5, 8, 7 and 9.

It should be observed in the above table that the sum of the deviations from the mean is zero, which can serve as a computational check. However, if absolute deviations were considered, the median would be used.

Applying a similar procedure to the universe of 50 accounts receivable listed in Figure 11, the tabular format in Figure 16 is obtained.

Obviously, an auditor is not expected to manually compute the standard deviation. The above illustrations are for the purpose of providing a better understanding of its structure, since it is one of the most significant statistical notions as will be evident in subsequent discussions of the normal curve and statistical inference.

The effect of squaring the deviations, even though the square root of their mean is later extracted, gives more relative weight to extreme values than the mean. The special consideration to extreme variations in the sample is significant since the limits will probably be even wider in the universe from which the sample was drawn. Thus, this added dimension to the variations in the sample tends to fill the gap caused by the exclusion of some items in the universe. However, a few very extreme items can produce an unusually large standard deviation which may be misleading. If this is a serious problem, the deviations may be measured from the median instead of the mean; however, in sampling, extreme variations are usually best handled through stratification.

The following is a summary of some of the properties of the standard deviation since it is so important in sampling theory and practice:

a. The standard deviation has a unique relationship with the mean of a normal distribution. In such cases, limits prescribed by the mean plus and minus one standard deviation contains 68 percent of the data. Likewise, 95 percent of the data will be within plus and minus two standard deviations, and practically all data are within three standard deviations of the mean.

ITEM X		ITEM MINUS MEAN (X-5,000)		ITEM MINUS MEAN SQUARED $(X-5{,}000)^2$	
2900	5000	-2100	0	4,410,000	0
3100	5000	-1900	0	3,610,000	0
3300	5000	-1700	0	2,890,000	0
3400	5100	-1600	100	2,560,000	10,000
3600	5200	-1400	200	1,960,000	40,000
3700	5300	-1300	300	1,690,000	90,000
3800	5300	-1200	300	1,440,000	90,000
4000	5300	-1000	300	1,000,000	90,000
4000	5300	-1000	300	1,000,000	90,000
4000	5300	-1000	300	1,000,000	90,000
4300	5400	- 700	400	490,000	160,000
4300	5500	- 700	500	490,000	250,000
4400	5600	- 600	600	360,000	360,000
4500	5700	- 500	700	250,000	490,000
4600	5700	- 400	700	160,000	490,000
4700	5900	- 300	900	90,000	810,000
4700	6000	- 300	1000	90,000	1,000,000
4700	6000	- 300	1000	90,000	1,000,000
4700	6300	- 300	1300	90,000	1,690,000
4700	6300	- 300	1300	90,000	1,690,000
4800	6400	- 200	1400	40,000	1,960,000
4900	6600	- 100	1600	10,000	2,560,000
5000	6700	0	1700	0	2,890,000
5000	6900	0	1900	0	3,610,000
5000	7100	0	2100	0	4,410,000
TOTAL	250,000		0		47,680,000

$$\text{Mean} = \frac{250{,}000}{50} = 5{,}000; \quad \text{Standard Deviation} = \sqrt{\frac{47{,}680{,}000}{50}} = \sqrt{953{,}600} = 976.52$$

Figure 16. Computation of standard deviation for 50 accounts receivable.

b. The standard deviation can be mathematically manipulated for use in statistical calculations.

c. A constant (either positive or negative) may be added to items in a distribution without changing the values of the variance or standard deviation. Obviously the mean value would change.

Example:

	Before adding constant		After adding 20 to each item	
	X	$(X - \mu)^2$	X	$(X - \mu)^2$
	20	$(-20)^2$	40	$(-20)^2$
	40	$(0)^2$	60	$(0)^2$
	60	$(20)^2$	80	$(20)^2$
Total	120	800	180	800

$$\text{Mean} = \frac{120}{3} = 40 \qquad \text{Mean} = \frac{180}{3} = 60$$

$$\text{Variance} = \frac{800}{3} = 266.67 \qquad \text{Variance} = \frac{800}{3} = 266.67$$

Standard deviation = $\sqrt{266.67} = 16.3$ Standard deviation = $\sqrt{266.67} = 16.3$

This property is very significant when dealing with a series of very large values since a constant can be subtracted from each value thereby reducing the series to manageable values. Then the standard deviation can be computed from the transformed series. This process is referred to as a linear transformation. Even when computers are used, such transformations reduce the data input effort. However, it should be remembered that the value of the mean of the transformed series compared with the original series is decreased by the amount of the constant. Hence, if this amount is added to the new mean the result will be the mean of the original series.

d. The standard deviation permits the conversion of individual values into standard scores, or Z scores. Z scores are absolute numbers in that they are not related to any units. The conversion formula is:

$$\text{standard score} = Z = \frac{\text{item value} - \text{distribution mean}}{\text{distribution standard deviation}}$$

$$\text{or in symbols} \qquad Z = \frac{X - \mu}{\sigma}$$

Example:

Analyst A found \$15,000 in undocumented costs. Other analysts on the same project averaged \$13,000 in undocumented costs with a standard deviation of \$1,000.

Analyst B on another project found \$6,250 in undocumented costs. Other analysts on this project averaged \$5,000 in undocumented costs with a standard deviation of \$500.

$$\text{The standard score for Analyst A} = \frac{15{,}000 - 13{,}000}{1{,}000} = 2.0$$

$$\text{The standard score for Analyst B} = \frac{6{,}250 - 5{,}000}{500} = 2.5$$

Thus, Analyst B did a better job of detecting undocumented costs since he was 2.5 standard deviations above his project peers, whereas Analyst A was only 2 standard deviations above.

In computing the variance from grouped data, a small consistent upward bias is introduced since the midpoints of the class intervals tend to be farther from the mean than the actual class interval averages. Thus, the deviations of the midpoints from the mean are more than the deviations of the actual class interval averages. The correction for this bias is known as Sheppard's correction, and it is accomplished by subtracting $k^2/12$ from the computed variance, where k is the size of the class interval. However, this poses no problem in most audit activities, especially where computers are available, since the original ungrouped data can be used; thereby avoiding the need for a correction.

One of the major properties of the standard deviation is that within particular intervals of a normal distribution, a known proportion of the observations are expected to fall. However, for extremely skewed distributions, there is no way of knowing what precise proportions of the observations would be expected within specified intervals, but there are theorems which provide tolerance limits. See Figure 23 for comparisons of tolerance limits for the normal distribution and others. Again, for the auditor, the most feasible solution is stratification. An example of a U or J shaped distribution is a mixed universe of eligible and ineligible participants in a health service program. One end of the distribution could represent costs associated with ineligible participants. All these costs are unallowable. The other end of the distribution could be a clustering of costs associated with eligible participants, where most of the costs are allowable, and therefore, are acceptable charges to the program. If this type of universe distribution could be stratified into two strata—one of the eligibles and the other of ineligibles—both the efficiency and precision of the sample would improve. A clue to how to effectively perform this kind of stratification may be gained through analysis of the nature and sources of the clustering from prior audits, audit surveys or a preliminary sample.

4. *Coefficient of Variation*

Since the standard deviation is in the original units of a series, it does not permit comparisons of series in different units. The coefficient of variation, an abstract measure of dispersion, is designed for comparative purposes. It is defined as:

$$\text{coefficient of variation} = V = \frac{\text{standard deviation}}{\text{mean}} \times 100$$

This gives the standard deviation as a percentage of the mean. The greater the dispersion, the higher the ratio of the standard deviation to the mean. Thus, the relative dispersion of distributions may be determined by comparing their coefficients of variation.

5. *Kurtosis*

Kurtosis is a descriptive characteristic of distributions which indicates the general form of the concentration about the mean and in the tails of the distributions. It relates to the peakness or flatness of a distribution.

The basis of reference for comparing kurtosis is the bell-shaped normal curve. For example, a leptokurtic curve has a narrower central portion and higher tails than the normal curve as shown in Figure 17. Properties of the normal curve will be discussed in a subsequent section.

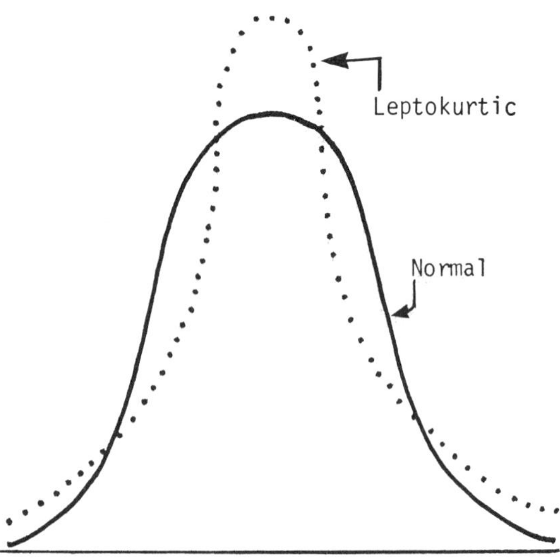

Figure 17. Comparison of normal and leptokurtic curves.

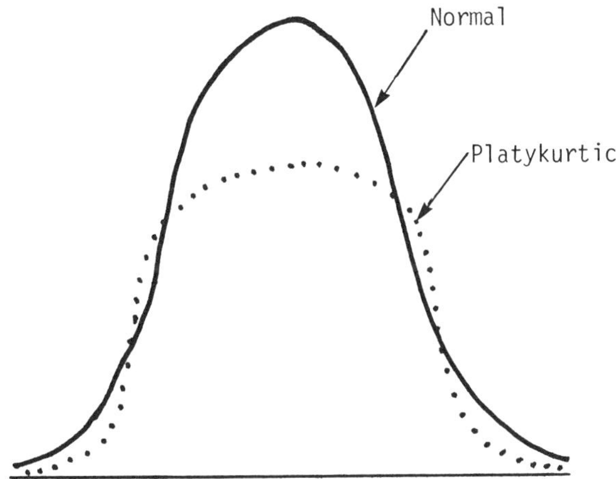

Figure 18. Comparison of normal and platykurtic curves.

A platykurtic curve has a broader central portion and lower tails than the normal curve as shown in Figure 18.

There are formulas for measuring relative kurtosis and others for measuring absolute kurtosis. When a relative measure is used, a factor of 3 indicates a normal distribution. If the factor is less than 3, the distribution is platykurtic, while greater than 3 indicates a leptokurtic distribution. When an absolute measure is used, zero indicates a normal distribution.

When a computer program is used to evaluate a sample result, a measure of kurtosis for the distribution of sample items can aid in deciding where further stratification is desired.

6. *Skewness*

Another aid to deciding whether additional stratification is desirable is the degree of skewness, or lack of symmetry, of the distribution of sample items. If a distribution is symmetrical, its mean, median and mode are equal. However, all symmetrical distributions are not normal.

Most frequency distributions show some skewness. Generally accounting distributions are skewed to the right as shown in Figure 15. Also, two sets of data can have the same mean and the same standard deviation but have different shaped distributions due to different degrees of skewness.

As in the case of kurtosis, there are absolute and relative measures of skewness. An absolute measure will be in terms of the units of the distribution, while relative measures are pure numbers, which are independent for any scale. For most indices, a zero factor indicates symmetry. Also, the larger the value of the factor, the greater the degree of skewness. Positive factors indicate right skewness, while negative factors indicate left skewness. If the skewness factor is a large value, further stratification should be one of the first options.

The following index of skewness reflects the fact that the greater the skewness the larger the difference between the mean and the median:

$$\text{skewness} = \frac{3(\text{mean} - \text{median})}{\text{standard deviation}}$$

D. Misuse of Graphs

Graphic methods display data in pictorial formats. Graphs are used to portray trends and relationships because it is easier to analyze and interpret a picture than a large set of numbers. Although graphs can convey a concise description of a set of data, they can also convey false impressions.

Perhaps the most prevalent of the misleading graphic practices is the omission of the base value, which is usually zero. When this occurs, significant information is visually surpressed.

The primary function of most graphs and charts is to provide the reader with a quick appreciation of the general trend of a data set. A proper interpretation of the general trend and relationship requires a correct perspective against which to evaluate the variations. Thus, the base line is of prime significance in assuring that the reader has the proper perspective.

For example, Figure 19A is drawn with a vertical scale from the zero base through the range of the data; whereas, Figure 19B omits about three-fourths of the vertical scale. Obviously, Figure 19B conveys the impression of a greater degree of variation in the data set than actually exists.

In other words, whenever the scale—vertical or horizontal—does extend to the base value, which is usually zero, the pictorial impression is out of perspective. Unfortunately, these kinds of misleading graphs and charts appear frequently in newspapers and magazines.

Although graphs and charts can be very informative, they require a careful analysis to prevent misinterpretations. Therefore, as a minimum, one should be very cautious in reading and interpreting graphs where the scales do not show the base values.

For additional discussions and illustrations of misleading graphic presentations, the reader should see Haack [21].

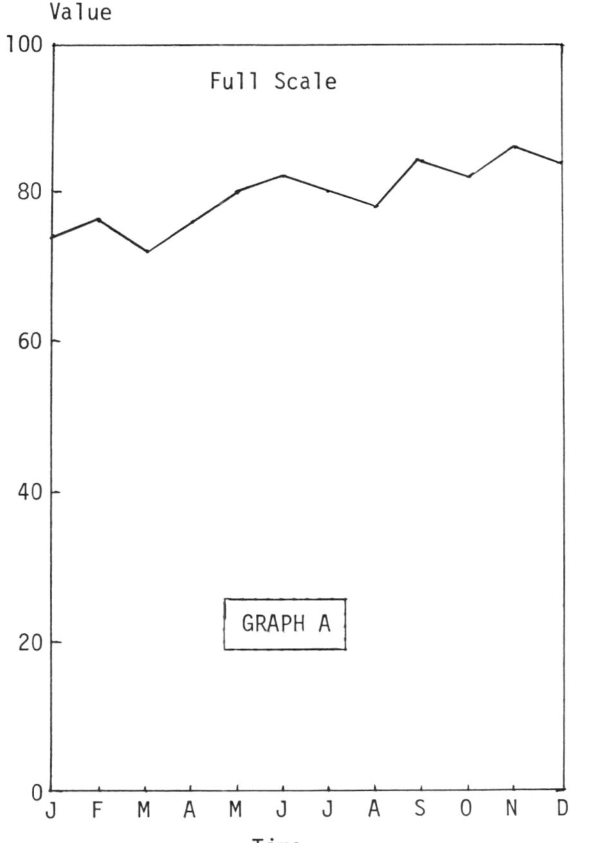

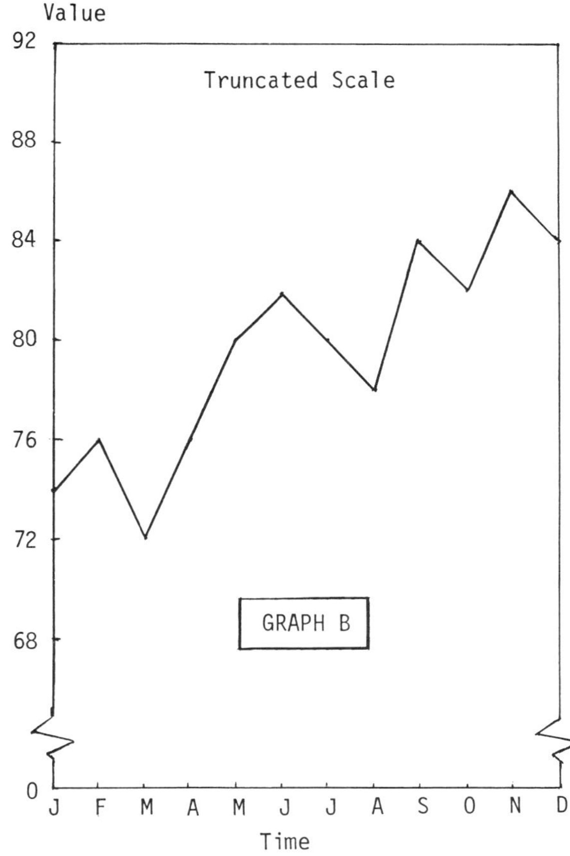

Figure 19. Misuse of graphs.

III. SOME STATISTICAL UNIVERSES OR DISTRIBUTIONS

Much of this section is more theoretical than previous sections. It is intended primarily for those readers who desire to know more about the theoretical underpinnings of statistical sampling. Except for the concepts of the normal curve, sampling distribution and Central Limit Theorem, the audit practitioner could skim this section initially and use it for future reference.

Obviously, one could make investments without any understanding of the investment market. Likewise, auditors may use packaged computer sampling programs without any knowledge of statistical inference or revelant concepts like a normally distributed universe or distribution. In both cases, however, the ultimate risks of making incorrect decisions are greater without an understanding and appreciation of the relevant concepts and basic notions.

The distributions discussed under descriptive statistics are empirical, since they are not constructed from the totality of all possible values. Conceptual distributions which consider the totality of all possible values—as in the case of a pair of dice—are referred to as theoretical, or statistical, universes or distributions.

The reason for studying theoretical universes is to become more alert to the kinds of universes approximated by accounting data. Most of these are highly skewed, which means that without effective stratification and/or larger sample sizes, the estimated sampling risk may be grossly underestimated.

In actual situations, the desired or expected characteristics of a universe are stipulated, and then one searches for statistical distributions which approximate such characteristics. Although the derivation of statistical relationships and properties involve complex mathematics, one can benefit from the efforts of statisticians without becoming involved with formulas by using statistical tables and packaged computer programs.

A. The Normal Curve

Several individuals independently discovered the normal curve. However, two of them—Pierre Simon de Laplace and Carl Friedrich Gauss—did more of the developmental research. They studied the distribution of errors observed in astronomical measurements. It was observed that repeated measurements of some variables seemed to follow predictable patterns. From this research a theory known as the normal curve of errors was developed. Later, a mathematician named Quetelet discovered that human characteristics and traits tend to follow the same curve as physical measurements. It is now recognized that many natural phenomena tend to be normally distributed.

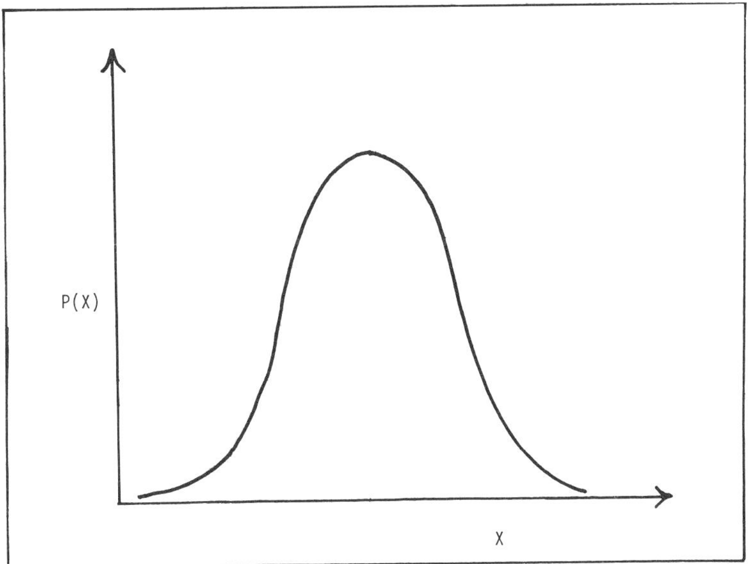

Figure 20. Curve of normal distribution.

The normal distribution is perhaps the most widely used theoretical distribution because the sampling distributions of many statistics are normal or nearly normal, especially for large sample sizes. The graph of the normal distribution is the bell-shaped curve depicted in Figure 20. It is completely specified by its mean and standard deviation. It is a continuous distribution, which is symmetric about its mean. The curve is asymptotic; that is, the graph extends infinity in both directions from the mean without touching the horizontal axis. However, for all practical purposes the graph can be considered to extend only three standard deviations from the mean in each direction, since less than 0.0015 of the area under the curve lies beyond three standard deviations in each direction.

B. The Standard Normal Distribution

To avoid the problem that every pair of values of the mean and standard deviation would require a different curve, one uses the standard form of the normal distribution. The standard normal distribution has a mean of zero, a standard deviation of one or unity, and an area of unity. Any normal distribution can be transformed or converted to the standard form by transforming the original X-values to something called Z-values. This is a simple scheme for expressing a variable as a multiple of its standard deviation from its mean value. The particular multiple usually is called the Z-value. The conversion or transformation formula is:

$$Z = \frac{\text{particular value of variable minus mean value}}{\text{standard deviation of variable}}$$

To illustrate, if the mean = 10 and the standard deviation = 5, then a value of X = 20 would be +2 standard deviations from its mean, or the equivalent of Z = + 2, computed as follows:

$$Z = \frac{X - \text{mean}}{\text{standard deviation}} = \frac{20 - 10}{5} = \frac{10}{5} = 2$$

In other words, a Z-value transformation divides the difference between a variable and the mean of its distribution by its standard deviation.

Obviously, if the mean is larger than the variable the resultant Z-value will be negative, denoting that it is to the left of the mean. With this conversion formula, one curve can be made to fit all combinations of means and standard deviations as shown in Figure 21.

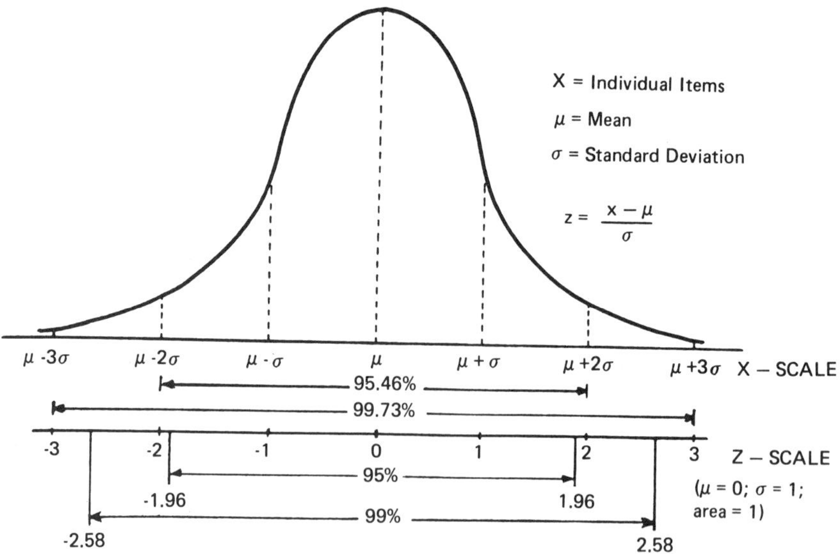

Figure 21. Normal and standard normal (Z) curves.

Tables of standard normal probabilities are constructed in various manners. Thus, one must ascertain the manner in which a particular table is constructed before it is used. The presentation format in Figure 22 is useful for a number of purposes.

The table is constructed with Z-values in the indicated columns. The corresponding probabilities are shown in columns headed by the letter A. The other columns headed by a Y, provide a means of constructing an approximation of a normal curve fitted to specific data. The table values include both the negative and positive values for the indicated Z.

If one wishes to know the Z-value associated with a particular probability, say .95 or 95 percent, he would search through the body of the table until he finds .95 (or the closest value to it) under the A columns, and then read the Z-value associated with it. In this case, the Z-value is 1.96.

The values in the Y columns are generally not used in sampling unless one desires a quick approximation of how much the sample deviates from a normal distribution, using the sample mean and standard deviation to construct a normal curve. However, the reader is reminded that the distribution of a single sample is not expected to be normal, nor is it required to be normal under sampling theory. It is the sampling distribution with which the sample is associated which approximates a normal curve if the sample is sufficiently large. On the other hand, the auditor may desire a clue as to how effective his audit stratification has been or as an indication of a need for further stratification. Fitting a normal curve to the data may assist in making a decision.

In any case, for those desiring to fit an approximately normal curve to a set of data, the values in the Y columns in Figure 22 indicate the relative height of the curve ordinate, compared to the width of the base at the specified Z-value distance from the mean.

There are a number of statistical problems which are solved with the aid of the standard normal distribution. The Z tables of interest in sampling are those constructed in terms of the area or probability between the mean and a specified Z-value. Since a normal distribution is completely defined by its mean and standard deviation, fixed proportions of a normal universe will fall within various ranges. The more important of these ranges are summarized in Table 14.1, while Figure 22 shows an extensive range of Z-values from zero to 5.0, and the corresponding probabilities represented by the values in the A columns.

The Z-values shown in Figure 22 are related to a theoretical normal distribution. These relationships also closely approximate those for actual distributions which do not deviate significantly from normality. Tchebycheff's and related theorems and the Central Limit

ORDINATES AND AREAS OF THE NORMAL CURVE

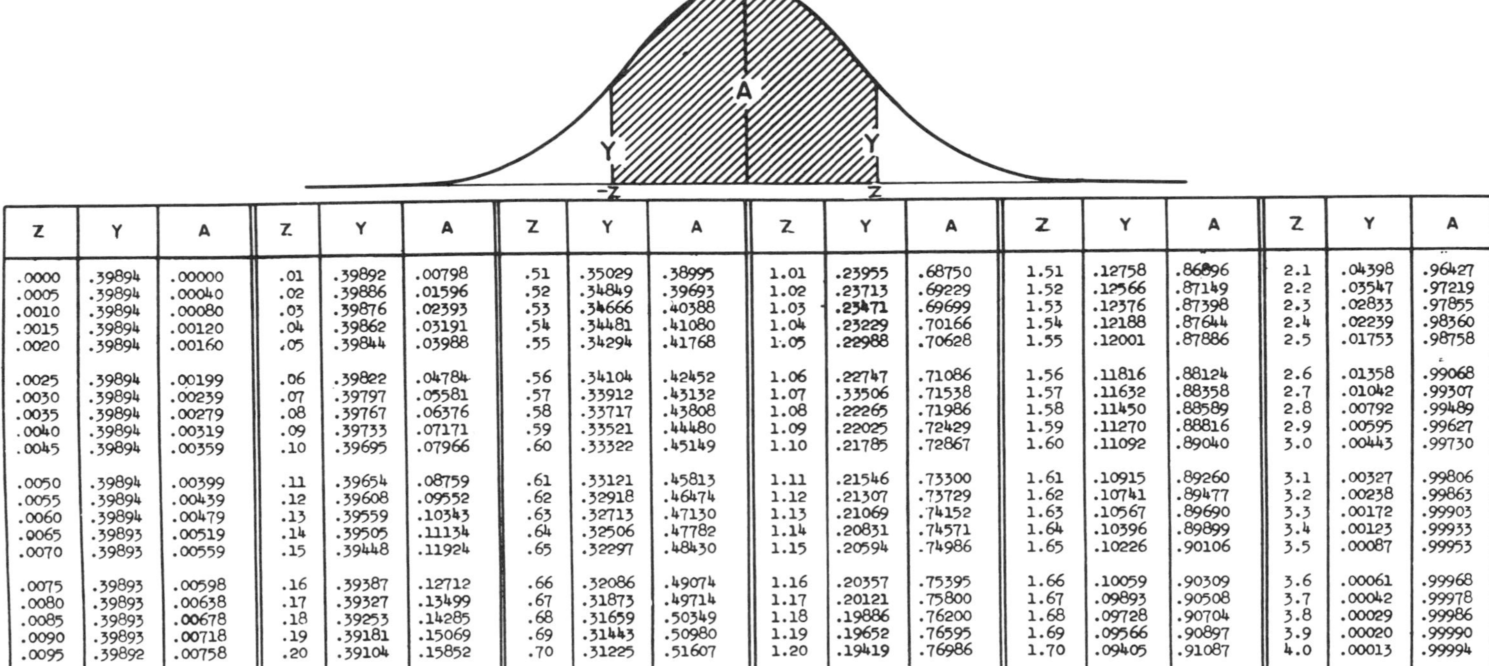

Z	Y	A	Z	Y	A	Z	Y	A	Z	Y	A	Z	Y	A	Z	Y	A
.0000	.39894	.00000	.01	.39892	.00798	.51	.35029	.38995	1.01	.23955	.68750	1.51	.12758	.86896	2.1	.04398	.96427
.0005	.39894	.00040	.02	.39886	.01596	.52	.34849	.39693	1.02	.23713	.69229	1.52	.12566	.87149	2.2	.03547	.97219
.0010	.39894	.00080	.03	.39876	.02393	.53	.34666	.40388	1.03	.23471	.69699	1.53	.12376	.87398	2.3	.02833	.97855
.0015	.39894	.00120	.04	.39862	.03191	.54	.34481	.41080	1.04	.23229	.70166	1.54	.12188	.87644	2.4	.02239	.98360
.0020	.39894	.00160	.05	.39844	.03988	.55	.34294	.41768	1.05	.22988	.70628	1.55	.12001	.87886	2.5	.01753	.98758
.0025	.39894	.00199	.06	.39822	.04784	.56	.34104	.42452	1.06	.22747	.71086	1.56	.11816	.88124	2.6	.01358	.99068
.0030	.39894	.00239	.07	.39797	.05581	.57	.33912	.43132	1.07	.33506	.71538	1.57	.11632	.88358	2.7	.01042	.99307
.0035	.39894	.00279	.08	.39767	.06376	.58	.33717	.43808	1.08	.22265	.71986	1.58	.11450	.88589	2.8	.00792	.99489
.0040	.39894	.00319	.09	.39733	.07171	.59	.33521	.44480	1.09	.22025	.72429	1.59	.11270	.88816	2.9	.00595	.99627
.0045	.39894	.00359	.10	.39695	.07966	.60	.33322	.45149	1.10	.21785	.72867	1.60	.11092	.89040	3.0	.00443	.99730
.0050	.39894	.00399	.11	.39654	.08759	.61	.33121	.45813	1.11	.21546	.73300	1.61	.10915	.89260	3.1	.00327	.99806
.0055	.39894	.00439	.12	.39608	.09552	.62	.32918	.46474	1.12	.21307	.73729	1.62	.10741	.89477	3.2	.00238	.99863
.0060	.39894	.00479	.13	.39559	.10343	.63	.32713	.47130	1.13	.21069	.74152	1.63	.10567	.89690	3.3	.00172	.99903
.0065	.39893	.00519	.14	.39505	.11134	.64	.32506	.47782	1.14	.20831	.74571	1.64	.10396	.89899	3.4	.00123	.99933
.0070	.39893	.00559	.15	.39448	.11924	.65	.32297	.48430	1.15	.20594	.74986	1.65	.10226	.90106	3.5	.00087	.99953
.0075	.39893	.00598	.16	.39387	.12712	.66	.32086	.49074	1.16	.20357	.75395	1.66	.10059	.90309	3.6	.00061	.99968
.0080	.39893	.00638	.17	.39327	.13499	.67	.31873	.49714	1.17	.20121	.75800	1.67	.09893	.90508	3.7	.00042	.99978
.0085	.39893	.00678	.18	.39253	.14285	.68	.31659	.50349	1.18	.19886	.76200	1.68	.09728	.90704	3.8	.00029	.99986
.0090	.39893	.00718	.19	.39181	.15069	.69	.31443	.50980	1.19	.19652	.76595	1.69	.09566	.90897	3.9	.00020	.99990
.0095	.39892	.00758	.20	.39104	.15852	.70	.31225	.51607	1.20	.19419	.76986	1.70	.09405	.91087	4.0	.00013	.99994

.0100	.39892	.00798	.21	.39024	.16633	.71	.31006	.52230	1.21	.19186	.77372	1.71	.09246	.91273	4.1	.00009	.99996
			.22	.38940	.17413	.72	.30785	.57848	1.22	.18954	.77754	1.72	.09089	.91457	4.2	.00006	.99997
			.23	.38853	.18191	.73	.30563	.53461	1.23	.18724	.78130	1.73	.08933	.91637	4.3	.00004	.99998
			.24	.38762	.18967	.74	.30339	.54070	1.24	.18494	.78502	1.74	.08780	.91814	4.4	.00002	.99999
			.25	.38667	.19741	.75	.30114	.54675	1.25	.18265	.78870	1.75	.08628	.91988	4.5	.00001	.99999
			.26	.38568	.20513	.76	.29887	.55275	1.26	.18037	.79233	1.76	.08478	.92159	5.0	.000001	.99999
			.27	.38466	.21284	.77	.29659	.55870	1.27	.17810	.79592	1.77	.08329	.92327			
			.28	.38361	.22052	.78	.29431	.56461	1.28	.17585	.79945	1.78	.08183	.92492			
			.29	.38251	.22818	.79	.29200	.57047	1.29	.17360	.80295	1.79	.08038	.92655			
			.30	.38139	.23582	.80	.28969	.57629	1.30	.17137	.80640	1.80	.07895	.92814			
			.31	.38023	.24344	.81	.28737	.58206	1.31	.16915	.80980	1.81	.07754	.92970			
			.32	.37903	.25103	.82	.28504	.58778	1.32	.16694	.81316	1.82	.07614	.93124			
			.33	.37780	.25860	.83	.28269	.59346	1.33	.16474	.81648	1.83	.07477	.93275			
			.34	.31654	.26614	.84	.28034	.59909	1.34	.16256	.81975	1.84	.07341	.93423			
			.35	.37524	.27366	.85	.27798	.60467	1.35	.16038	.82298	1.85	.07206	.93569			
			.36	.37391	.28115	.86	.27562	.61021	1.36	.15822	.82617	1.86	.07074	.93711			
			.37	.37255	.28862	.87	.27324	.61570	1.37	.15608	.82931	1.87	.06943	.93852			
			.38	.37115	.29605	.88	.27086	.62114	1.38	.15395	.83241	1.88	.06814	.93989			
			.39	.36973	.30346	.89	.26848	.62653	1.39	.15183	.83549	1.89	.06687	.94124			
			.40	.36827	.31084	.90	.26609	.63188	1.40	.14973	.83849	1.90	.06562	.94257			
			.41	.36678	.31819	.91	.26369	.63718	1.41	.14764	.84146	1.91	.06438	.94387			
			.42	.36526	.32551	.92	.26129	.64243	1.42	.14556	.84439	1.92	.06316	.94514			
			.43	.36371	.33280	.93	.25888	.64763	1.43	.14350	.84728	1.93	.06195	.94639			
			.44	.36213	.34006	.94	.25647	.65278	1.44	.14146	.85013	1.94	.06076	.94762			
			.45	.36052	.34729	.95	.25406	.65789	1.45	.13943	.85294	1.95	.05959	.94882			
			.46	.35889	.35448	.96	.25164	.66294	1.46	.13741	.85571	1.96	.05844	.9500			
			.47	.35723	.36164	.97	.24923	.66795	1.47	.13542	.85844	1.97	.05730	.95116			
			.48	.35553	.36877	.98	.24681	.67291	1.48	.13344	.86113	1.98	.05618	.95230			
			.49	.35381	.37587	.99	.24439	.67782	1.49	.13147	.86378	1.99	.05508	.95341			
			.50	.35207	.38292	1.00	.24197	.68269	1.50	.12952	.86639	2.00	.05399	.95450			

Figure 22. Probability and standard deviation (Z) factors.

Tolerance limits around the mean	If the distribution is approximately normal, the following percentages of units or area will be within the tolerance limits	If distribution has one mode, and mode equals mean, and frequencies decline continuously on both sides of mode, the following percentages will be within tolerance limits	Under any circumstances at least the following percentages will be within the tolerance limits
Mean $\pm$ 1 standard deviation	68.27 %	55%	—
Mean $\pm$ 2 standard deviations	95.45 %	89%	75%
Mean $\pm$ 3 standard deviations	99.73 %	95%	89%
Mean $\pm$ 4 standard deviations	99.994%	97%	94%

Figure 23. Probabilities associated with normal and skewed distributions.

Table 14.1. Percentages of Normal Curve Area between Selected Z-Values

Range of Z-value	Percent of area or units within range
±1.0000	68.27
±1.6449	90.00
±1.9600	95.00
±2.0000	95.45
±2.5759	99.00
±3.0000	99.73

Theorem, discussed in a subsequent section, offer solutions when the shape of the universe distribution is unknown or highly skewed.

The table in Figure 23 shows the areas or probabilities associated with selected multiples of the standard deviation for both normal and skewed distributions. These percentages reflect the reduced probabilities associated with slightly skewed and badly distorted distributions.

It is observed from Figure 23 that the auditor has at least a 75 percent probability associated with the mean plus and minus two standard deviations for the most distorted distribution or universe.

C. The t–Distribution

The t-distribution, like the normal distribution, is a continuous, symmetric, asymptotic distribution. However, unlike the normal distribution, it is completely specified by a single parameter, the degrees of freedom, which reflect the number of values in a distribution which are free to vary after certain constraints have been imposed. In other words, the degrees of freedom are the number of independent values in the distribution, while the constraints are the number of relations specified among the values. For example, one might impose the constraint that the sum of a set of values is fixed. In this case, for a sample of n values, only the first n-1 of the values are independent, since the last value must be the difference between the fixed sum and the sum of the first n-1 values. In general, the rationale for degrees of freedom will vary among statistical techniques; however, in simple random sampling involving a single statistic, the number of degrees of freedom is the sample size minus one, that is, n-1.

	Confidence Level (Probability) for One-sided Confidence Interval				
	.55	.75	.90	.975	.995
	Confidence Level (Probability) for Two-sided Confidence Interval				
DF*	.10	.50	.80	.95	.99
1	0.158	1.000	3.078	12.706	63.657
2	0.142	0.816	1.886	4.303	9.925
3	0.137	0.765	1.638	3.182	5.841
4	0.134	0.741	1.533	2.776	4.604
5	0.132	0.727	1.476	2.571	4.032
6	0.131	0.718	1.440	2.447	3.707
7	0.130	0.711	1.415	2.365	3.499
8	0.130	0.706	1.397	2.306	3.355
9	0.129	0.703	1.383	2.262	3.250
10	0.129	0.700	1.372	2.228	3.169
11	0.129	0.697	1.363	2.201	3.106
12	0.128	0.695	1.356	2.179	3.055
13	0.128	0.694	1.350	2.160	3.012
14	0.128	0.692	1.345	2.145	2.977
15	0.128	0.691	1.341	2.131	2.947
16	0.128	0.690	1.337	2.120	2.921
17	0.128	0.689	1.333	2.110	2.898
18	0.127	0.688	1.330	2.101	2.878
19	0.127	0.688	1.328	2.093	2.861
20	0.127	0.687	1.325	2.086	2.845
21	0.127	0.686	1.323	2.080	2.831
22	0.127	0.686	1.321	2.074	2.819
23	0.127	0.685	1.319	2.069	2.807
24	0.127	0.685	1.318	2.064	2.797
25	0.127	0.684	1.316	2.060	2.787
26	0.127	0.684	1.315	2.056	2.779
27	0.127	0.684	1.314	2.052	2.771
28	0.127	0.683	1.313	2.048	2.763
29	0.127	0.683	1.311	2.045	2.756
30	0.127	0.683	1.310	2.042	2.750
35	0.127	0.682	1.306	2.030	2.724
40	0.126	0.681	1.303	2.021	2.704
45	0.126	0.680	1.301	2.014	2.690
50	0.126	0.679	1.299	2.009	2.678
60	0.126	0.679	1.296	2.000	2.660
70	0.126	0.678	1.294	1.994	2.648
80	0.126	0.678	1.292	1.990	2.639
90	0.126	0.677	1.291	1.987	2.632
100	0.126	0.677	1.290	1.984	2.626
120	0.126	0.677	1.289	1.980	2.617

Figure 24. t-Factors for specified probabilities and various degrees of freedom.

*DF = degrees of freedom. For the mean estimate, the number of DF is one less than the sample size; for the difference and ratio estimates, DF may be set as one less than the number of differences in the sample; for the regression estimate, DF may be set as two less than the number of differences in the sample; and for stratified samples, the number of DF is reduced from the corresponding unstratified sample by one less than the number of strata. When the number of DF is greater than 120, use Z factors shown in Figure 22.

Compared with the shape of the normal distribution, the t-distribution, for degrees of freedom less than about 120, contains more area in the tails of the distribution. For more than 120 degrees of freedom, the normal distribution closely approximates the t-distribution and may be used in its place. The t-value is the ratio of a statistic to its standard error.

The t-value differs from the Z-value only in that the denominator of the t-value is an estimate of the universe standard deviation while the denominator of the Z-value is the universe standard deviation. Tables for the t-distribution differ from those for the normal distribution since the t-distribution's shape changes for each degree of freedom specified. In the case of sampling, the shape changes for different sample sizes until the sample becomes large.

Since the t-distribution is dependent on degrees of freedom, there is a problem in constructing tables of probabilities, since each degree of freedom (or sample size) considered would require a separate table such as the one for the standard normal distribution of Zs. To cope with this problem, tables of the t-distribution are constructed so that only certain probability levels are presented. The values shown in the body of these tables are for specified values of t, and the corresponding probabilities. An abbreviated example of such a table is Figure 24. This figure shows t factors for selected one-sided, and corresponding two-sided, confidence intervals for various degrees of freedom (DF) to 120; above 120, use Z factors in Figure 22.

For example, if one wishes to find the t-value associated with a two-sided confidence level of 95 percent with 60 degrees of freedom (for mean estimate from sample of 61) he would go to the row in Figure 24 for 60 degrees of freedom, read across to the column headed .95, and then read the required t-value, which, in this case, is 2.000.

Where computer programs are available for evaluating sample results, these programs automatically determine the appropriate t-value for the specified confidence level, sample size and sample design. The Z and t distributions have been discussed so that the auditor will have a better appreciation and understanding of how these sampling distributions are involved in the statistical evaluation of samples. This insight should permit a better interpretation of the data provided from computer applications.

D. Binomial Universe or Distribution

A binomial universe or distribution is one in which events may be categorized as belonging to either of two mutually exclusive classes: right or wrong, eligible or ineligible, present or absent, and the like.

The probabilities of occurrences of events of these kinds in repeated trails, where the selection is with replacement, are given by the binomial distribution.

To develop the notion of the binomial distribution, consider a record file which contains correct vouchers, R, and incorrect vouchers, W. Let P be the proportion of correct vouchers and, $Q = 1 - P$, be the proportion of incorrect vouchers. The vouchers are thoroughly mixed, one voucher is selected from the file, its quality (correct or incorrect), is noted, and then it is replaced in the file. The vouchers are again thoroughly mixed, and a second voucher is selected, its quality noted, and it is then replaced in the file. This process is repeated until a total of n successive selections with replacement has been obtained.

The objective is to determine the probability that a specified number, X, of incorrect vouchers will be observed in the n successive selections with replacement. Fortunately, tables are available which provide this information. Also, in actual practice seldom does an auditor sample with replacement, which reflects a constant probability for each item selected. However, where the universe is very large and the sample is relatively small, the results of sampling without replacement approximate sampling with replacement. Otherwise—if the universe is small and the selection is without replacement—a different statistical universe or distribution is required. It is known as the hypergeometric probability distribution.

E. The Hypergeometric Universe or Distribution

Suppose an auditor intends to conduct an eligibility study of participants in a program, and experience indicates that the error rate (proportion of ineligible participants in the program) is $P = \frac{1}{4}$. To simplify the computations, also assume that there are only 40 participants at a specified location, and once a participant is selected, he is not selected again, which is sampling without replacement. In this case, the outcome of each selection is not independent, and thus the binomial is not appropriate. If there are 40 participants and $P = \frac{1}{4}$, then there are expected to be 10 ineligible participants at this location. But once the record of the first participant is selected for audit, the value of P changes. For instance, if the first record were for an ineligible participant, then P equals 9/39 on the second selection. On the other hand, if the first record is for an eligible participant, then P equals 10/39 on the second selection. That is, the outcome of each selection changes the probability of the outcome of the next selection.

Fortunately, tables of hypergeometric probabilities are available thereby precluding the need for one to perform the mathematical computations. The detection sampling probability table, Appendix C, and the flexible sampling probability table, Appendix D, are examples of probabilities based on the hypergeometric, which reflects sampling without replacement.

For a detailed discussion of the hypergeometric distribution, see Cochran [8] and Deming [11].

F. The Poisson Universe or Distribution

The Poisson distribution, in addition to serving as a convenient way of approximating binomial probabilities, is a statistical universe model in its own right. The Poisson distribution is used to estimate probabilities for situations such as the number of defects in products. The primary concern in this book, however, is for its use in monetary-unit sampling applications which are described in Chapter 12.

If each of an occurrence of interest is called a "success," then in an audit sense, a success could be a finding. In the binomial distribution each occurrence can happen in one of two ways, as either a success or a failure. In the Poisson distribution, however, an occurrence can happen in only one way. In the use of Poisson probabilities an occurrence of interest is a success. Also, a failure can be defined as an occasion on which a success could have occurred, but did not.

The essential difference between the binomial and Poisson distributions is that in the binomial, successes occur in relation to some specific number of events; whereas, in the Poisson successes are considered to be continuous. In both distributions, successes are assumed to occur at random and to be independent of each other. However, the Poisson distribution indicates the probability of there being a specified number of successes in some segment of a continuum, such as a production-line operation. It is the appropriate universe model when one is dealing with the number of elements per unit of time or space and the average number of elements per unit is relatively small.

G. Sampling Distributions

A "parameter" is a numerical expression that summarizes the values of some characteristic for all items of the universe, such as the mean, proportion or total. Although such a value is usually unknown (which is the reason sampling is resorted to), it is a constant, independent of sampling.

The sample value, or statistic, is an estimate of the corresponding universe value, computed from the items in the sample, such as the sample mean. It is a random variable which depends on the sample design and on the particular combination of items which happen to be selected. Thus, the particular sample estimate is only one among many possible estimates which could have been obtained by the same sample design.

Statistics are subject to both sampling and nonsampling errors. Sampling errors arise because only part of the total universe is observed in the sample. Nonsampling errors are associated with imperfect observation and measurement procedures. Although both types of errors should be considered in designing the sample, most of the discussions in this book deal with the variability of sample statistics around the true universe, or with sampling errors.

A specific sample statistic, such as the sample mean, is of little practical value, per se; its value lies in what it may reveal about the universe mean. One is interested in knowing how much the sample mean may differ from the universe mean, which is a value of interest. However, the sample mean depends on which combination of items happened to be selected, since different samples of the same size and same design taken from the same universe could have different mean values. Actually the deviation of any particular sample mean from the universe mean is unknown. There is a statistical way of expressing sampling variability in terms of probability. The array of all possible values for a statistic, and the relative frequency, or probability, with which the values would occur, in the long run for a particular universe, specified sample design and given sample size, is known as a sampling distribution.

One of the basic concepts in statistical inference is the notion of a sampling distribution. It combines the concepts of a random variable and a probability distribution. To illustrate, assume that a sample design (selection technique and estimation procedure) has been specified, that it has been applied to a particular universe, selecting sample after sample of the same size; and that the mean of each sample has been calculated, and plotted as the distribution of the relative frequencies. As the number of plotted means increases, the shape of the distribution approaches the normal distribution. A sampling distribution is a theoretical probability distribution which represents all possible values for a statistic, each with its probability of occurrence.

Under the assumptions of an appropriate statistical, or probability, sampling design and a specified universe, statistics calculated from individual samples may be regarded as random variables. An appropriate sampling procedure allows one to treat sample statistics as random variables. It is the manner in which the chance factors operate in the sampling process that permits one to derive a sampling distribution.

An important characteristic of a sampling distribution is its relation to the universe. A sampling distribution represents all possible values of a statistic, such as a mean, which can occur on the basis of assumptions about the universe. Such a distribution gives the relative frequency, or probability, with which various values of a statistic will occur in the long run for a specified sample design.

Although sampling distributions are obtained through mathematical derivations, based on assumptions about the manner in which chance factors operate, an intuitive appreciation of the information contained in a sampling distribution can be obtained by considering a simple experiment. For instance, one could roll a pair of dice a large number of times, record the outcome of each roll, and tabulate or graph the results. Since there are 11 different possible values, see Figure 30, each sample taken would have to include a fairly large number of rolls. Each sample would produce a graph of a frequency distribution. However, each graph would be somewhat different, since the short run effects of chance factors will only approximate the long run outcome. But if one were to repeat the experiment by rolling the dice a very large number of times, the resulting relative frequency distribution would approximate a normal distribution. Thus, a sampling distribution represents the relative frequency with which the various possible values of a statistic will occur in the long run.

H. Central Limit Theorem

The Central Limit Theorem, one of the most significant statistical theorems, states that regardless of the shape of the distribution of the original universe, a sampling distribution of means will approach a normal distribution as the sample size becomes large, which is at least 100 in most monetary audit situations. Even though an individual random sample selected from a positively skewed universe tends to be positively skewed too, especially if extreme and unusual items have not been separated or stratified, a sampling distribution of means tends to form a normal distribution when samples are not too small.

The usefulness of the Central Limit Theorem lies in the intuitive assurance it provides that various data sets will behave in a similar manner. The implication is that the approximation to the normal distribution improves continuously as the sample size increases. However, for highly skewed data sets the improvement may be slow. For this reason, the auditor should investigate each occurrence individually, searching for ways to reduce the skewness through additional stratification.

Another fundamental theorem indicates that the mean of a sampling distribution of means will always be the same as the mean of the original universe. Also, the standard deviation of a sampling distribution (called the standard error) is equal to the standard deviation of the universe divided by the square root of the sample size if the sample size is less than 5 percent of the universe or the sampling is with replacement. Sampling with replacement is not advocated in auditing because the use of a sample item more than once for the same purpose is difficult to explain and rationalize to the auditee. If the sample is greater than 5 percent of the universe, the finite correction factor improves the estimate of sample precision. Ignoring the finite correction factor under these circumstances produces a more conservative, larger estimate of the confidence interval. The finite correction factor and its use is discussed in the chapter on statistical evaluation of sample results.

I. Illustrations of Sampling Distributions

The fact that the mean of a sampling distribution is equal to the mean of the universe can be best illustrated by a simple example.

Assume that a universe consists of the following five numbers:

2, 3, 6, 8, 11.

The mean of this universe, a value one would endeavor to estimate from a sample, is exactly 6. No matter what selection procedure one uses, only 10 samples of two items each may be formed from this universe of five numbers if selected numbers are not replaced. These 10 samples and their mean values are listed below:

Sample number	1	2	3	4	5	6	7	8	9	10
Sample values	2	2	2	2	3	3	3	6	6	8
	3	6	8	11	6	8	11	8	11	11
Sample mean	2.5	4	5	6.5	4.5	5.5	7	7	8.5	9.5

The numbers on the line immediately above constitute the sampling distribution, without replacement, of sample means for samples of two items. The mean of this sampling distribution is 60/10 = 6, which is exactly the same value as the universe mean. Also, the same value would have resulted if sampling with replacement had been employed.

The same example may be extended to illustrate the relation between the standard deviation of the sampling distribution (the standard error) and the universe standard deviation.

The standard deviation is computed as follows:

	Item	Item minus mean	Item minus mean squared
	2	2 - 6 = -4	$(-4)^2 = 16$
	3	3 - 6 = -3	$(-3)^2 = 9$
	6	6 - 6 = 0	$(0)^2 = 0$
	8	8 - 6 = 2	$(2)^2 = 4$
	11	11 - 6 = 5	$(5)^2 = 25$
Total	30	0	54
Mean	30/5 = 6		54/5 = 10.8

The mean of the last column, 10.8, is the variance of the universe of 2, 3, 6, 8 and 11; $\sqrt{10.8} = 3.28634$, is the standard deviation of the universe.

In a similar manner, the standard error is obtained as follows:

	(X)*	$X - \mu$	$(x - \mu)^2$
	2.5	2.5 - 6 = -3.5	12.25
	4.0	4.0 - 6 = -2.0	4.00
	5.0	5.0 - 6 = -1.0	1.00
	6.5	6.5 - 6 = 0.5	0.25
	4.5	4.5 - 6 = -1.5	2.25
	5.5	5.5 - 6 = -0.5	0.25
	7.0	7.0 - 6 = 1.0	1.00
	7.0	7.0 - 6 = 1.0	1.00
	8.5	8.5 - 6 = 2.5	6.25
	9.5	9.5 - 6 = 3.5	12.25
Total	60.0	0	40.50
Mean	60/10 = 6		40.5/10 = 4.5

*These are all possible sample means for samples of two items.

The mean of the last column, 4.05, is the variance of the sampling distribution. The square root of 4.05 = 2.01246 is the standard deviation (the standard error) of the sampling distribution of means.

Now to illustrate that the standard error is equal to the product of the universe standard deviation divided by the square root of the sample size and the finite correction factor (since the sample is more than 5 percent of the universe size). A formula for the finite correction factor is:

$$\sqrt{\frac{\text{universe size} - \text{sample size}}{\text{universe size} - \text{one}}}$$

$$\text{Standard error} = \frac{\text{universe standard deviation} \times \text{finite correction factor}}{\sqrt{\text{sample size}}}$$

Substituting values from the above illustration gives:

$$2.01246 = \frac{3.28638}{\sqrt{2}} \sqrt{\frac{5 - 2}{5 - 1}}$$

Squaring both sides of the above equation, and completing the computation produces:

$$4.05 = \frac{10.8}{2} \left|\frac{5 - 2}{5 - 1}\right|$$

$$4.05 = 5.4 \times 0.75$$

$$4.05 = 4.05.$$

This relation between the standard error (sampling distribution standard deviation) and the universe standard deviation is the basis for sample evaluations which are discussed in the chapter on statistical evaluation of sample results.

As mentioned, mathematical theory has demonstrated that the distribution of the means of all possible samples of the same size is (or approximates) a normal distribution, except for very small samples and samples from extremely irregular universes. The close similarity between the distribution of all possible sample means and the normal curve can best be illustrated by a simple example.

Assume a universe of 12 accounts receivable as shown in Figure 25. Also suppose that the mean value of the 12 accounts receivable is estimated by selecting samples of 7 different accounts. The total combination of 12 things taken 7 at a time is 792; therefore, there

$40,000	$ 26,000	$36,000	$18,000
72,000	126,000	54,000	96,000
44,000	62,000	30,000	38,000

Figure 25. Universe of 12 accounts receivable.

would be 792 possible samples. Each of these samples would have a mean which could be considered an estimate of the mean of the 12 accounts receivable. These sample means would range from $33,143 to $70,571 with an overall mean of $53,500. Using a $4,000 class interval, the frequency distribution table in Figure 26 would result.

Figure 27 shows a graph of a frequency polygon plotted from the frequency distribution table in Figure 26, together with a fitted normal curve. As can be observed, the distribution of 792 sample means does not deviate much from a normal curve despite the small size of the samples. Sampling theory indicates that sampling distributions

Mean Class Boundaries	Frequency (No. of Samples)
$32,000 - $35,999	6
36,000 - 39,999	27
40,000 - 43,999	61
44,000 - 47,999	98
48,000 - 51,999	136
52,000 - 55,999	150
56,000 - 59,000	130
60,000 - 63,999	108
64,000 - 67,999	62
68,000 - 71,999	14
Total	792

Figure 26. Frequency distribution table of 792 sample means.

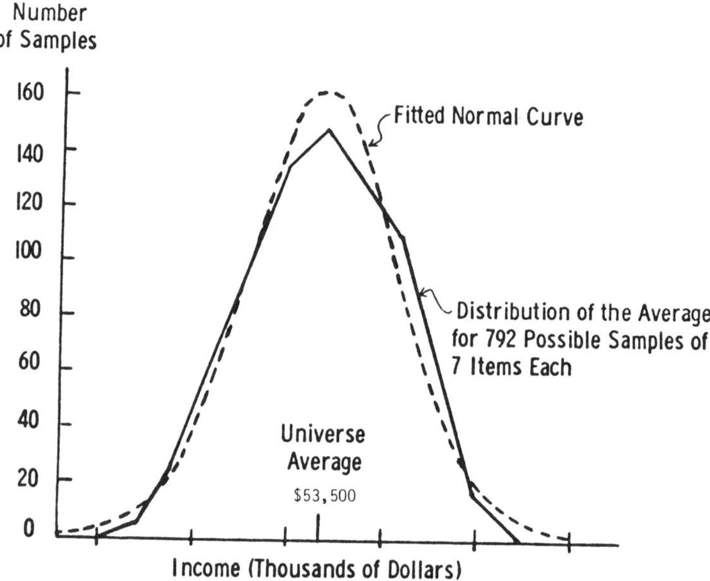

Figure 27. Frequency polygon and normal curve fitted to 792 sample means. Figure 27 adapted from AT&T booklet, "What is Scientific Sampling."

for sample sizes over 30 tend to approximate a normal distribution. However, for many auditing situations where the underlying distribution is extremely skewed and the occurrence rate is low, a minimum sample of 100 is suggested, including the items audited 100 percent.

A significant feature of statistical, or probability, sampling is that one can estimate from sample statistics a distribution which closely approximates the distribution of the results that would be obtained from all possible samples of the same size from the same universe. Also, the properties of the underlying distribution can be used to compute the precision of the sample results.

The significance of the fact that a sampling distribution approaches a normal distribution as the sample size increases is that the characteristics of a normal curve apply. For instance, if one were to consider all possible samples of 100 invoices from a universe of invoices, about 68 percent of the samples would have mean values between the sampling distribution mean and plus and minus one standard error. For example, if the sampling distribution mean were $50.00 and the standard error were $7.50, then about 68 percent of the sample means would be expected to fall between $42.50 and $57.50 ($50 ± $7.50). Thus, the probability is about 68 percent (68 times out of 100) that a random sample of 100 invoices in this instance would not differ from the true universe mean by more than \$7.50. Also, about 95 percent of the sample means would be expected to be between the mean of the sampling distribution and plus and minus two standard errors or between \$35.00 and \$65.00 (\$50 ± 2 × \$7.50), and almost all (99.7 percent) of the sample means would be expected to be between the mean of the sampling distribution and plus and minus three standard errors or between \$27.50 and \$72.50 (\$50 ± 3 × \$7.50).

J. The Three Distributions

In sampling, there is concern for three different distributions, their characteristics and relations. The three distributions are (i) the universe distribution, (ii) a particular sample distribution, and (iii) the sampling distribution of the statistic of interest, as shown in Figure 28. As indicated in the figure, both the sample mean and the sample standard deviation are computed from actual sample data. The sample mean, in turn, is used as an estimate of both the universe mean and the sampling distribution mean. The standard error, which is the standard deviation of the sampling distribution, is estimated from the sample standard deviation.

SAMPLE MEAN	MEAN OF SAMPLING DISTRIBUTION	UNIVERSE MEAN
Computed from sample data	Estimated from sample mean	Estimated from sample mean
SAMPLE STANDARD DEVIATION	STANDARD ERROR	UNIVERSE STANDARD DEVIATION
Computed from sample data	Estimated by dividing sample standard deviation by square root of sample size	Estimated from sample standard deviation

Figure 28. The relation among the mean and standard deviation of a sample, sampling distribution and universe.

IV. INFERENTIAL STATISTICS AND PROBABILITY

Statistical sampling is more precisely described as probability sampling since it is based on probability theory. Therefore, some basic idea of probability is essential to understanding the statistical inference aspects of sampling. Also, most data which are used in the formulation of opinions and decisions reflect some degree of uncertainty. When appropriate sampling procedures are used, probability theory provides a mathematical basis for expressing the sampling risk.

Statistical inference is a process by which one makes generalizations about universe parameters based on sample statistics. A parameter is a value characterizing a universe, and a sample statistic is a value for an equivalent characteristic of a sample. Sample statistics are usually considered as estimates of the corresponding universe characteristics or parameters.

A. Meaning of Probability

Since statistical inference is based on probability theory, one should have at least a general idea of what is meant by probability. There are several definitions of probability, however, in this book probability is related to the notion of relative frequency. This concept defines the probability of a particular outcome as the limit approached as the number of trials increases indefinitely.

Probability is usually expressed as a proportion, a fraction, or a percentage. The numerator of the ratio represents the number of ways a particular event may occur and the denominator is the total number of possible occurrences.

Since probability is a proportion, it cannot exceed 100 percent or unity. Also, probabilities can never be negative. Thus, probabilities must be in the range of zero to one. See the probability scale depicted in Figure 29. If the probability of a particular outcome is established on the basis of observations, it is called an empirical probability. If the probability of a particular outcome is established on the basis of assumptions and rules for combining probabilities, it is called a mathematical or theoretical probability. Empirical probabilities are usually tentative and subject to revision based upon additional observations or subjective probability. This concept of probability does not require the notion of repeated trials and the stabilizing of relative frequencies. Subjective probability is the total acceptance of the personal assessment by an individual or group without any way to "prove" whether it is right or wrong. However, there are procedures called Bayes' rule or Bayes' theorem which combine subjective probabilities with current, objective information. In this book the formulation of most statistical concepts will be within the framework of the relative frequency theory

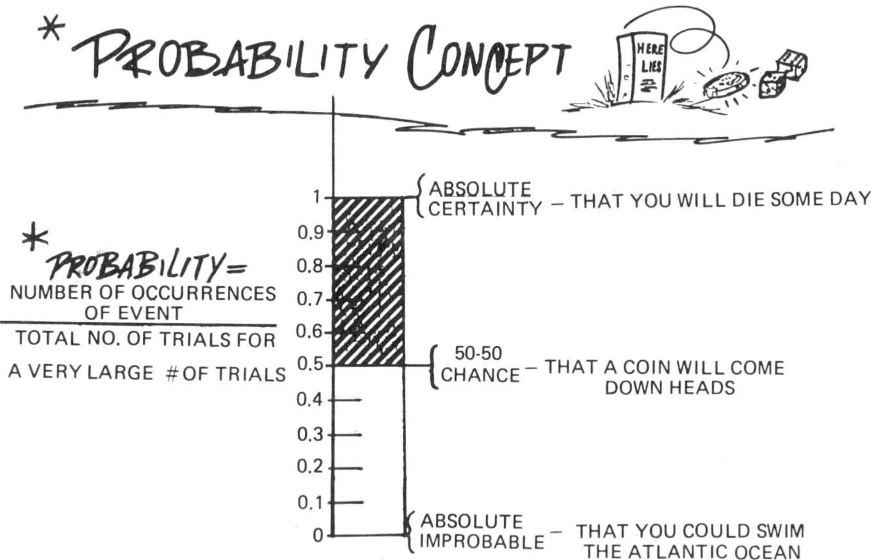

Figure 29. The probability scale.

of probability. Those readers desiring more information on subjective probability should see Kyburg and Smokler [34]. Discussions and applications of Bayes theorem are covered by Schlaifer [61].

The probability of a fair coin coming up heads certainly cannot be based on a single toss of the coin. A statement about the frequency with which the event "heads" will occur is related to a large number of repeated tosses, which statistically is referred to as "in the long run." Obviously, any single toss of a coin could come up either head or tail. These two possible events are said to be mutually exclusive; that is, the occurrence of one precludes the occurrence of the other. Also, the events—heads and tails—are collectively exhaustive; that is, together they account for all possible outcomes of a coin.

As shown in Figure 29, probability is expressed as a proportion from zero to unity (one). Since it is a proportion, it cannot exceed unity. Also, negative probabilities are meaningless.

Probability measures a degree of regularity to be expected in events. For instance, it is the relative frequency to be expected in the appearance of "heads" in a large number of tosses of the coin, that is, in the long run. In other words, probability may be viewed as the limit approached as the number of trails, or tosses, increases indefinitely. Thus, a probability statement can be demonstrable only in terms of the relative frequency of an event, or occurrence, in the long run.

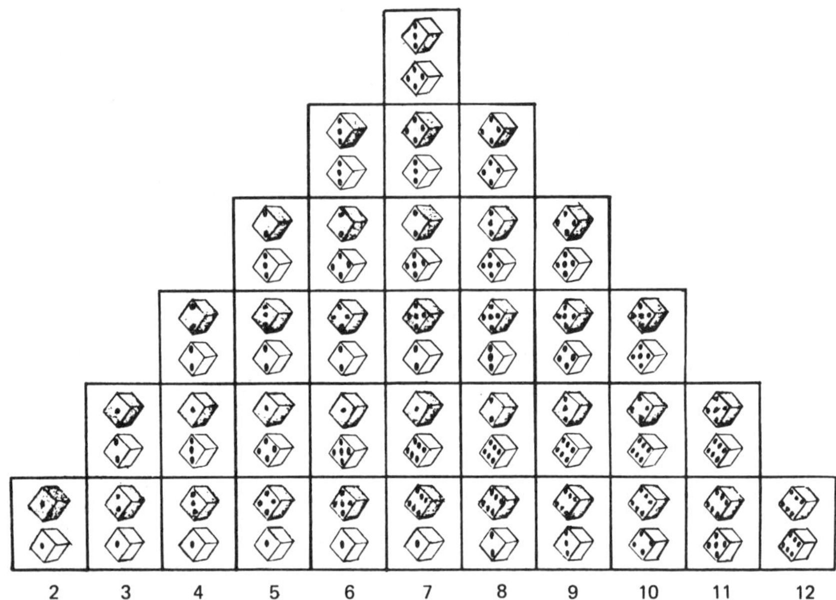

Figure 30. Graphic probability distribution of combinations of pair of dice.

Table 14.2. Probability Distribution for the Outcomes of a Single Roll of a Pair of Dice

Outcome (score)	Frequency	Probability
2	1	1/36 = 0.02778
3	2	2/36 = 0.05556
4	3	3/36 = 0.08333
5	4	4/36 = 0.11111
6	5	5/36 = 0.13889
7	6	6/36 = 0.16667
8	5	5/36 = 0.13889
9	4	4/36 = 0.11111
10	3	3/36 = 0.08333
11	2	2/36 = 0.05556
12	1	1/36 = 0.02778
77	36	36/36 = 1.00000

Expected value = mean = $\frac{77}{11}$ = 7

Inasmuch as the interpretation of probability in most statistical applications is the relative frequency notion, an important concept is a sample space, which is the set, or complete listing, of all possible outcomes of some event. This relates to the frequency interpretation of probability since the probability of any particular outcome is obtained by counting. Thus, the probability of a success can be obtained by computing the ratio of the number of successes to the total number of outcomes.

B. Probability Distribution

Certain probability concepts can be easily illustrated with a pair of dice. For example, Figure 30 shows all the possible 36 combinations or outcomes of a pair of dice, and Table 14.2 is the probability distribution of the outcome of a single roll of a pair of dice. The probability for each of the scores (outcomes), 2 through 12, is shown. It should be noted that the total of all these probabilities sum to unity or one. Also, the score 7 has the highest probability of occurrence (6/36 or 1/6) since it has the largest number of possible combinations. Likewise, the scores 2 and 12 have the lowest probability of occurrence, since there is only one combination of each.

To facilitate discussions, several concepts which are important in probability theory will be introduced.

A random variable is a variable for which one knows the probability distribution, and whose observed values are the outcome of chance. Although one cannot predict the value of the next observation of a random variable, one knows the relative frequency with which each possible value will occur in the long run.

The relative frequency distribution for the outcomes of an event is called a probability distribution as illustrated in Table 14.2 for a pair of dice. The significant characteristics of a probability distribution are that (i) it represents all the theoretically possible values of the random variable; (ii) it indicates the probability of each of the theoretically possible values occurring on any trial; and (iii) the sum of all probabilities in the distribution is unity or one.

An expected value is the mean of the random variable. Thus, the expected value for the probability distribution of a pair of dice, as shown in Table 14.2 is 7. That is, in a long series of rolls of a pair of dice, one would expect an average, or mean, score (outcome) of 7.

C. Principles or Rules of Probability

Now for a brief discussion, with illustrations, of several commonly used rules for combining probabilities. These rules permit the calculation of unknown probabilities from other known or more easiliy computed probabilities.

1. The Addition Rule

The addition rule for mutually exclusive events states that if there are several mutually exclusive events, that is, only one of them can occur at a time, the probability of occurrence of an event is the sum of the individual probabilities. To illustrate, suppose one desires to compute the probability of rolling a 7 or an 11 on one toss of a pair of dice. Referring to Figure 30, it is noted that there are 36 equally likely combinations of a pair of dice. Since there are 6 of the 36 combinations which result in a 7, the probability of rolling a 7 is 6/36 or 1/6. Also, the probability of rolling an 11 is 2/36 or 1/18. Applying the addition rule results in the following:

$$\begin{aligned}\text{Probability}(7 \textit{ and } 11) &= \text{probability } (7) + \text{probability } (11)\\ &= 1/6 + 1/18 = 3/18 + 1/18 = 4/18\\ &= 2/9.\end{aligned}$$

Thus, the probability of rolling *either* a 7 *or* an 11 on *one toss* is 2/9.

2. The Multiplication Rule

The multiplication rule states that the probability of the occurrence of several independent outcomes—those outcomes which in no way affect or influence each other—either together or in succession is the product of the probabilities of the individual events. To illustrate, one may wish to compute the probability of rolling a 7 followed by an 11 on two rolls of a pair of dice. Remembering that the probability of rolling a 7 is 1/16 and the probability of rolling an 11 is 1/18, and applying the multiplication rule, results in the following:

$$\begin{aligned}\text{Probability } (7 \textit{ and } 11) &= \text{probability } (7) \times \text{probability } (11)\\ &= 1/6 \times 1/18 = 3/18 \times 1/18 = 3/324\\ &= 1/108.\end{aligned}$$

Thus, the probability of rolling a 7 followed by an 11 on two rolls of a pair of dice is 1/108.

3. The General Addition Rule

When the probabilities of events are not mutually exclusive, the general rule for addition applies. This rule adjusts the sum to compensate for double counting. To illustrate, consider computing the probability of rolling a 7 on the first roll <u>or</u> an 11 on the second roll. Obviously, the multiplication rule does not apply since it is not a joint occurrence. Also, the addition rule for mutually exclusive events does not apply since the events are independent but not mutually exclusive. The general rule for addition, resulting from combining the two above rules, states that if there are several events, the probability of occurrence of an event is the sum of the individual probabilities minus the product of the individual probabilities. If the events are mutually exclusive, then the product of the probabilities is zero, resulting in the initial addition rule. The general addition rule may be illustrated as follows for rolling a 7 on the first roll or an 11 on the second roll:

$$\begin{aligned}\text{Probability (7 or 11)} &= \text{probability (7)} + \text{probability (11) minus} \\ &\quad \text{probability (7)} \times \text{probability (11)} \\ &= 1/6 + 1/18 - (1/6 \times 1/18) \\ &= 2/9 + 1/108 = 24/108 - 1/108 \\ &= 23/108.\end{aligned}$$

Thus, the probability of rolling a 7 on the *first roll or* an *11* on the *second roll* is 23/108.

4. The Conditional Rule

When the probabilities of events are not independent, the conditional rule applies. A deck of regular playing cards provides a means to illustrate conditional probabilities. For example, before any card is exposed, each denomination has a 4/52 probability of selection. Thus, if A is the event that the first card dealt is an ace, then the probability (A) is 4/52. However, if the first card is an ace, then the probability of the second card being an ace is 3/51, since the first card was not replaced, and there are 3 aces in the 51 remaining cards. If B is the event that the second card is an ace after the first card has been exposed, then the probability (B if A) is 3/51. This situation is referred to as the conditional probability of event B assuming that A occurs. Where A and B are not independent events, the rule for conditional probabilities may be illustrated with the drawing of aces described above as follows:

$$\begin{aligned}\text{Probability (A and B)} &= \text{probability (B if A)} \times \text{probability (A)} \\ &= 3/51 \times 4/52 = 12/2{,}652 \\ &= 1/221\end{aligned}$$

Thus, the probability of both the first card and the second card being aces is 1/221. It is interesting to note that the probability of neither card being an ace is:

$$\text{Probability (not A and not B)} = 47/52 \times 48/52 = \frac{2{,}256}{2{,}652} = \frac{188}{221}$$

This brief discussion of probability concept is to provide some background on the basic ideas of probability for understanding the conceptual framework of statistical, or probability, sampling. Those readers desiring more rigorous and detailed discussions should refer to Feller [15], Hogg and Craig [22], Huntsberger [23] or Mosteller, Rourke and Thomas [45].

V. SUMMARY

The combination of frequencies for all observed values of a variable is a frequency distribution. The frequency distribution may be either in absolute or relative terms (percentages).

Two common ways of presenting a frequency distribution are the frequency histogram and polygon. In a histogram the frequency is in the form of a bar whose height corresponds to the frequency of a particular value. Connecting the midpoints of the top of the bars with lines form a polygon.

Indices of central tendency seek to describe the clustering of data. A number of such indices are available; however, the three most popular indices of central tendency are the arithmetic mean, the median and the mode. An index of central tendency provides no information about the dispersion.

The range is the simplest index of dispersion, since it is merely the difference between the highest and the lowest values of a data set.

The average deviation is the mean of the absolute deviation of each value in a set from either the mean or median of the set.

The most frequently used index of dispersion is the standard deviation because of its usefulness in inferential statistics. It is the square root of the mean of the squared deviations of individual values from their mean.

The coefficient of variation (the ratio of the standard deviation to the mean) is used for comparative purposes; it indicates relative dispersion.

Kurtosis refers to the peakness or flatness of a curve relative to the normal curve. Whereas, skewness refers to the lack of symmetry of a curve or distribution.

Graphs are informative, but they can be misleading, especially if the base line is not shown. Interpret graphs critically and carefully.

Conceptually, one can construct a distribution for the totality of all possible values of a specified universe. Such conceptual distributions are referred to as theoretical or statistical universes.

A binomial universe is one in which the units may be classified as belonging to either of two groups: right or wrong, defective or non-defective, and the like. The two classifications are generalized as "success" or "failure." The probabilities associated with various possible outcomes of a sample from such a universe are indicated by the terms of the binomial distribution.

However, as the sample size increases the calculation of binomial probabilities becomes more complicated because of the increased number of terms to be calculated and summed. The use of the table of areas for the standard normal distribution provides a simple method for obtaining approximations to the binomial probabilities. The normal distribution, therefore, is a good approximation for the binomial as long as the product of the sample size and success rate is equal to or greater than 5, regardless of the individual values of the sample size and success (occurrence) rate.

The hypergeometric is the appropriate model when samples are selected without replacement from a finite universe consisting of two kinds of items, successes and failures. However, when the universe size is large in comparison with the sample size, the binomial and hypergeometric values are nearly the same. Appendices C and D are based on the hypergeometric distribution.

If the proportion of the occurrence of an event in a universe is very small and the sample is not sufficiently large, the normal distribution does not provide a satisfactory approximation of the discrete binomial probabilities. Under these circumstances the Poisson distribution provides a more accurate approximation of the binomial.

Note that most distributions of accounting data would be expected to be positive skewed, that is, with a tail towards the higher values. It is not necessary, however, for the universe distribution to be normally distributed for probability sampling procedures to apply according to the Central Limit Theorem, provided the sample is large enough.

Suppose one takes a random sample of 100 vouchers from the universe of all expense vouchers and examines them. A particular sample distribution may take any shape, but is can be summarized by its mean and standard deviation. Although it is not done in practice, assume that a large number of random samples of 100 vouchers each has been selected and examined. The mean of each sample would be expected to be somewhat different.

If a frequency distribution of the mean values of a large number of samples of 100 vouchers were constructed, the resulting distribution would be called a sampling distribution of means. According to the Central Limit Theorem, such a sampling distribution of means closely approximates a normal distribution even when the universe distribution is not normal, if the samples are large enough. Also, the mean of the sampling distribution will be equal to the mean of the universe from which the samples were selected. Thus, the mean of a sampling distribution is usually denoted by the symbol for mu, which is used to designate the universe mean. However, the standard deviation of a sampling distribution is given a special name, the standard error.

The standard error of the estimate is a significant concept in sampling theory. It has been proven that the standard error equals the standard deviation of the universe divided by the square root of the sample size, but since the standard deviation of the universe is seldom known, in practice the standard deviation of a sample is used as an estimate of the corresponding universe value.

The concept of chance or probability is a familiar one. Except for the certainty of death, few aspects of daily living elude chance. In fact, probability affects us even before we are born in the chance combination of genes which affects inheritable characteristics. Often, occupations are not the ones originally planned. Also, daily conversations contain such words as probably, likely, usually, perhaps, maybe, seldom, and the like. These words reflect aspects of uncertainty, and hence are associated with probability.

When a weather forecaster indicates that there is a 70 percent chance of rain, he is saying that on other days when there were similar climatic conditions, it rained on about seven of every ten such days.

There are two ways of viewing a probability statement. Any initial probability statement is based on assumptions. Additional evidence may suggest that the assumptions be modified or changed. Also, as assumptions change, the related probabilities change. The classical approach views a probability as a given unchangeable fact. The modern approach views it as a deducted and tentative assumption subject to change on the basis of evidence and experience.

Inferential statistics is the procedure of reasoning from the sample (partial) to the universe (general). In other words, generalizations and conclusions are based on observations (samples) supported by relevant experience and knowledge.

The primary task of inferential statistics is the development of methods which enable one to calculate and minimize the sampling risk associated with predictions. The foundation of modern statistics and sampling theory is the concept of probability.

There are, in general, two different bases for estimating the probability that a certain event will occur at a given trial. In the first, estimates are based on a logical analysis of the different ways in which the event may occur or fail to occur; the result is called mathematical probability. The second, called statistical or empirical probability, is based on a knowledge of what has occurred on similar occasions in the past, and a belief that probably similar results will be obtained in the future.

Mathematical probability is useful in the theory of sampling for calculating relationships between universes and samples. Probability interpretations relate to the universe and not to any specific event or item in it. For example, insurance rates are based on the average life expectancy.

Probability is expressed as a proportion, fraction or a percentage from zero to unity (one). The numerator is the number of ways a particular event may occur and the denominator is the total possible outcomes. The reasons for those limits is that a probability is a proportion, and a proportion cannot exceed 100 percent or unity (one); also the notion of a "negative success" is meaningless.

Probability interpretations are used to evaluate the precision of estimations and predictions. When an auditor assigns a 95 percent confidence or assurance to his projection (estimation) what he is really saying is that he is using a method of estimation which he expects to provide estimates within a specified range or interval 95 percent of the time if the procedures were repeated. The appropriate estimation method is dependent upon the objective. This is especially true in auditing where there may be several objectives, and a single method of estimation may not be equally efficient and economical in satisfying all objectives.

15

Estimation and Evaluation

I. GENERAL

This chapter provides guidance for the statistical estimation and evaluation of universe characteristics derived from sample data. Despite any assumptions which the auditor might make prior to the selection of the sample items, a sample can be statistically evaluated only after it has been examined and analyzed. For this reason, findings from preliminary samples may often suffice for some audit purposes.

Briefly, the application of statistical sampling procedures to audit activities involves the choice of (i) sample selection methods, (ii) audit review and analysis procedures, (iii) estimating methods, and (iv) corresponding methods for statistically evaluating the reliability of the estimated universe characteristic or value.

As previously mentioned, the random selection of sample items distinguishes statistical sampling from judgment sampling. If appropriate random selection procedures are used, the sample results can be statistically evaluated. The evaluation approach commonly used in audit applications involves the calculation of confidence intervals, which are expected to contain the universe characteristic or value a specified proportion of times if repeated samples of the same size were drawn in the same manner.

It should be noted that various estimating procedures were used in auditing long before the introduction of statistical sampling; however, unless the sample is randomly selected, there is no valid basis for constructing confidence intervals. For example, in judgment (non-statistical) sampling, the examination of 200 transactions may reveal $1,000 in expenses which are not documented, resulting in an average of $5 of undocumented expenses per transaction. Projecting this average to a universe of 2,000 transactions would give an estimated $10,000 of undocumented expenses; however, there would be no valid basis for constructing confidence limits. Also, a judgment sample could have likely been bias either intentionally or unintentionally.

It will be recalled that any sample selected by a chance mechanism, such as a random digit table, with known, non-zero probabilities of selection is referred to as a random or statistical sample. The chance mechanism guarantees freedom from selection bias. Also, a simple random sample is one which is selected by a procedure which gives every sample of the same size from the same universe an equal probability of being selected. However, the probabilities of selection need not be equal for all samples, as long as the probabilities are known and are non-zero.

If one infers findings from a sample to the universe, the random selection procedure is an integral part of the inference. Nevertheless, it is impossible to ascertain from the examination of a specific sample whether it is free from selection bias. The theoretical aspect of a random sampling procedure is not associated with a specific sample on a particular occasion but with its promise to be impartial among samples in the long run. This aspect emphasizes the virtues of replicated sampling.

Statistical sampling involves an inference. The aim is to obtain, with optimal reliability, unknown information about a universe on the basis of a review and analysis of a sample from that universe. The method of selecting and analyzing the sample is known as the sampling operation. The objective of any sampling operation is either to estimate some unknown characteristic or value of a universe or to test the reasonableness of some supposition about the nature of a universe. The evaluation objective is accomplished through the computation of confidence intervals that attempt to measure the effect of random sampling variations which cause the sample estimate of the universe value to deviate from the actual universe value.

The probability (expressed as a percentage) that each of these confidence intervals will contain the actual universe value, if a very large number of samples of the same design and size were drawn from a universe, is known as the confidence level. The confidence level indicates the degree of reliability that may be associated with the corresponding confidence interval.

Statistical significance is determined by constructing mathematically ranges of acceptance and rejection at a specified confidence level. Since auditors have an obligation to ascertain the nature and cause of findings, and to make financial adjustments, as well as recommend procedural changes, the accept-or-reject concept seems to have little usefulness in achieving audit goals. Moreover, the types of errors found will influence the kind of action taken by the auditor, and therefore, it seems unrealistic to expect him to make a final choice of his decision parameters before the sample is collected. Those readers who are interested in hypothesis testing should see Elliott and Rogers [13].

Most accounting distributions tend to be skewed to the right because of relatively few credits and large upper limits. Despite the presence of skewness in the universe distribution, for practical purposes the procedures based on normal distributional theory remain useful if the sample size is moderately large, the extreme values are stratified and audited 100 percent, and the occurrence rate is not too low. Otherwise, the sample results may be misleading. On the other hand, one of the virtues of monetary-unit sampling is that it is not based on normal distributional theory, and it is optimal with low error rates. However, it is hampered by conservatively high upper confidence limits.

A statistical inference, however, does not constitute an audit decision. An audit decision involves a critical analysis and an expert judgment of the nature, merit, quality, significance, magnitude, degree and value of the totality of all available audit evidence, not merely the estimates obtained from the sample.

The monetary-unit sampling strategy of constructing an upper bound on the total monetary error is discussed in Chapter 12, and therefore, will not be repeated in this chapter. Also, the concept and guidance on the use of upper limit occurrence rates (attributes) are discussed in Chapter 13. Thus, the attribute (occurrence rate) strategy, as discussed in this chapter, is concerned with two-sided confidence interval determinations.

II. SAMPLING RISK

Statistical sampling is associated with only one aspect of the overall audit risk. Statistical sampling is a method by which the auditor obtains results from a randomly selected portion of a total number of accounts or transactions that will be the same, within calculable limits of risk, as the results that would have been produced by a complete (100 percent) review of the same sampling frame with the same criteria, procedures and thoroughness, and over the same time period, as for the sample. Therefore, sampling risk relates to the probability that the sampling procedures for both compliance and substantive testing might produce different results, if the same procedures were applied in a similar manner to all accounts or transactions.

The sampling risk can be objectively computed and controlled when statistical sampling is used. In fact, the special feature of statistical sampling is that it permits the use of the theory of probability for the computation, from the sample data, of the margin of sampling error (precision) with the associated risk that has been introduced by using a sample in lieu of a complete review.

The other aspect of overall audit risk is associated with the possibility that the auditing procedures, even if applied to a complete review, might fail to detect material monetary errors or fail to reveal compliance deviations which would influence the auditor's decision. This aspect of audit risk is referred to as the nonsampling risk, and it is attributable to the nature of the auditing procedures, the timing of the application of the procedures, the system or operation being reviewed, and the maturity, skill and thoroughness of the auditor.

Controlling the nonsampling audit risk is important since the total audit risk cannot be less than the nonsampling risk. However, control over nonsampling risk involves well designed audit programs and adequate supervision, topics which are not covered in this book. Therefore, references to audit risk in this chapter and elsewhere relate only to the risk associated with sampling.

In summary, there are three principal sources of risk in audit sampling applications: (i) sampling variability, generally called sam pling error or precision, which depends on the design and, to a lesser extent, on the sample size; (ii) sample selection biases which reflect how accurately the design is executed; and (iii) the effects of erroneous criteria, assumptions and observations which cause differences between observed and actual characteristics or variables. The latter two sources are usually called nonsampling risks and are not covered by the statistical evaluation aspect of the sample results.

III. SAMPLING PRECISION AND ASSURANCE

When the auditor uses statistical sampling to estimate a universe characteristic, the Central Limit Theorem provides a basis for calculating the impreciseness of the estimate for a specified sampling risk. Two interrelated concepts are used for this purpose. One is the precision, which estimates the probable proximity of the estimate from the sample to the corresponding unknown universe characteristic or value. The other is the reliability (confidence, assurance), which measures the frequency with which the difference between the sample estimate and the actual universe characteristic or value does not exceed the estimated precision, if repeated samples of the same size were taken.

For example, a calculated precision of $25,000 at a confidence level of 95 percent indicates that, among all possible samples of the same size from the same universe, 95 percent would be expected to have a difference between the sample estimate and the actual universe value less than or equal to $25,000. Conversely, the auditor has a 5 percent sampling risk that he may have a sample where the difference is greater than $25,000.

Replication provides an intuitive way of minimizing the acceptance of higher sampling risks through comparison of the results of individual replicates. Any significant differences noted among the replicates should be analyzed for cause. If there seems to be a systematic pattern, the auditor might subjectively examine additional items which fit the pattern, and thus, augment the sample results with judgmental testing in support of an audit opinion.

In auditing, one often proceeds from an effect (finding) to the cause. In doing this, there is an assumption that the occurrences of a characteristic in a sample will follow a similar pattern as the universe, provided the sample is large enough. For example, if replicated samples reveal an occurrence (error) proportion of about 0.05 or a ratio of audited values to corresponding book values of 0.97, then on the basis of the sample observations, the auditor assumes that the total universe exhibits nearly a corresponding proportion and ratio unless he has reasons to believe otherwise.

Since the sample proportion and ratio are random variables, they can be defined with confidence limits. As mentioned, these limits can be controlled by making them approach each other as closely as desired through stratification and by increasing the number of observations (sample size).

IV. RELIABILITY STATEMENT

In statistical sampling, a reliability statement consists of two parts: (i) a confidence or precision interval and (ii) a confidence or reliability (or assurance) level.

The terms confidence interval and confidence level are used in the consideration of the following, based on information obtained from the sample:

a. The probability α (or confidence level 100 α percent) that the unknown universe characteristic or value will lie within specified limits (or confidence intervals) and
b. The confidence interval within which the unknown universe characteristic or value may be expected to lie with a specified confidence level of 100 α percent.

It should be noted that in specifying a precision amount for a certain confidence level, the auditor is asserting that at the confidence level, he expects the actual universe characteristic or value will not differ from the sample estimate by more than the precision amount. The sample estimate of the universe characteristic, plus and minus the amount of the achieved precision forms the corresponding confidence interval. For example, if the sample estimate of the total audited amount is \$500,000 with a precision of \$20,000 at the 95 percent two-sided confidence level, then the confidence interval would be from \$480,000 (\$500,000 – \$20,000) to \$520,000 (\$500,000 + \$20,000). That is, if repeated samples of the same size were taken from this universe, using the same sampling and audit procedures, the confidence intervals of 95 out of 100 of these samples would be expected to contain the universe value. In a technical sense, the reliability statement does not relate to any particular sample. The concept, per se, provides no assurance that a particular sample is not one of the five percent extremes; however, replication provides an intuitive means of identifying "odd balls."

Thus, for most purposes, the auditor desires to know how likely the universe estimate derived from the sample deviates from the actual universe characteristic or value. Knowledge from the Central Limit Theorem that the sampling distribution of means approaches the normal distribution as the sample size increases, and that the mean of the sampling distribution is equal to the mean of the universe, together with properties of the normal distribution, enables the auditor to make statistical evaluation or reliability statements.

The concept of the normal distribution is significant in stating the reliability of sample estimates of universe means and aggregates. The sample means of distributions of any shape tend to form normal distributions as the sample sizes increase. This tendency is more rapid for those universes which are only moderately skewed. Sample size, however, is not the sole determinant of the reliability of the results. Smaller samples from carefully designed and executed samples provide more useful information than loosely improvised larger samples.

If the sample is effectively stratified and large enough, the underlying sampling distribution is approximately normal, and therefore, the chances are 68 out of 100 (68 percent assurance) that the universe estimate derived from the sample does not deviate from the actual universe characteristic or value by more than one standard error (the standard deviation of the sampling distribution), and the chances are 95 out of 100 that the deviation will not exceed two standard errors. Similarly, statistical assurance factors for other multiples of the standard error are available in tables and in computer programs.

There is also the concept of one-sided and two-sided confidence levels. Suppose one wishes to determine the confidence interval corresponding to a confidence level of 95 percent. As stated, this problem does not have a unique solution, since there are a number of different intervals which would cover 95 percent of the area under the normal probability distribution, depending upon how one proposes to exclude areas from the upper and lower tails of the distribution. Often the interval is assumed to be such that equal areas are excluded from each tail, resulting in what is referred to as a two-sided confidence interval. In the case of a 95 percent confidence level, this would mean the exclusion of 2.5 percent of the area from each tail of the distribution. On the other hand, a one-sided confidence interval excludes only the 2.5 percent of the area from one tail, depending upon whether an upper precision limit or a lower precision limit is of interest. Thus, the confidence level is increased to 97.5 percent.

If in the example cited in this section, the same upper precision limit of $520,000 is referred to as a one-sided confidence level, by specifying that the universe value is expected to be less than this amount, then the 2.5 percent of the area associated with the lower tail of the distribution is added to the two-sided confidence. In this case, the 95 percent plus 2.5 percent, results in a corresponding one-sided confidence level of 97.5 percent. Lower one-sided confidence levels are computed in a similar manner. See the comparisons of one-sided and two-sided confidence levels shown in Figure 24, Chapter 14.

The conversions from two-sided to one-sided probabilities, and vice-versa, may be accomplished by applying the following relationships:

$$C = 2C_1 - 1, \text{ for conversion to two-sided}$$

$$C_1 = \frac{1}{2} + \frac{C}{1}, \text{ for conversion to one-sided}$$

where

C = probability for two-sided interval corresponding to a specified value

C_1 = probability that either the mean is greater than the lower limit or that it is less than the upper limit.

For example, to convert from one-sided probability 0.95, one has:

$$C = 2(0.95) - 1 = 1.9 - 1 = 0.90, \text{ for two-sided}$$

To perform the reverse process, one has:

$$C_1 = \frac{1}{2} + \frac{0.90}{2} = \frac{1.90}{2} = 0.95, \text{ for one-sided.}$$

V. STATISTICAL ESTIMATORS

After the sample has been selected, examined and analyzed, the results are used to estimate universe characteristics or values. The statistical procedures for deriving universe estimates from sample data are referred to as estimators. For example, the sample mean is a random variable because it varies from sample to sample, and thus has a sampling distribution. The sample mean is also an estimator because it can be used to estimate the universe mean.

Statistically, a best estimator is unbiased, consistent, efficient, and sufficient. The bias of an estimator is not in the same sense as a selection bias, since an absence of selection bias does not guarantee the avoidance of procedural or estimation bias.

Any selection, observation, measurement, recording or estimation procedure which systematically distorts an estimate from the actual universe value is considered bias. Although there may be distortions for individual samples, they average out to zero in the long run if the method is unbiased. This is one of the reasons replication is so significant in audit sampling.

An unbiased sample is one that is free from selection and procedural bias, such as incorrect criteria, erroneous recordings, nonresponses, unavailable data, auditor bias, and so on. However, an unbiased sample selection procedure should be distinguished from an unbiased estimator, which may be used with any data, even a judgment sample. Although unbiasedness is a desirable virtue, it is not necessarily an indispensable requirement of a good estimator, especially if the amount of bias is small compared with the precision of the estimate.

An estimator is said to be consistent when the mean of its sampling distribution approaches the actual universe mean as the sample size increases. Also, an estimator is said to be most efficient when its sampling distribution reflects the smallest standard deviation.

An estimator is said to be sufficient if it is useful for small samples, exhausts all information in the sample, and no other estimator can provide additional information. For many universe estimates, however, there is no estimator which satisfies all these requirements.

Although the sample mean is an unbiased estimator of the universe mean, it is not unique in this respect since there are other unbiased estimators. Also, when the mean estimate is close to the book value, there is a tendency to accept universes with material errors more often than the specified sampling risk because of the strong correlation between estimates of the mean and the standard error.

There are other estimators which are more efficient than the sample mean. For example, the ratio estimator, though slightly bias, is more favorable than the sample mean in auditing. The slight bias poses no problem since statisticians have been able to measure its extent as being equal to no more than the reciprocal of the sample size.

If the sample size is at least 50, this bias is considered negligible and is usually ignored. Because of its many valuable uses in financial and operational management, the ratio estimator will be discussed in more detail in a subsequent section.

With the knowledge that an estimator is unbiased or that the extent of the bias can be measured, then a best estimator is the one with its sampling distribution concentrated most closely about its mean. In other words, a best estimator may be the one that is most efficient.

This leads to the standard deviation of the sampling distribution—the standard error—as the best measure of dispension of sample estimates, and consequently, the precision of the results. Hence, for comparative purposes, the best estimator for large sample sizes (50 or more) may be considered the estimator with the smallest standard error.

Therefore, the standard error, which is discussed in Chapter 14, is used to measure the margin of sampling error; that is, the difference between the estimate from the sample and the result that would have been obtained from a complete review of the universe. However, it neither measures the accuracy nor the usefulness of the estimate. Also, a small standard error merely indicates that a complete review would have produced a result nearly the same as the estimate from the sample. In addition, it does not indicate successful coverage nor adequate audit work, and it does not measure any biases that may afflict complete reviews as well as samples.

Confidence limits are calculated as multiples of the standard error. The difference between upper and lower limits is referred to as the confidence interval. The confidence interval is also a variable since it varies from sample to sample of the same design and selection scheme. The estimate of a universe characteristic or value derived from a sample is much more likely, on the average (from replications), to be closer to the actual universe characteristic or value than either of the confidence limits.

Classical statistical estimating procedures, based on large-sample theory, are useful in some accounting applications such as the division of revenues for services performed by more than one company. However, some of the procedures involving the application of statistical sampling to auditing have failed to recognize the special structure of auditing universes and unique aspects of the auditing environment. Generally, the auditor has much more information about the accounting universe and the related environment than is available to those who conduct sample surveys, but the auditor usually has tighter precision requirements and uses relatively smaller samples.

According to Kaplan [27], auditors must use auxiliary information estimators which explicitly use all available information. The difference, ratio and regression estimators belong to a class of auxiliary information estimators which are being used in more audit applications. Each of these is discussed in subsequent sections.

There is much guidance available about selecting the most appropriate estimator which is consistent with the audit objectives, the characteristics of the universe and the sample design. Some of this guidance seems to imply that the auditor must make an unrevocable decision concerning the use of a particular estimator. Practice, however, reveals that useful comparative information can be obtained quickly from packaged computer programs which provide computations for various estimators. For example, if several estimators provide comparable information, the auditor naturally has more intuitive assurance in his estimates. On the other hand, if the various estimators provide significantly different information, he would be obliged to investigate to determine cause or to ascertain a basis for further stratification.

There is also a misconception that a sample can be used either to estimate occurrence rates (attributes) or to estimate dollar amounts (variables), but not both attributes and variables from the same sample. This misconception may be associated with the fact that auditors have two separate and distinct testing procedures: compliance testing with an interest in occurrence rates and substantive testing where the primary concern is dollar balances. Although the sample size for compliance testing may be smaller, the same sample could be the initial replicate for substantive testing. The point to remember is that a sample is useful for any kind of relevant audit information that it may provide.

Aside from the statistical considerations of a best estimator, one must consider the timeliness, simplicity, cost and other administrative and policy considerations. Generally, auditors tend to choose the estimator which provides the smallest confidence interval for the most critical universe estimate. Also, the choice of an estimator may depend upon the availability and efficiency of a computer capability to assist in the sampling procedure. Furthermore, it is better to use simple estimating procedures which are readily understood by the staff and the auditee than to use more elaborate estimating schemes which may be misapplied and misunderstood.

Most of the remainder of this chapter is devoted to discussions of several widely used statistical estimators which may be used with stratified and unstratified sampling.

VI. MEAN-PER-UNIT ESTIMATOR

Although it is seldom used with audit sampling, the mean-per-unit estimator may be used in projecting (estimating) a universe characteristic or value from a sample. After a simple random sample is selected, and an audited amount determined for each sample unit, a mean estimation may be computed as follows: the sample mean of audited amounts is multiplied by the total number of audit units in the universe. The product is an estimate of the universe total. Even though the use of the mean-per-unit estimate in auditing is limited to making estimates of universe totals when the recorded amount of each sampling unit is known, an understanding of this concept will facilitate comprehending other statistical estimators.

As mentioned, confidence limits can be computed for universe estimates from knowledge of sample estimates. Based on the Central Limit Theorem, cumulative frequency distributions of sample means will approximate a normal distribution, even if the universe distribution is not normal, provided the universe is not highly skewed, the sample size and the rate of occurrence of the universe characteristic being estimated are not small. In auditing, however, the non-normality of the universe distribution can be further diminished through audit stratification.

Unfortunately, there is no uniform notation in publications or sample audits except a tendency to use capital letters to denote universe totals. In this chapter, X is used to denote universe book value or recorded amount and Y denotes the estimated audited value or amount, since X is generally employed to denote an auxiliary variable used to estimate some characteristic of another variable Y. Sample estimates follow a similar form except lower case letters are used. Also, for a sample of n units, y_j refers to the y characteristic of the jth unit. The symbol Σ, the Greek capital letter sigma means "the sum of" or total and indicates that items following it are to be added together. For example, Σy_j for j from 1 to 3 means add $y_1 + y_2 + y_3$. Also, sample summations or totals are for $j = 1$ to n, where n is the sample size.

In order to permit comparison among the results for the various estimators discussed in this chapter, the data in Table 15.1 will be used in each illustration. In a practical situation, the sample size may be more than the 50 items used for illustrative purposes.

The mean estimation of the total audited amount Y is computed from the following relationship:

$$Y = N\bar{y}$$

$$\begin{bmatrix}\text{Mean estimation} \\ \text{of total} \\ \text{audited amount}\end{bmatrix} = \begin{bmatrix}\text{Number of} \\ \text{audit units} \\ \text{in universe}\end{bmatrix} \times \begin{bmatrix}\text{Mean of} \\ \text{sample} \\ \text{audited amount}\end{bmatrix}$$

Substituting data from Table 15.1 gives the following:

$$\text{Mean estimation of audited total} = 2576 \times 109.86 = 282{,}999$$

Before the standard error can be derived, the estimated standard deviation of the sample result must be calculated. Using the formula for computing the standard deviation for a sample and data from Table 15.1, the estimated standard deviation of the sample audited amount is 132.457.

The computer printout in Figure 31, under MEAN METHOD, reveals the same values for these estimates based on computations performed by a computer. That is, the total dollars audited is shown as 282,299 and the standard deviation for the sample values is 132.457.

With an estimate of the standard deviation, one can compute an estimate of the standard error of the estimated audited total using the following:

$$\hat{\sigma}_{\bar{y}} = \frac{Sy}{\sqrt{n}} N \sqrt{\frac{N - n}{N - 1}}$$

$$\begin{bmatrix}\text{Estimated} \\ \text{standard} \\ \text{error of} \\ \text{audited} \\ \text{total}\end{bmatrix} = \begin{bmatrix}\text{Product of estimated} \\ \text{standard deviation of} \\ \text{audited amounts, and} \\ \text{universe size, and} \\ \text{finite universe} \\ \text{correction factor}\end{bmatrix} \div \begin{bmatrix}\text{Square root} \\ \text{of} \\ \text{sample size}\end{bmatrix}$$

Note: the expression $\sqrt{\frac{N - n}{N - 1}}$ is the factor which adjusts for sampling without replacement. If the sample size is less than 5 percent of the universe, this factor may be ignored.

Substituting 132.457 for Sy, 2576 for N, and 50 for n in the relationship for computing the standard error of the estimated audited total, gives:

$$\text{Estimated standard error} = \frac{132.457}{\sqrt{50}} \times 2576 \sqrt{\frac{2576 - 50}{2576 - 1}}$$

$= 47{,}793$ (shown as 47,792.8 in Figure 31).

Table 15.1. Universe and Sample Data for Illustrative Computations (From a universe of 2576 inventory records with a book value of \$312,089, a simple random sample of 50 records, without replacement, was selected with the following results:)

Sample item	(x) Recorded sample amount	(y) Audited sample amount	(d) Audited − recorded difference	(x^2) Recorded amount squared	(y^2) Audited amount squared	(d^2) Difference squared	$(x)(y)$ Product of recorded amount and audited amount	$(x)(d)$ Product of recorded amount and difference
j	(1)	(2)	(3) = (2) - (1)	(4) = (1) × (1)	(5) = (2) × (2)	(6) = (3) × (3)	(7) = (1) × (2)	(8) = (1) × (3)
1	483	483	0	233,289	233,289	0	233,289	0
2	441	392	-49	194,481	153,664	2,401	172,872	-21,609
3	346	346	0	119,716	119,716	0	119,716	0
4	57	48	-9	3,249	2,304	81	2,736	-513
5	49	0	-49	2,401	0	2,401	0	-2,401
6	167	167	0	27,889	27,889	0	27,889	0
7	94	94	0	8,836	8,836	0	8,836	0
8	406	406	0	164,836	164,836	0	164,836	0
9	57	57	0	3,249	3,249	0	3,249	0
10	77	74	-3	5,929	5,476	9	5,698	-231
11	12	0	-12	144	0	144	0	-144
12	134	134	0	17,956	17,956	0	17,956	0
13	144	144	0	12,996	12,996	0	12,996	0
14	41	41	0	1,681	1,681	0	1,681	0
15	73	70	-3	5,329	4,900	9	5,110	-219

16	95	95	0	9,025	9,025	0	9,025	0
17	338	338	0	114,244	114,244	0	114,244	0
18	12	12	0	144	144	0	144	0
19	395	388	-7	156,025	150,544	49	153,260	-2,765
20	12	10	-2	144	100	4	120	-24
21	87	79	-8	7,569	6,241	64	6,873	-696
22	7	0	-7	49	0	49	0	-49
23	210	204	-6	44,100	41,616	36	42,840	-1,260
24	2	0	-2	4	0	4	0	-4
25	6	6	0	36	36	0	36	0
26	137	133	-4	18,769	17,689	16	18,221	-548
27	75	75	0	5,625	5,625	0	5,625	0
28	146	142	-4	21,316	20,164	16	20,732	-584
29	2	2	0	4	4	0	4	0
30	90	85	-5	8,100	7,225	25	7,650	-450
31	109	109	0	11,881	11,881	0	11,881	0
32	414	409	-5	171,396	167,281	25	169,326	-2,070
33	54	54	0	2,916	2,916	0	2,916	0
34	16	16	0	256	256	0	256	0
35	44	44	0	1,936	1,936	0	1,936	0
36	51	51	0	2,601	2,601	0	2,601	0
37	46	46	0	2,116	2,116	0	2,116	0
38	6	6	0	36	36	0	36	0
39	164	164	0	26,896	26,896	0	26,896	0
40	32	22	-10	1,024	484	100	700	-320

Table 15.1. (Continued)

Sample item	(x) Recorded sample amount	(y) Audited sample amount	(d) Audited − recorded difference	(x^2) Recorded amount squared	(y^2) Audited amount squared	(d^2) Difference squared	$(x)(y)$ Product of recorded amount and audited amount	$(x)(d)$ Product of recorded amount and difference
j	(1)	(2)	(3) = (2) - (1)	(4) = (1) × (1)	(5) = (2) × (2)	(6) = (3) × (3)	(7) = (1) × (2)	(8) = (1) × (3)
41	47	47	0	2,209	2,209	0	2,209	0
42	49	49	0	2,401	2,401	0	2,401	0
43	326	326	0	106,276	106,276	0	106,276	0
44	1	1	0	1	1	0	1	0
45	46	0	-46	2,116	0	2,116	0	-2,116
46	5	0	-5	25	0	25	0	-25
47	40	40	0	1,600	1,600	0	1,600	0
48	22	22	0	484	484	0	484	0
49	53	53	0	2,809	2,809	0	2,809	0
50	39	39	0	1,521	1,521	0	1,521	0
Total	5,729	5,493	-236	1,527,635	1,463,153	7,574	1,491,607	-36,028
Mean	114.58	109.86	-4.72	30,552.70	29,263.06	151.48	29,832.14	-720.56

```
This program evaluates the results of sampling for variables (dollars)
with or without stratification, for the mean and ratio estimation methods,
using book values and audited values.  First enter the number of strata
examined followed by the number of items examined in each stratum.  Then
enter the following data for each stratum:  the total number of items in
the stratum, and the book amount and the audited amount for each item in
the sample.  At the end of each stratum enter '3E33'.
THE FILE MAY BE IN ONE OF THE FOLLOWING TWO FORMATS:
      100 DATA 12,3,14,2,18,0,19,0,16,4,3E33
      200 DATA 85,14,95,23,88,0,3E33
      (WITH THIS KIND OF FILE, USE LINE NUMBERS 1 - 1999 ONLY)
                         OR
      100 12 3 14 2 18 0 19 0 16 4 3E33
      200 85 14 95 23 88 0 3E33
      (WITH THIS KIND OF FILE ANY LINE NUMBERS 1 - 99999 MAY BE USED)

IF SUCH A FILE ALREADY EXISTS, ENTER 'GO', ELSE ENTER 'STOP' GO


1DATA1,5Ø
2DATA2576,312Ø89
3DATA483,483,441,392,346,346,57,48,49,Ø,167,167,94,94,4Ø6,4Ø5
4DATA57,57,77,74,12,Ø,134,134,114,114,41,41,73,7Ø,95,95,338,338
5DATA12,12,395,388,12,1Ø,87,79,7,Ø,21Ø,2Ø4,2,Ø,6,6,137,133
6DATA75,75,146,142,2,2,9Ø,85,1Ø9,1Ø9,414,4Ø9,54,54,16,16,44,44
7DATA51,51,46,46,6,6,164,164,32,22,47,47,49,49,326,326,1,1,46,Ø
8DATA5,Ø,4Ø,4Ø,22,22,53,53,39,39
RUN
```

```
APPRAISAL OF SAMPLE RESULTS
                                              MEAN             RATIO
                                              METHOD           METHOD
STRATUM 1 OF 1
 UNIT COST AUDITED--5493/5Ø                   1Ø9.86
 RATIO AUDITED--5493/5729                                      .958806
 TOTAL COST AUDITED
                 1Ø9.86 X 2576                282999.
                 .95 X 312Ø89                                  299233.
 STANDARD DEVIATION                           132.457          12.12Ø2
 STANDARD ERROR OF ESTIMATE                   47792.8          4373.19
 SAMPLING ERROR OR PRECISION FOR:
  7Ø PC LEVEL--1.Ø48 X STD ERROR              5ØØ86.8          4583.1
  8Ø PC LEVEL--1.299 X STD ERROR              62Ø82.8          568Ø.77
  9Ø PC LEVEL--1.677 X STD ERROR              8Ø148.5          7333.83
  95 PC LEVEL--2.Ø1  X STD ERROR              96Ø63.5          879Ø.1
```

Figure 31. Computer printout of mean and ratio evaluations of audited amount.

With the standard error, one can compute confidence limits and intervals for the universe estimate. A description of how confidence limits and intervals are computed follows.

If the sample size is very large, Z factors in Figure 22, Chapter 14, may be used to compute confidence intervals. In auditing, however, t factors, which reflect the impact of sample size, are generally used in lieu of Z factors. Figure 24, Chapter 14, shows t factors for simple random samples for both one-sided and two-sided confidence intervals for selected probabilities. For comparison, the Z factor of 1.96 for the 95 percent two-sided confidence interval would be replaced by a t factor of 1.984 for a simple random sample of 100, resulting in a slightly larger interval. For very large sample sizes, corresponding Z and t factors are equal.

Using the t factor of 2.01 for a sample of 50, and the data from the above example, the 95 percent two-sided confidence limits and interval are derived as follows:

95% LOWER CONFIDENCE LIMIT

[Universe estimate] − [t factor for 95% confidence for sample size and design × Standard error of estimate]

or 282,999 − (2.01) (47,793)

resulting in 282,999 − 96,064

or 186,935

95% UPPER CONFIDENCE LIMIT

[Universe estimate] + [t factor for 95% confidence for sample size and design × Standard error of estimate]

or 282,999 + (2.01) (47,793)

resulting in 282,999 + 96,064

or 379,063

Thus, the two-sided 95 percent confidence interval would be from $186,935 to $379,063. Figure 31 shows a computer computation of 282,999 as the estimate and 96,063.5 as the 95 percent precision.

Often the result of a confidence interval is misinterpreted. An auditor would be unwarranted in affirming that there is a 95 percent probability that the actual audited total is between $186,935 and $379,062. From a frequency concept of probability, the proper assertion is that if he were to draw repeatedly simple random samples of size 50 and calculated the confidence intervals at the 95 percent two-sided confidence level, about 95 percent of the intervals would be expected to encompass the actual universe value. In other words, if the auditor were to compute a confidence interval for every possible sample of the same size using the same proceures with the same criteria and thoroughness, the proportion of the intervals containing the universe characteristic or value would equal the specified confidence level.

Actually, the mean-per-unit estimator surpresses the most significant information furnished by the sample: when there are observed differences between audited values and recorded values. When at least several differences are observed between the sample audited values and corresponding book values, the auditor may use auxiliary information estimators, such as the difference, ratio and regression estimators, to estimate and evaluate universe characteristics and values.

These more efficient estimators do not alter the designs of the more commonly used sampling plans. The same basic sample data are obtained; only the computations of universe estimates and their precisions are affected. When properly used, auxiliary information estimators result in more precise estimates of universe characteristics and values than the mean estimation. Although these estimators are somewhat more complex, they are discussed in their more elementary form in subsequent sections. The purpose of the illustrative computations is to emphasize similarities and differences among them. In practice, manual computations are not necessary, since there are many packaged computer programs to perform this chore.

VII. DIFFERENCE ESTIMATOR

When the universe standard deviation is large, as it is for many accounting universes, the sample size required to achieve a reasonable precision with a mean-per-unit estimate is much too large to be practical. Since the auditor usually has available information about sample items as well as total universe book values, he may use auxiliary information estimators, with or without stratification. As mentioned, auxiliary information estimators provide more precise estimates than the mean-per-unit estimator which uses only data from the sample audited amounts. The difference, ratio and regression estimators use both the sample recorded values and audited values, but each in a different manner.

Difference estimation uses the mean and standard deviation of the individual differences between each sample unit's audited amount and the corresponding recorded amount instead of merely the mean and standard deviation of the sample audited amounts as in mean-per-unit estimation. The difference method, therefore, has an intuitive appeal since the amount of the correction in the book value is emphasized.

An audited difference is defined as the audited amount minus the corresponding recorded amount. The mean of the differences is computed by dividing the sum of the sample differences by the sample size, and it is extended (projected) by multiplying the result by the number of audit units in the universe. This estimate of the total universe difference is algebraically added to the total recorded amount to compute an estimate of the total audited amount for the universe. When the recorded amount exceeds the audited amount, the algebraic sign of the difference is negative. Similarly, when the recorded amount is less than the audited amount, the sign of the difference is positive.

The mean-per-unit estimate involves a univariate characteristic: the audited amount. In applying auxiliary information estimators, however, bivariate characteristics are involved, and the use of some additional notations may be helpful. As in the case of mean-per-unit estimate, Y_j refers to the audited amount associated with the jth audit unit in the universe; that is, the amount the auditor would find to be correct if he were to verify the book amount X_j associated with the jth audit unit in the universe. Sample estimates follow a similar form except lower case letters are used.

The estimated total difference D between the audited amounts and the recorded amounts is computed from the following relationship:

$$D = N(\bar{y} - \bar{x})$$

$$\begin{bmatrix}\text{Estimated}\\ \text{total}\\ \text{difference}\end{bmatrix} = \begin{bmatrix}\text{Number}\\ \text{of audit}\\ \text{units in}\\ \text{universe}\end{bmatrix} \times \begin{bmatrix}\text{Mean of} & & \text{Mean of}\\ \text{sample} & - & \text{sample}\\ \text{audited} & & \text{recorded}\\ \text{amounts} & & \text{amounts}\end{bmatrix}$$

Substituting data from Table 15.1, gives the following:

$$\begin{aligned}\text{Estimated total difference} &= 2576\,(109.86 - 114.58)\\ &= -12{,}158\end{aligned}$$

Adding the total difference estimate to the total recorded amount gives an estimated total audited amount as follows:

$$\begin{aligned}\text{Difference estimation of total audited amount} &= 312{,}089 - 12{,}158\\ &= 299{,}931\end{aligned}$$

```
THIS PROGRAM ACCEPTS A FILE OF SAMPLE VALUES FOR A VARIABLES APPRAISAL.
THE PROGRAM WILL ACCEPT AN UNSTRATIFIED OR STRATIFIED SAMPLE.
DO YOU NEED INSTRUCTIONS FOR CREATING THE FILE OF VALUES (YES OR NO)?YES
FOR EACH SAMPLING UNIT, YOU MUST ENTER THE AMOUNT EXAMINED AND THE
AMOUNT OF 'DIFFERENCE'.  THE 'DIFFERENCE' AMOUNT IS THE AMOUNT TO BE APPRAISED
USING THE DIFFERENCE AND RATIO ESTIMATORS.  IN ONE AUDIT SITUATION THE
'DIFFERENCES' COULD BE ERRORS; IN ANOTHER SITUATION THEY COULD REPRESENT THE
VALUE OF ASSETS, AND SO ON.  THE PROGRAM WILL PROVIDE APPRAISALS FOR THE
EXAMINED (RECORDED) VALUES, THE DIFFERENCE VALUES AND THE 'ADJUSTED' VALUES
(DIFFERENCE VALUE SUBTRACTED FROM EXAMINED VALUE).
AT THE END OF EACH STRATUM, ENTER '3E33'.
ENTER NUMBER OF STRATA (IF UNSTRATIFIED, ENTER '1')?1
ENTER SAMPLE SIZE?50
ENTER UNIVERSE SIZE?2576
ENTER UNIVERSE VALUE?312089
ENTER SAMPLE VALUES?
```

```
1DATA1,50
2DATA2576,312089
3DATA483,0,441,49,346,0,57,9,49,49,167,0,94,0,406,0,57,0,77,3
4DATA12,12,134,0,114,0,41,0,73,3,95,0,338,0,12,0,395,7,12,2
5DATA87,8,7,7,210,6,2,2,6,0,137,4,75,0,146,4,2,0,90,5,109,0,
6DATA414,5,54,0,16,0,44,0,51,0,46,0,6,0,164,0,32,10,47,0,49,0
7DATA326,0,1,0,46,46,5,5,40,0,22,0,53,0,39,0
RUN
    SUMMARY OF INPUT DATA:
    STRA- NO. EX-  TOTAL NUMBER   TOTAL DOLLARS  AMOUNT        AMOUNT
    TUM   AMINED   IN STRATUM     IN STRATUM     EXAMINED      QUESTIONED
     1     50         2576           312089        5729           236

APPRAISAL OF SAMPLE RESULTS
                                                DIFFERENCE     RATIO
                                                METHOD         METHOD
STRATUM 1     OF 1
 UNIT COST QUESTIONED-- 236  / 50               4.72
 RATIO QUESTIONED-- 236  / 5729                                .041194
 TOTAL COST QUESTIONED
                 4.72    X 2576                 12158.0
                 .041194 X 312089                              12856.2
 STANDARD DEVIATION                             11.4821        12.1202
 STANDARD ERROR OF ESTIMATE                     4142.95        4373.19
 SAMPLING ERROR OR PRECISION FOR:
  70 PC LEVEL-- 1.048   X  STD ERROR            4341.81        4583.1
  80 PC LEVEL-- 1.299   X  STD ERROR            5381.69        5680.77
  90 PC LEVEL-- 1.677   X  STD ERROR            6947.73        7333.83
  95 PC LEVEL-- 2.01    X  STD ERROR            8327.33        8790.1
```

Figure 32. Computer printout of difference and ratio evaluations.

In the evaluation of the sample results, the differences are treated mathematically exactly as the audited values were treated in the mean-per-unit method. In Figure 32, under DIFFERENCE METHOD, is a computer printout of the desired estimates. Thus, the 95 percent confidence interval of the differences (total cost questioned) is from \$3,831 (\$12,158 - \$8,327) to \$20,485 (\$12,158 + \$8,327).

The minimum number of differences depends upon the variability of the differences, the sample design and the method of estimation. This is the kind of situation where audit stratification, error analysis and audit appraisal rein supremely, for the differences should be reasonable and typical for the particular circumstances. In such cases, Roberts [57] suggests a minimum of 15 or 20 differences. Others, such as McCray [42] suggest at least 30 differences are necessary. Roberts also suggests that if non-zero differences are small and both understatements and overstatements exist, the unstratified difference estimator should not be used, because the required sample size may be impractically large. Under these circumstances, he suggests using some form of stratification or probability-proportional-to-size (PPS) selection. For many of these schemes, each sampling unit has probability of selection which is proportional to its book value, such as dollar-unit sampling. On the other hand, if the few differences are concentrated in a limit operation, the audit significance should not be ignored.

If all the differences have the same algebraic sign, then the difference universe distribution is skewed, and the degree of skewness reflects the magnitude and proportion of individual differences. This is one of the reasons why error analysis and audit appraisal of each non-zero difference is significant. Also, audit stratification will reduce the degree of skewness.

VIII. RATIO ESTIMATOR

A. Use of Ratios

Another very useful method of estimation which yields more precise results than the mean-per-unit is the ratio estimate. A limitation on the use of the ratio estimate is that the total value of the denominator variable must be known; however, this poses no problem in most accounting situations, since the denominator of the ratio usually involves book or recorded values.

Ratios are often encountered in accounting and managerial activities, since financial ratios are key operational indicators. By casting data from financial statements as comparisons among various operations of a company, ratios assist in identifying areas and operations which may require management attention.

In business activities, ratios are used in many ways. For example, management can ascertain the company's position as a ratio of return on equity. Ratios can also be used for comparisons with historical data, industry standards, and in trend analyses. Additionally, ratios can be used to set performance goals and appraise accomplishments.

Other examples of financial ratios include the ratio of current assets to current liabilities; ratio of total liabilities to total equity; ratio of dollars of accounts receivable in a particular age category to total dollars of accounts receivable; ratio of unsupported dollars in account balances to total dollar account balances, and so on. Thus, it is obvious that ratios are common and important indicators in accounting and managerial operations.

B. Comparison of Ratios and Proportions

Since a proportion, which is associated with sampling for attributes, is similar in many respects to a ratio, which involves variables, it is worthwhile to understand the difference.

An estimate of a proportion, such as an error rate, is defined as the number of errors in a sample divided by the sample size. Thus, both a proportion and a ratio reflect a division of one number by another; however, the ratio has one random variable in its numerator and another random variable in its denominator. A proportion has a random variable in the numerator and a fixed number in its denominator. Thus, the denominator may be fixed in advance such as the size of a sample.

For illustrative purposes, suppose a simple random sample of 200 accounts is selected. In determining the proportion of accounts in error, it is immediately known before any of the sample accounts are examined that the denominator of the proportion will be 200. Hence, if eight sample accounts are found to be in error, the error proportion is 8 ÷ 200 = 0.04 or 4 percent. Suppose it is also desirable to use the same sample to estimate the ratio of audited values to book values. In this case, neither the numerator nor the denominator is known before the accounts are examined, and the cumulative amounts for each will vary after each sample account is examined. Thus, if the audited amount for the 200 accounts is $107,200 and the recorded amount for those accounts is $119,000, then the ratio of audited values to recorded values is 107,200 ÷ 119,000 = 0.9. Incidentally, if both the numerator and denominator are divided by the same number, such as the sample size, the value of the ratio is not changed. Thus, one could determine the same ratio by dividing the mean of the audited amounts for the sample by the mean of its recorded amount. In this case, 107,200 ÷ 200 = 536 (mean of audited amounts) and 119,000 ÷ 200 = 595 (mean of recorded amounts), and hence, 536 ÷ 595 = 0.9.

C. Advantages of Ratio Estimation

One of the aims of ratio estimation is to obtain increased precision by taking advantage of the correlation between the audited amounts and the recorded amounts. There is usually a high correlation, since for most universe units the two values are the same; that is, there is a zero difference.

Despite a slight bias, the ratio estimate has an intuitive appeal for auditors because it has been one of the traditional methods of analysis and comparison. Ratios from several subsamples and prior samples are often intuitively compared.

Another feature of ratio estimation which appeals to auditors is the fact that the standard error associated with the undocumented amounts (differences) is numerically identical to the standard error for the audited amounts. For example, an auditor is interested in determining whether a submitted inventory valued at $1,000,000 is reasonable. Using ratio estimation to project his sample result to the universe, he may question $70,000 with a standard error of, say $20,000. He is intrigued to learn that the audited amount of $930,000, under ratio estimation, has exactly the same standard error of $20,000. However, the relative precision of the latter is 2 percent compared with over 28 percent for the former.

Also, in many audit situations, the ratio estimate is more feasible than other estimates because of the kind of available data. For instance, if the total number of sampling units is unknown, such as the number of applicable line items on a very large computer listing, it is difficult to determine the total audited amount using the mean-per-unit or difference estimate. By contrast, a ratio estimate of the universe total can be computed simply by multiplying the ratio derived from the sample by the controlled recorded amount of the account balances being audited, without a need to know the number of relevant line items. In computing the precision, an estimate of this number would suffice. It is assumed, however, that the recorded values are all positive. If there are negative book values, they may be assigned to a separate stratum and considered separately through audit stratification.

D. Coping with Slight Bias

The ratio estimate and its standard error are slightly biased, but they are consistent. An estimate is said to be biased if the mean of the sampling distribution of the estimate is not equal to the actual universe value being estimated. The ratio estimate bias is equivalent to the reciprocal of the sample size, and the ratio of the bias in the standard error is of the order of the reciprocal of the square root of the sample size, and thus, becomes negligible for large sample sizes. According to Cochran [8], the bias in practice is usually insignificant even in samples of moderate size. As a working rule, he suggests sample sizes over 30, such that the coefficients of variation (standard deviations divided by means) of the two variables (numerator and denominator) are both less than 10 percent. If the coefficients of variation are unknown, a minimum sample of 100 is suggested.

Deming [12] describes how the detection and correction of an estimation bias can be accomplished through replication. If there is any estimation bias, it involves the reciprocal of the sample size, and hence, decreases rapidly as the sample size increases. For example, two replicates provide two estimates of the ratio: one estimate from the mean of the two individual replicates and another estimate by combining the two replicates. Since Deming favors 10 replicates, separate ratios could be estimated a number of ways, such as the mean of the 10 individual replicates, the mean of two groups of five replicates each, the mean of five groups of two replicates each, and then the mean from combining the 10 replicates. These estimates are then plotted against the reciprocal of the number of replicates, and the result is fitted with a linear-regression line. By extrapolating this line to zero, one obtains an estimate of the ratio which one would obtain if the sample size were increased indefinitely. When the line goes through the origin, the ratio estimate will be unbiased.

Also, a suggestion by Quenouille [52], when applied to two replicates, provides a calculation which is equivalent to an extrapolation of the ratio regression line through zero. Suppose, r denotes the ratio derived from the two replicates combined, and r_1 and r_2 the ratios derived from the two replicates individually, then the

$$\text{Extrapolated ratio} = 2r - \tfrac{1}{2}(r_1 + r_2).$$

In addition, Lahiri [35] derived a sample selection scheme which provides unbiased ratio estimators. The procedure is as follows:

a. Select n sampling units without replacement between 1 and the universe size N. This provides each sample of size n an equal probability of being selected.
b. Sum the sample values for the variable in the ratio denominator.
c. Select a random number k between 1 and T, where T is as large or larger than the largest possible denominator variable of the ratio. The sum of the n largest values of the denominator variable may be used for T.
d. Retain the sample selected in Step a if k falls between 1 and the value of the denominator. Otherwise, reject it, and repeat the steps until the random number k falls between 1 and value of the denominator variable.
e. The first sample retained by this scheme will provide an unbiased estimate of the ratio of the x variable to the y variable.

This scheme requires advanced knowledge of the denominator variable. In many audit applications, the denominator of the ratio estimator consists of book values, which are usually available. Hence, there should be no problem in using the Lahiri scheme to obtain an unbiased ratio estimate in accounting universes.

E. Illustration of Ratio Estimation

In the use of ratio estimation, the auditor uses two sample values (audited amount and the corresponding recorded amount) and one known universe value (total recorded amount) to compute an estimated universe value (total audited amount). To be more specific, for each sample item the auditor observes the recorded value x_j and the corresponding audited value y_j. Then the ratio of the sum of the audited sample amounts to the sum of the corresponding recorded amounts is multiplied by the total universe recorded amount, resulting in a ratio estimate of the total audited amount Y_r in the universe. (The ratio may be differences to recorded amounts; see Figure 32.)

The ratio estimate of the universe audited amount Y_r is computed as follows:

$$Y_r = rX$$

$$\begin{bmatrix}\text{Ratio estimate} \\ \text{of total} \\ \text{audited amount}\end{bmatrix} = \begin{bmatrix}\text{Ratio of sample} \\ \text{audited amounts to} \\ \text{recorded amounts}\end{bmatrix} \times \begin{bmatrix}\text{Universe} \\ \text{recorded} \\ \text{amount}\end{bmatrix}$$

Substituting data from Table 15.1, gives:

$$\text{Ratio estimate of audited total} = \frac{5493}{2729} \times 312{,}089$$

$$= 299{,}233 \text{ (see Figure 31)}$$

Using appropriate ratio formulas in Roberts [57], the standard error of the estimated total audited amount is 4373. Note that this is the same estimate for the RATIO METHOD shown in Figure 32. Hence, the 95 percent two-sided confidence limits are:

$$299{,}233 \pm (2.01)\ (4373)$$

or

$$299{,}233 \pm 8{,}790$$

producing a 95 percent confidence interval from \$290,443 to \$308,023.

Figure 31, under the RATIO METHOD, shows computer computations using the data in Table 15.1 and the ratio of audited amounts to recorded amounts. By contrast, the RATIO METHOD computations in Figure 32 involve the ratio of differences (questioned cost) to recorded values. As mentioned before, it is observed that the values of the standard error and the corresponding precisions are exactly the same for the ratio of audited values to recorded values as for the ratio of differences (questioned cost) to recorded values. This is illustrated in Figures 31 and 32, where the standard error is 4373.19 for both ratio estimates.

F. Separate and Combined Ratios

There are two methods for computing a ratio estimate of the universe total for a stratified sample. One method is to compute a separate ratio estimate for the total of each stratum, and then add these estimates. For this kind of sampling procedure, however, most of the formulas for computing the estimated variance are valid only if the sample in each stratum is large enough to permit the formula to apply separately to each stratum. Otherwise, the cumulative estimation bias may not be negligible relative to the standard error.

The other method of ratio estimation uses a single combined ratio from the various strata. The combined ratio estimate is subject to less bias than the separate ratio estimate. In the combined ratio estimate procedure, the bias relative to the standard error is negligible if the coefficients of variation (standard deviations divided by the means) are less than 10 percent.

Thus, when only a small sample is selected in each stratum, the combined ratio estimate is suggested unless there is a good audit reason for doing otherwise.

G. Other Considerations

McCray [42] cites the following reasons for using the difference and ratio estimators in auditing instead of the mean-per-unit estimator (i) corresponding confidence intervals are usually significantly smaller; (ii) they provide a way of measuring the internal control by focusing on causes of non-zero differences; (iii) the varying degrees of internal control could provide a basis for stratification; and (iv) they permit explicit incorporation of prior information into the estimation procedure by relating reported book amounts to audited amounts. However, since differences cannot be assumed to be normally distributed, McCray contends that an audit sample should contain at least 30 non-zero differences before the difference or ratio estimate is used. This contention apparently assumes that the auditor's decision will be based solely upon the sample result. With other sources of evidence and experience information, a smaller number of non-zero differences may suffice, or at least, provide a basis for stratification, especially if the differences are restricted to a limited operation. For example, if five audited differences are all caused by a single computer programming error, the auditor's actions might be different from those when audited differences are all caused by unrelated keypunching errors. As mentioned before, the types and nature of errors found influence the auditor's actions and decisions.

IX. REGRESSION ESTIMATOR

When book or recorded values are available and their distribution is not very skewed, linear-regression estimation may be used in lieu of or in addition to difference estimation or ratio estimation.

When the correlation between the audited amount and the recorded amount is not zero, the regression estimate is more efficient than the mean-per-unit estimate. In fact, the regression estimate is usually more precise than the mean-per-unit, difference and ratio estimates. Also, the book values do not have to be positive as in the case of the ratio. The precision of the ratio estimate approaches the regression precision only when each audited value is nearly proportional to the corresponding book value. Additionally, when the relationship between x_j and y_j is approximately linear, but the line does not pass through the origin, the regression estimate is more precise than the ratio estimate. Nevertheless, the ratio estimator has a major advantage over the other estimators discussed in this chapter; it provides an estimate of the total audited amount without a need to know the number of audit units in the universe.

Like the ratio estimate, the regression estimate is consistent but slightly biased; however, the bias is trivial for large sample sizes. The tendency towards normality of the sampling distribution of the regression estimate is not only affected by the distribution of book values, but also by the distribution and proportion of non-zero differences. There should be enough non-zero differences to be considered typical or indicative of a trend; however, a single difference may have audit significance.

The linear regression estimator is somewhat similar to the difference estimator except the adjustment is multiplied by b, where b is the regression in the simple regression equation. The linear-regression estimate of the total audited amount Yg is computed from the following relationship:

$$Yg = N\bar{y} + b\ (X - N\bar{x})$$

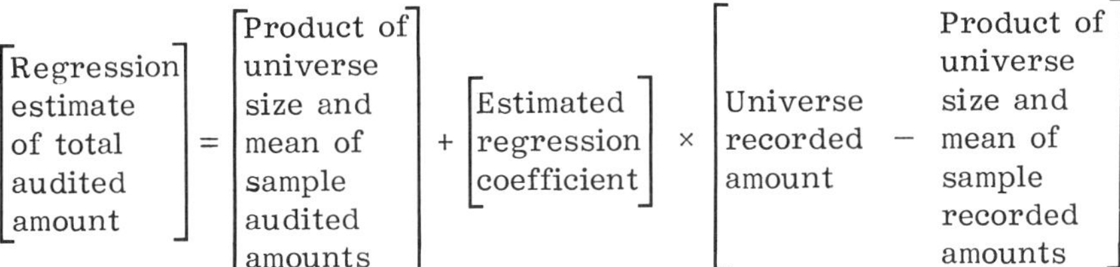

By suitable choices of b in the above relationsip, the mean-per-unit, difference and ratio estimators may be considered particular cases of the regression estimator. For instance, when b is zero, the relationsip becomes the mean-per-unit estimator for the total audited amount; when b is one, it becomes the difference estimator; and when b is $\bar{y}/\bar{x}$, it becomes the ratio estimator.

All the data needed to compute the regression estimate are available from the sample result or the universe records except the estimated regression coefficient b, which is an estimate of the change in y when x changes by unity. The coefficient may be preassigned when repeated previous audits have shown that the sample value of b remains somewhat constant or it may be estimated from the sample data.

In most applications, however, b is estimated from the sample result, using a formula similar to the one found in Roberts [57].

Substituting data from Table 15.1 in the unstratified regression formula in Roberts [57, p. 87] gives an estimated regression coefficient equal to 0.989684.

Using 0.989684 as the regression coefficient b, 2576 as the universe size N, 109.86 as the mean of the sample audited values $\bar{y}$, 312,089 as the total book value X, and 114.58 as the mean of the sample book values $\bar{x}$, the linear regression estimate of the total audited amount Y_g is determined as follows:

$$\begin{aligned} Y_g &= (2576)(109.86) + 0.989684\,[(312{,}089) - (2576)(114.58)] \\ &= 299{,}755 \end{aligned}$$

The necessary computations to determine standard errors and confidence intervals for regression estimates are somewhat complicated and involved. Interested readers may obtain appropriate formulas from Roberts [57]. Also, there are computer programs which perform the necessary computations. Figures 31, 32 and 33 are examples of such a program, using data from Table 15.1. Figure 33 contains estimates based on regression computations.

To compute confidence limits for the linear regression estimate it is necessary to estimate the standard deviation of the differences between sample audited amounts and the "true" linear regression. Afterwards, the remaining procedures are similar to those used for the mean, difference and ratio estimators.

This program evaluates the results of sampling for variables (dollars) with or without stratification, for the regression estimation method, using book values and audited values. First enter the number of strata examined followed by the number of items examined in each stratum. Then enter the following data for each stratum: the total number of items in the stratum, and the book amount and the audited amount for each item in the sample. At the end of each stratum enter '3E33'.

```
THE FILE MAY BE IN ONE OF THE FOLLOWING TWO FORMATS:
      100 DATA 12,3,14,2,18,0,19,0,16,4,3E33
      200 DATA 85,14,95,23,88,0,3E33
      (WITH THIS KIND OF FILE, USE LINE NUMBERS 1 - 1999 ONLY)
                         OR
      100 12 3 14 2 18 0 19 0 16 4 3E33
      200 85 14 95 23 88 0 3E33
      (WITH THIS KIND OF FILE ANY LINE NUMBERS 1 - 99999 MAY BE USED)

IF SUCH A FILE ALREADY EXISTS, ENTER 'GO', ELSE ENTER 'STOP' GO


1DATA1,5Ø
2DATA2576,312Ø89
3DATA483,483,441,392,346,346,57,48,49,Ø,167,167,94,94,4Ø6,4Ø5
4DATA57,57,77,74,12,Ø,134,134,114,114,41,41,73,7Ø,95,95,338,338
5DATA12,12,395,388,12,1Ø,87,79,7,Ø,21Ø,2Ø4,2,Ø,6,6,137,133
6DATA75,75,146,142,2,2,9Ø,85,1Ø9,1Ø9,414,4Ø9,54,54,16,16,44,44
7DATA51,51,46,46,6,6,164,164,32,22,47,47,49,49,326,326,1,1,46,Ø
8DATA5,Ø,4Ø,4Ø,22,22,53,53,39,39
RUN




APPRAISAL OF SAMPLE RESULTS

                                                     REGRESSION
STRATUM 1 OF 1                                         METHOD
 UNIT COST AUDITED                                   1Ø9.86
 REGRESSION COEFFICIENT                              .989684
 TOTAL COST AUDITED                                  299755


 STANDARD DEVIATION                                  11.5175
 STANDARD ERROR OF ESTIMATE                          4154.93
 SAMPLING ERROR OR PRECISION FOR:
  7Ø PC LEVEL--1.Ø48 X STD ERROR                     4354.4
  8Ø PC LEVEL--1.299 X STD ERROR                     5397.3
  9Ø PC LEVEL--1.677 X STD ERROR                     6967.8
  95 PC LEVEL--2.Ø1  X STD ERROR                     8351.4
```

Figure 33. Computer printout of regression evaluation of audit amount.

From Figure 33, it is observed that the 95 percent confidence limits are:

299,755 ± 8351

Thus, producing a 95 percent confidence interval from $291,404 to $308,106.

It is evident from the results that the three auxiliary information estimators exhibit much better precision than the mean-per-unit estimator. However, the interpretation of the results to auditees may be more difficult when complicated procedures are used. On the other hand, there is an intuitive appeal when various estimators provide similar results. In such cases, the choice among estimators becomes secondary.

Those readers who are interested in formulas for stratified and more complex sampling applications should see Roberts [57], Cochran [8] or Deming [12].

X. ATTRIBUTE SAMPLING EVALUATION

In using the flexible sampling strategy (Chapter 13), the auditor desires to estimate the universe occurrence rate. He also wishes to state at a specified confidence level that the universe occurrence rate does not exceed his maximum tolerable occurrence rate. Thus, as used in this book, flexible sampling is concerned with a one-sided confidence level. For example, if a simple random sample of 180 transactions reveals four occurrences of interest, then there is an achieved upper precision limit of 5 percent at the 95 percent one-sided confidence level (see Appendix D). Technically, from the frequency concept of probability, one should not assert from this evaluation that there is a 95 percent assurance that the universe occurrence rate does not exceed 5 percent. The proper assertion would be that if repeatedly simple random samples of size 180 were selected and the upper precision limits computed at the 95 percent one-sided confidence level, about 95 out of 100 of the computer upper limits would exceed the actual universe occurrence rate.

Attribute sampling, as used in this book, involves the estimation of an overall occurrence (error) rate and a related confidence interval. That is, attribute sampling is concerned with a two-sided confidence level, such as 5 percent plus or minus 2 percent, corresponding to an interval from 3 percent to 7 percent, at a specified confidence level. In other words, it is the qualitative counterpart of sampling for variables or dollars. Except for the sampling units and the formulas for the computation of the standard error, all other aspects are the same.

Unfortunately, the attribute sampling strategy requires the auditor to specify an expected occurrence rate in order to determine the sample size. If he overestimates this rate, he will oversample. By contrast, the flexible sampling strategy prevents oversampling by using minimum sample sizes. Thus, no prior estimate of the universe occurrence rate is necessary in flexible sampling. To determine the minimum sample under flexible sampling, the auditor only needs to specify his maximum tolerable occurrence rate, his desired confidence level, and a very rough estimate of the universe size. To be conservative, when he is not sure of the universe size, he can use the table in Appendix D for a universe size over 2,000. Incidentally, the estimates of universe characteristics for both attribute and flexible sampling strategies are computed the same way and the overall upper occurrence limits are the same; however, in flexible sampling the auditor breaks down the overall rate into individual components of sources and causes. The estimated universe occurrence rate of a characteristic is assumed to be the same as the sample rate; that is, the number of occurrences of the characteristic in the sample divided by the sample size.

The flexible sampling strategy, described in Chapter 13, is adequate for most audit compliance testing, and where dollar differences are observed, for substantive testing, too, since both can be estimated from the same sample. There are, however, some limited occasions, such as conducting or updating a survey or in an initial audit of an account, when more precision in the estimation of occurrence rates may be desirable. When these circumstances prevail, the auditor may decide to use the attribute sampling strategy. However, a more precise estimation of an occurrence rate requires more audit effort for the same upper precision limit or maximum tolerable rate. For example, if the audit objective can be satisfied with an upper precision of 6 percent at a specified confidence level, much less effort is required than if the requirement is to squeeze the estimate into a more precise interval such as 5 percent plus or minus 1 percent. In both cases, the upper precision limit (worse probable condition) is a 6 percent occurrence.

For those readers who desire to pursue the attribute sampling strategy, the following discussion is provided; however, a flexible sample may be converted into an attribute sample whenever desirable. Thus, to conserve audit resources, the flexible sampling strategy is suggested as the initial approach even when attribute sampling is an objective.

Usually sampling for attributes is considered prior to discussions of sampling for variables, since the concepts and the computations are easier. On the other hand, the theory of sampling for attributes may be considered a special case of the general sampling theory for variables, where the variable can assume only two values: zero or unity. Thus, the universe may be represented by the symbols "0" to mean "does not possess the characteristic" and "1" to mean "does possess the characteristic."

If the universe proportion of the occurrence of a characteristic is P, then the proportion of the absence of the characteristic is 1-P. Then P is the universe rate of occurrence and the universe variance is P (1-P).

The most efficient unbiased estimator of a universe proportion is the corresponding sample proportion, which is designated p. Also, the standard error of a proportion p is the square root of the variance p (1-p) divided by the sample size less one, or

$$\text{Standard error of proportion} = \sqrt{\frac{(\text{proportion})\ (1\text{-proportion})}{\text{sample size} - 1}}$$

if one ignores the finite universe correction factor. Confidence intervals for the estimate of a universe proportion can be calculated in a similar manner as for the universe mean.

Due to the simplicity of proportions or attributes as compared with variables, there is no need to rely on the large sample convergence to the normal distribution, since the sampling distribution of p for any size sample can be found in tables or calculated. Three such sampling distributions are the binomial, Poisson and hypergeometric. The roles of these distributions in sampling for attributes are discussed in Chapter 14. However, some general summary remarks will serve to relate those discussions to the estimation and evaluation of universe proportions or attributes.

In sampling for attributes, that is, ascertaining the presence or absence of characteristics, the hypergeometric distribution is fundamental to sampling without replacement. In such a procedure, each item drawn into the sample is withheld so that it does not have another opportunity of being selected for the sample. The tables in Appendix C and Appendix D are based on the hypergeometric distribution, and therefore, are useful in sampling without replacement, the usual procedure in auditing.

The binomial distribution is fundamental to the sampling of items with replacement, that is, where each of the items drawn into the sample is returned to the sampling frame after each drawing so that it has an opportunity of being selected again. Thus, the universe remains unchanged for each drawing of a sample item. Sometimes when the sample size is relatively small compared to the size of the universe, the binomial distribution is used to approximate the hypergeometric distribution because of computational simplicity.

When the occurrence rates are small, say less than five percent, and the sample size is large, the Poisson distribution becomes a good approximation of the binomial. The Poisson distribution is used in most monetary-unit sampling procedures as discussed in Chapter 12.

The normal distribution is a good approximation of the hypergeometric, binomial and Poisson distributions for reliability statements when the sample size is not too small and the occurrence rate is not too rare. Both the normal and Poisson distributions are used when feasible because of their relative computational simplicity.

In sampling for attributes, the estimated proportion or occurrence rate P completely specifies the universe, since the universe variance is a function of P, being P (1-P). Also, neither the universe variance nor the sample variance can exceed 0.25. To illustrate, if P equals 0.5, then 1-P also equals 0.5, and the product (0.5 × 0.5) equals 0.25. Any P less than or greater than 0.5 will have a variance less than 0.25. This also controls the maximum value of the standard error for attributes, since the variance of the sampling distribution is equal to the square of the standard error.

To avoid oversampling when using attribute sampling, it is suggested that a preliminary sample, at least the size of a minimum sample, be selected as in flexible sampling (Chapter 13), and then be evaluated using a formula similar to the one used in this section or a computer program to evaluate the results with a two-sided confidence level. If the desired precision is not achieved, use a computer program to estimate the additional sampling required or use the following rule-of-thumb:

$$\text{Total sample} = \left[\frac{\text{Achieved precision}}{\text{Desired precision}}\right]^2 \times \text{initial sample size.}$$

Suppose, for illustrative purposes, an initial sample of 100 is selected and the error rate is 5 percent with an achieved precision of ±4 percent when ±2 percent is desired at the same confidence level. Using the above relationship, one gets:

$$\text{Total sample} = \left(\frac{4}{2}\right)^2 \times 100 = 400.$$

Since 100 items have already been examined, the additional sample would be 300. If the supplementary sampling is large, replication should be considered with a cumulative evaluation after each replication to avoid oversampling.

To illustrate the computation of confidence limits and intervals for attribute sampling, suppose a simple random sample of 100 accounts is selected from a universe of 10,000 acounts, and 6 accounts out of the 100 are in error, resulting in a sample error rate of 0.06 or 6 percent. Since the universe error rate is not likely to be exactly 6 percent, the auditor uses the concept of confidence limits. Instead of saying that the error rate for all accounts is 6 percent, he can compute a range or interval of probable error rates.

The relationship for calculating the estimated standard error Sp of a proportion (or attribute) is as follows:

$$Sp = \sqrt{\frac{N-n}{N\ (n-1)}\ p\ (1-p)}$$

$$\text{Estimated standard error} = \sqrt{\frac{\begin{bmatrix}\text{(Universe size} - \\ \text{sample size)}\end{bmatrix}\begin{bmatrix}\text{Sample} \\ \text{error rate}\end{bmatrix}[1 - \text{Sample error rate}]}{\text{(Universe size) (Sample size} - 1)}}$$

Substituting the above data in this relationship and performing the indicated mathematical operations result in the following:

$$\text{Estimated standard error} = \sqrt{\frac{(10{,}000 - 100)\ (0.06)\ (1 - 0.06)}{(10.000)\ (100 - 1)}}$$

$$= 0.023748$$

After the standard error is obtained, the computation of confidence limits is easy and straightforward, as follows, using the sample example at the 90 percent two-sided confidence level for a simple random sample of 100:

90% LOWER CONFIDENCE LIMIT

$$\begin{bmatrix}\text{Sample} \\ \text{error} \\ \text{rate}\end{bmatrix} - \begin{bmatrix}\text{Confidence} \\ \text{factor, 90\%}\end{bmatrix} \times \begin{bmatrix}\text{Standard} \\ \text{error}\end{bmatrix}$$

or 0.06 − [(1.66) (0.023748)]

resulting in 0.06 − 0.039

or 0.021 or 2.1%

90% UPPER CONFIDENCE LIMIT

$$\begin{bmatrix}\text{Sample} \\ \text{error} \\ \text{rate}\end{bmatrix} + \begin{bmatrix}\text{Confidence} \\ \text{factor, 90\%}\end{bmatrix} \times \begin{bmatrix}\text{Standard} \\ \text{error}\end{bmatrix}$$

or 0.06 + [(1.66) (0.023748)]

resulting in 0.06 + 0.039

or 0.099 or 9.9%

Based on the above upper and lower confidence limits, there is only one chance in 10 that the universe error rate is less than 2.1 percent or more than 9.9 percent. Also, the range from 2.1 percent to 9.9 percent is the 90 percent confidence interval. Other confidence limits and intervals are computed in a similar manner, using the desired confidence coefficient (factor) for the specified two-sided confidence level.

However, in cases where both the occurrence rate and the sampling size are small, care should be exercised, since the means of the underlying sampling distributions may deviate from normality.

Although the computation of confidence intervals for attributes is relatively simple compared to those for variables, there are computer programs to perform the computations. However, flexible sampling is preferable. The above manual computations serve to clarify conceptional and procedural considerations, and do not have to be performed in actual audit practice.

The usual manner of improving the precision of the result (reducing the confidence interval) is through stratification and increasing the size of the sample. The former is usually more efficient than the latter, since the improvement is only in terms of the square root of the increased sample size.

The precision may also be improved by using a one-sided confidence interval in lieu of a two-sided interval, such as those in Appendix D, and the upper precision limits for flexible sampling discussed in Chapter 13.

It is important that the auditor understands the statistical significance of occurrence or error rates in compliance testing. As explained in Chapter 13, an overall error rate for an accounting universe is usually not very useful in error analysis and audit appraisal. Also, a statistical sample should not be the primary source of reliance on internal controls. However, it is additional corroborating evidence which should be consistent with the results of all other auditing procedures involved in the evaluation of internal controls. In addition, there is no necessary correlation between error rates and monetary impact. There could be very few errors but each monetarily significant. By contrast, there could be many errors which reflect no monetary impact, such as in some inventory records.

XI. EVALUATION OF REPLICATED SAMPLES

Replicated sampling, described in Chapter 13, is the procedure of selecting from the entire universe two or more independent random samples of the same or about the same size. Replication provides an easy way to estimate and evaluate both variables and attributes, such as means, aggregates, ratios, frequencies, proportions, and so on.

Table 15.2. Summary Data for Replication Example

Employees in area k		Random minutes (replicates) 1	2	3	4	Cents per minute
Engaged in activity A	1	1	0	0	1	5
	2	0	1	1	0	6
	3	0	0	0	1	5
	4	0	1	0	0	7
	.					
	.					
	.					
	12	0	0	0	0	
	Sum	1	2	1	2	
Estimated portion of 12 employees		1/12	2/12	1/12	2/12	
Estimated dollars w = 480 per replicate		\$24.00	\$62.40	\$28.80	\$48.00	

Replicates provide a simple and direct method of calculating standard errors of any kind of estimate. Also, as described previously, the procedure can be used to detect and localize abnormal errors where an unusual frequency or value occurs in a replicate and to adjust for bias in ratio estimations.

To illustrate the estimation and evaluation of an attribute Pa and a dollar amount Ya using replication, the following work sampling exercise from Rosander [58] is used. Those readers desiring more details should consult this reference.

The objective is to obtain estimates of the proportion of work and total dollar wages devoted to an activity A, and to compute standard errors of these estimates. Activity data and the wage rate in cents per minute are recorded for individual employees.

The four random minutes shown in Table 15.2, being selected independently, constitute the four replicates. Obviously, more replicates could be used. To simplify computations, only one randomly selected work area k is used; similar computations would be made for each randomly selected area.

Using data from Table 15.2, the estimates and their standard errors are computed as follows, using the range method:

a. For proportion (attribute) estimate:

$$P_a = \frac{\text{Activity A minutes for area } k \text{ employees}}{(\text{Number of replicates}) \times (\text{number of employees in activity})}$$

$$= \frac{6}{4 \times 12} = 0.125 \text{ or } 12.5 \text{ percent}$$

$$\text{Estimated standard error } P_a = [p_{max} - p_{min} \div m$$

$$= \left[\begin{matrix}\text{Largest estimate} \\ \text{from a replicate}\end{matrix} - \begin{matrix}\text{Smallest estimate} \\ \text{from a replicate}\end{matrix}\right] \div \left[\begin{matrix}\text{Number} \\ \text{of} \\ \text{replicates}\end{matrix}\right]$$

$$= \left[\frac{2}{12} - \frac{1}{12}\right] \div 4$$

$$= 0.02083 \text{ or } 2.1 \text{ percent}$$

This estimate of the standard error of P_a could be used to calculate confidence intervals as in that described for attribute sampling in the preceding section.

b. For dollar (variable) estimate:

$$Y_a = \left[\frac{\text{Minutes per work day}}{\text{Number of replicates}}\right] \times \left[\begin{matrix}\text{Sum of products cents} \\ \text{per minute and the} \\ \text{number of employee minutes}\end{matrix}\right]$$

$$= \frac{480}{4} [0.05\ (3) + 0.06\ (2) + 0.07\ (1)]$$

$$= \$40.80$$

$$\text{Estimated standard error } Y_a = [y_{max} - y_{min}] \div m$$

$$= \frac{\$62.40 - \$24.00}{4}$$

$$= \$9.60$$

For the same sample data, using the method of sums of squares, the estimates are as follows:

a. For proportion (attribute):

$$P_a = \frac{\text{Number of sample occurrences for activity A}}{\text{Total number of sample occurrences}}$$

$$= \frac{6}{48} = 0.125 \text{ or } 12.5 \text{ percent}$$

The following relationship is used to estimate the proportion variance S^2:

$$S^2 = \frac{1}{m\,(m-1)}\,\Sigma\,(p_j - p)^2$$

$$= \left[\frac{1}{\text{Number of replicates} \times \text{Number of replicates minus one}}\right] \times \left[\begin{array}{l}\text{Sum of the squares of}\\ \text{the differences between}\\ \text{individual replicate}\\ \text{proportions and overall}\\ \text{sample proportion}\end{array}\right]$$

Substituting in this relationship produces:

$$\text{Estimated proportion variance} = \frac{1}{4 \times 3}\,\Sigma\,\left(p_j - \frac{1}{8}\right)^2, \text{ where } p_j \text{ is estimate from } j\text{th replicate}$$

$$= 0.005786$$

$$\text{Estimated standard error} = \sqrt{0.00578} = 0.024 \text{ or } 2.4 \text{ percent.}$$

b. For the dollar aggregate estimate, the following relationship is used to compute the variance using sum of the squares method:

$$S^2 = \frac{1}{m\,(m-1)}\,\Sigma\,(y_j - \bar{y})^2$$

$$= \left[\frac{1}{\text{Number of replicates} \times \text{Number of replicates minus one}}\right] \times \left[\begin{array}{l}\text{Sum of the squares of}\\ \text{the differences between}\\ \text{individual replicate}\\ \text{estimates and the mean}\\ \text{estimate for the sample}\end{array}\right]$$

Using $\bar{y} = 40.80$, computed above, and summation over the four replicates results in:

$$S^2 = \frac{1}{4 \times 3} \Sigma (y_j - 40.80)^2 = 78.72$$

Thus, standard error $S = \sqrt{78.72} = \$8.87$.

Comparing the two methods of estimating the standard error of the proportion (attribute) of 12.5 percent, the range method gives 2.1 percent and the sum of the squares method 2.4 percent. For the dollar estimate (variable) of $40.80, the range method gives an estimated standard error of $9.60 while the sum of the square gives $8.87.

It should be obvious from this simple exercise that the range method of replication is a much quicker and easier way of calculating standard errors than the sum of the squares method. Also, the range method gives an unbiased estimate of universe characteristics and values. A computer program can calculate both methods quickly.

XII. RESEARCH ON AUDIT SAMPLING PROCEDURES

Many researchers including Stringer [66], Stephan [64], Meikle [44], Anderson and Teitlebaum [1], Kaplan [27], Neter and Loebbecke [47], Jones [26] and Cox and Snall [9] have recognized that classical survey sampling procedures are inadequate for applying statistical sampling to auditing, since they do not provide for the special structure of accounting universes nor the unique environment in which audit sampling occurs. Also, the auditor usually has much more information about his universe than is available to those performing survey samples. Kaplan and Cox and Snell suggest that the auditor should use statistical estimators which explicitly use all the available auxiliary information. Among such estimators are the difference, ratio and regression estimators. According to Kaplan the ratio and regression estimators perform as well as any auxiliary information estimators. Felix and Grimlund [14], however, indicate that the low error rates in most accounting universes cause auxiliary estimators—difference, ratio and regression—to suffer from a lack of normal distribution robustness of the sampling distribution. For this reason Roberts [57] suggests that there should be a minimum of 15 or 20 differences (errors) in an audit sample before the difference or ratio estimator is used. McCray [42] believes that there should be at least 30 non-zero differences since one cannot assume that the differences are normally distributed. Nevertheless, he cites several reasons for using the ratio and difference estimators in auditing in lieu of the mean-per-unit estimator; namely, (i) the corresponding confidence intervals are usually significantly smaller; (ii) the ratios and differences provide a means of measuring the extent of internal control by focusing on causes of differences; (iii) the different degrees of internal control could provide a basis for stratification; and (iv) they explicitly incorporate prior information into the estimation process by relating book values to audited values.

Unfortunately, if no differences (errors) are found in the sample, confidence intervals cannot be computed using these auxiliary information estimators. There are, however, other techniques for computing upper confidence limits of the total value of monetary errors as well as the occurrence rates for error-free samples. For upper limits on the occurrence rates, flexible sampling and tables in Appendix D can be used; for upper limits on total dollar errors, dollar-unit sampling may be used.

A study by Baker and Copeland [4], a supplement to the Neter and Loebbecke AICPA study, reveals that the precision of the regression estimator, in general, tends to be almost the same as the precision of the stratified difference and ratio estimates. Similar to the difference and ratio estimators, the reliability of the regression confidence coefficient is poor at low error rates.

A study by Beck [5] also supplements the Neter and Loebbecke AICPA study by examining the behavior of the stratified and unstratified regression estimators for the same accounting universes. The results were similar to those reported for the difference and ratio estimators in the original Neter and Loebbecke study.

Leitch, Neter, Plante and Sinka [36] present a modification of the multinomial bound which enables the auditor to obtain bounds for larger numbers of errors in audit samples than the original basic technique. The modification consists of clustering taintings found in the sample and obtaining a conservative bound by assuming that all taintings in a cluster are as large as the largest tainting in the cluster. This modified multinomial bound is usually tighter than the Stringer bound used in most dollar-unit sampling applications.

Loebbecke and Neter [38] suggest that the sampling procedures used in a particular audit situation reflect considerations of the audit objectives and anticipated environmental factors. Some of the considerations in choosing appropriate procedures should include provisions for sample enlargement, nature of the sampling frame and bias of the audit performance. The authors also suggest that provisions be made for a fall-back procedure in case the expected environmental factors differ from those actually encountered.

Research will continue in search for more efficient estimators which exploit all the unique auxiliary information provided by accounting universes, such as the following:

a. Information provided by a particular selected sampling unit will often furnish information about another sampling unit whether or not the other sampling unit is selected. For instance, an error in one sample account will often provide information that another account is also in error.
b. Some situations are such that it may be possible to estimate correct dollar balances so that the error in estimating one balance will be offset entirely or in part by an error in estimating another balance.

Undoubtedly, future research will improve the state-of-the-art by providing more efficient and effective audit sampling procedures. Meanwhile, the practical aspects of audit sampling, as discussed in this book, should not change appreciably since they reflect the integration of the auditor's judgment and knowledge with sample information and other audit evidence in the formulation of an informed, professional opinion, and therefore, are in consonance with provisions of the American Institute of Certified Public Accountants SAS No. 39, *Audit Sampling* [3].

XIII. SUMMARY

It is necessary not only to seek the most efficient sample selection procedure, but to select an effective and efficient estimating procedure to accomplish a specific audit objective.

Inherent in every statistical sample design is at least one method of estimation which provides an estimate of some universe characteristic or value. Every statistical estimate has a mean and standard error (the standard deviation of the underlying sampling distribution). Different estimators can be used to estimate the same universe characteristic or value, such as proportions, means, totals, differences, ratios and regressions; however, each has a different precision.

The mean-per-unit estimator provides an estimate of the total audited amount in the universe by multiplying the mean of audited values of the sampled units by the number of units in the universe or stratum. Thus, in using the mean-per-unit estimator, one needs only the audited value for each sampled unit; therefore, the procedure ignores information about the recorded values of the sampled units. Utilizing information about recorded values provides more efficient estimators. Such a class of estimators is referred to as auxiliary information estimators.

Auxiliary information estimators, such as difference, ratio and regression estimators, provide alternative methods of estimating universe characteristics and values. They are designated as auxiliary information estimators because they utilize auxiliary information in computing universe estimates. In auditing, the auxiliary information is the recorded value of audit units, which is used in estimating the universe audited value. In other words, auxiliary information estimators use the relationship between the audited values and the recorded values of sampled transactions to estimate universe characteristics and values.

The difference estimator utilizes the mean difference between the audited amounts and the corresponding recorded amounts in the sample to estimate the total audited amount in the universe. The ratio estimator uses the sample ratio of audited amounts to recorded amounts to estimate the total universe audited amount. The regression estimator uses linear regression techniques to compute the relationship between audited and recorded amounts in the sample, and then utilize the relationship to estimate the total audited amount in the universe.

If it is possible to relate the universe characteristic to be estimated to another universe characteristic for which the value is known and with which it varies proportionately (that is, a high value of the characteristic occurs with a high value of another characteristic and a low value of the characteristic occurs with a low value of the other), then the relationship between the two characteristics can be used to improve the sampling precision through ratio estimation. Thus, the ratio estimate is of considerable value in estimating the aggregate of some universe characteristic, such as audited value, when there is a high correlation between this characteristic and another universe characteristic, such as recorded values, and the aggregate of the recorded values is known. Also, there are occasions when the ratio of two variables is an end objective in itself, such as in comparative analysis, studies of interrelationships, trend analyses, and managerial indicators of performance.

In deciding on the best estimator to use, a graph in which each audited value y_i is plotted against the corresponding recorded value x_i is helpful. For instance, the ratio estimate is a best linear unbiased estimate if (i) the relationship of the audited and recorded values is a straight line through the origin and (ii) if the variance of the vertical audited values about the line seems to increase proportionally to the recorded values on the horizontal axis.

The comparative assumptions of the three auxiliary information estimators are that the ratio estimator assumes that the sample produces a constant percentage difference (error); the difference estimator assumes that the sample produces a constant absolute difference (error); and the linear regression estimator assumes that the sample produces some of both types of differences.

However, with the availability of computer programs to perform the necessary computations, it is suggested that the auditor project his sample result to the universe using the three auxiliary estimators; ratio, difference and linear regression. If the three estimators are close, he has additional intuitive support for the use of the results. Otherwise, he should perform more investigation, analysis and appraisal to ascertain the reasons for unusual differences.

A basic requirement of a statistical estimator is a measure of its precision, often in the form of confidence intervals for the estimates of universe characteristics. Such a precision, or confidence, interval may be two-sided, such as $100,000 plus and minus $10,000 for an interval of $90,000 to $110,000. However, for the same precision, a one-sided interval has a higher reliability than the corresponding two-sided confidence interval.

Confidence limits can be computed for universe estimates based on the Central Limit tendency of the distribution of sample means to approach the normal distribution as the sample sizes increase. This tendency is usually so strong that moderate departure from normality in the universe distribution will not materially affect the reliability of the universe estimates, provided there is adequate sampling and effective stratification. This tendency is also present in universes consisting largely of zero differences, but the approach to normality is too slow to be considered reliable until the sample is large enough to include a minimum of 10 non-zero differences, and preferably at least 30.

If a distribution is approximately normal, one can estimate from a sample both the mean and the standard deviation (standard error). With the standard error, one can calculate the range or interval about the mean which will contain any specified proportion of the units in the distribution.

In making inferences from a sample to the universe, precision and confidence level are interrelated competitors; the higher the confidence level for the evaluation of an estimate, the less the precision, and vice versa.

In the choice of an estimator, the auditor should consider not only the precision, but also simplicity, cost and other audit and administrative considerations. Obviously, a simple procedure is easier to apply and explain. Also, the cost of deriving rigorous or more precise estimates may negate their cost effectiveness. Another important consideration is the availability and efficiency of packaged computer programs to perform the computational chores.

An estimate made from a sample is subject to random and nonrandom variations. The variation due to random selection is expressed by the standard error, which is an objective measure and not a matter of opinion. The standard error measures how much sample means or other statistics vary because of chance differences arising from the random selection process. That is, the standard error of the mean gives a measure of variation among repeated samples of the same size drawn in the same manner from a universe.

It is important to realize that the standard error does not measure all the audit risks which occur in sampling. It measures only the difference associated with the particular sampling procedure. In audit sampling applications, there are also nonsampling risks arising from defects in the sampling frame, erroneous criteria, mistakes in the collection of data, and mistakes in processing and analyzing the results. When the nonsampling risks are large, the total risk will be reduced only slightly by decreasing the sampling risk in selecting larger samples. Under these circumstances, it is better to use minimum samples and utilize the resources saved to reduce the nonsampling risk. In other words, if the methods of obtaining information or of testing or of appraising findings are not satisfactory, no sample, not even a 100 percent examination, will provide valid and useful audit information.

In summary, among the three auxiliary estimators discussed, the difference estimator is most efficient when the audited amount is not proportional to the recorded amount. When the audited amount is nearly proportional to the recorded amount, the ratio estimator is most efficient. When the relationship between the audited amount and recorded amount is a mixture or is not obvious, the regression estimator is usually most efficient. These estimators, like the mean-per-unit rely on the normal distributional theory and a minimum frequency of non-zero differences between audited and recorded amounts, but they tend to be less reliable when the errors are primarily in one direction, such as all overstatements which emphasize skewness.

Although dollar-unit sampling is discussed in Chapter 12, its primary features are summarized here for comparative purposes. In dollar-unit sampling the universe consists of individual dollar units rather than audit units. The selection procedure automatically stratifies because larger book value transactions have a greater chance of being included in the sample. After the sample dollar units have been audited, attribute theory is used for estimating an upper universe monetary error limit based on the mean tainting ratio in the sample. Dollar-unit sampling relies neither upon the normal distributional theory nor a minimum error rate. However, it tends to become more conservative and less efficient as the error rate increases. It is used most often in estimations involving only overstatement errors.

In audit sampling, an important objective is to select and analyze the sample data in such a manner that they may be used to compute valid confidence limits, where valid indicates that the calculated statistical risk is substantially correct or not materially understated. In some sample surveys, there is no concern about validity because the methods are sufficiently robust to handle most situations. However, the usual classical sampling methods generally are not sufficiently robust for the computation of valid confidence limits for the total dollar error in skewed accounting universes consisting of very few monetary errors. In other words, the departure from normality in the distribution of the universe may be so severe that relying on the Central Limit tendency may not lead to valid confidence limits unless the sample is extremely large or effectively stratified. Much research is being directed towards a solution of this problem. Meanwhile, many auditors resort to a combination of flexible and dollar-unit sampling. Obviously, if the non-zero differences are not rare and the underlying distribution is not very skewed, auxiliary information estimators provide valid and useful estimates of universe characteristics and values if the sample is large enough.

References

1. Anderson, R. and A. Teitlebaum. "Dollar-Unit Sampling: A Solution to the Audit Sampling Dilemma," *CA Magazine*, April 1973.
2. Arkin, H. "Discovery Sampling in Auditing," *Journal of Accountancy*, February 1961.
3. *Audit Sampling*, Statement of Auditing Standards (SAS) No. 39, New York: American Institute of Certified Public Accountants, 1981.
4. Baker, R. and R. Copeland. "Evaluation of the Stratified Regression Estimator for Auditing Accounting Populations," *Journal of Accounting Research*, Autumn 1979.
5. Beck, P. "A Critical Analysis of the Regression Estimator in Audit Sampling," *Journal of Accounting Research*, Spring 1980.
6. Blalock, H. *Social Statistics*. New York: McGraw-Hill, 1972.
7. Brisley, C. "How You Can Put Work Sampling to Work," *Factory Management and Maintenance*, July 1952.
8. Cochran, W. *Sampling Techniques*, 3rd ed. New York: John Wiley & Sons, 1977.
9. Cox, D. and E. Snell. "On Sampling and the Estimation of Rare Errors," *Biometrika*, 1979.
10. Coxton, F., D. Cowden and B. Bolch. *Practical Business Statistics*, 4th ed. Englewood Cliffs, N.J.: Prentice-Hall, 1969.

11. Deming, W. *Some Theory of Sampling*. New York: John Wiley & Sons, 1950.
12. Deming, W. *Sample Design in Business Research*. New York: John Wiley & Sons, 1960.
13. Elliott, R. and J. Rogers. "Relating Statistical Sampling to Audit Objectives," *Journal of Accountancy*, July 1972.
14. Felix, W. and R. Grimlund. "A Sampling Model for Audit Tests of Composite Accounts," *Journal of Accounting Research*, Spring 1977.
15. Feller, W. *An Introduction to Probability Theory and Its Applications*, 3rd ed. New York: John Wiley & Sons, 1968.
16. Fienberg, S., J. Neter and R. Leitch. "Estimating the Total Overstatement Error in Accounting Populations," *Journal of the American Statistical Association*, June 1977.
17. Fisher, R. and F. Yates. *Statistical Tables for Use in Biological, Agricultural and Medical Research*. Edinburgh: Oliver and Boyd, 1938.
18. Garstka, S. "Models for Computing Upper Error Limits in Dollar-Unit Sampling," *Journal of Accounting Research*, Autumn 1977.
19. Goodfellow, L., J. Loebbecke and J. Neter. "Perspectives on CAV Sampling Plans," *CA Magazine*, Part I, October 1974, and Part II, November 1974.
20. Hansen, M., W. Hurwitz and W. Madow. *Sample Survey Methods and Theory*, Vols. 1 and 2. New York: John Wiley & Sons, 1953.
21. Haack, D. *Statistical Literacy—A Guide to Interpretation*. North Scituate, Massachusetts: Duxbury Press, 1979.
22. Hogg, V. and A. Craig. *Introduction to Mathematical Statistics*, 2nd ed. New York: Macmillian, 1965.
23. Huntsberger, D. *Elements of Statistical Inference*, 2nd ed. Boston: Allyn and Bacon, 1967.
24. Huntsberger, D., P. Billingsley and D. Croft. *Statistical Inference for Management and Economics*. Boston: Allyn and Bacon, 1975.
25. Ijiri, Y. and R. Kaplan. "The Four Roles of Sampling in Auditing: Representative, Corrective, Protective, and Preventive," Carnegie-Mellon University Working Paper WP64-68-9, June 1969.
26. Jones, H. "Some Problems in Sample Audits," Unpublished paper, 1976.
27. Kaplan, R. "Statistical Sampling in Auditing with Auxiliary Information Estimators," *Journal of Accounting Research*, Autumn 1973.
28. Kaufman, S. "Statistical Sampling in Bank Auditing," *Auditgram*, February 1965.

29. Kendall, M. and B. Smith. *Tables of Random Sampling Numbers*, Tracts for Computers No. 24. Cambridge: Cambridge University Press, 1939.
30. Kennedy, J. "Work Sampling Methods in Accounting," Unpublished paper presented at the Annual Meeting of American Statistical Association, August 11-14, 1980.
31. Kinney, W. "A Decision—Theory Approach to the Sampling Problem in Auditing," *Journal of Accounting Research*, Spring 1975.
32. Kish, L. *Survey Sampling*, New York: John Wiley & Sons, 1965.
33. Kraft, W. "Statistical Sampling for Auditors: A New Look," *Journal of Accountancy*, August 1968.
34. Kyburg, H. and H. Smokler. ed. *Studies in Subjective Probability*. New York: John Wiley & Sons, 1964.
35. Lahiri, D. "A Method for Sample Selection Providing Unbiased Ratio Estimates," *International Statistical Institute Bulletin*, Vol. 33/2, 1951.
36. Leitch, R., J. Neter, R. Plante and P. Sinka. "Modified Multinomial Bounds for Larger Numbers of Errors in Audits," Research Study 80-010, The University of Georgia, 1981.
37. Leslie, D., A. Teitlebaum and R. Anderson. *Dollar-Unit Sampling—A Practical Guide for Auditors*. Toronto: Copp Clark Pitman, 1979.
38. Loebbecke, J. and J. Neter. "Considerations in Choosing Statistical Sampling Procedures in Auditing," *Journal of Accounting Research*, Vol. 13 Supplement, 1975.
39. Madow, W. and L. Madow. "On The Theory of Systematic Sampling," *Annals of Mathematical Statistics*, Vol. 15, March 1944.
40. Mahalanobis, P. "Recent Experiments in Statistical Sampling in the Indian Statistical Institute," *Journal of Royal Statistical Society*, Vol. 109, 1946.
41. Mandel, B. "Work Sampling in Financial Management-Cost Distribution in Post Office Department," *Management Science*, February 1971.
42. McCray, J. "Ratio and Difference Estimation in Auditing," *Management Accounting*, Vol. 55, December 1973.
43. McMurdo, C. "Work Sampling for Distributing Machine and Clerical Time," *New Frontiers in Administrative and Engineering Quality Control*, American Society for Quality Control, 1962.
44. Meikle, G. *Statistical Sampling in an Audit Context*. Toronto: The Canadian Institute of Chartered Accountants, 1972.
45. Mosteller, F., R. Rourke and G. Thomas. *Probability With Statistical Applications*, 2nd ed. Reading Mass.: Addison-Wesley, 1970.

46. Neter, J. and W. Wasserman. *Fundamental Statistics for Business and Economics*, 2nd ed. Boston: Allyn and Bacon, 1961.
47. Neter, J. and J. Loebbecke. *Behavior of Major Statistical Estimators in Sampling Accounting Populations—An Empirical Study*. New York: American Institute of Certified Public Accountants, 1975.
48. Neter, J., R. Leitch and S. Fienberg. "Dollar Unit Sampling: Multinomial Bounds for Total Overstatement and Understatement Errors," *Accounting Review*, January 1978.
49. Newmark, J. *Statistics and Probability in Modern Life*. San Francisco: Holt, Rinehart and Winston, 1975.
50. Neyman, J. "On the Two Different Aspects of the Representative Method: The Method of Stratified Sampling and the Method of Purposive Selection," *Journal Royal Statistical Society*, Vol. 97, 1934.
51. Parish, W. "Work Sampling: Use With Caution," Unpublished paper presented at the 1979 Annual Meeting, American Statistical Association, August 13-16, 1979.
52. Quenouille, M. "Notes on Bias Estimation," *Biometrika*, Vol. 43, 1956.
53. RAND Corporation, *A Million Random Digits With 100,000 Normal Deviates*. Glencoe, Illinois: Free Press, 1955.
54. Ray, W. *Robust Statistical Methods*. New York: Springer-Verlog, 1978.
55. Reneau, J. "CAV Bounds in Dollar Unit Sampling: Some Simulation Results," *Accounting Review*, July 1978.
56. Richardson, W. "Work Sampling Today," *Factory Management and Maintenance*, September 1959.
57. Roberts, D. *Statistical Auditing*. New York: American Institute of Certified Public Accountants, 1978.
58. Rosander, A. *Case Studies in Sample Design*. New York: Marcel Dekker, 1977.
59. Rosander, A., H. Guterman and A. McKeon. "The Use of Random Work Sampling for Cost Analysis and Control," *Journal of the American Statistical Association*, June 1958.
60. Schlaifer, R. *Probability and Statistics for Business Decisions*. New York: McGraw-Hill, 1959.
61. Schlaifer, R. *Analysis of Decisions Under Uncertainty*. New York: McGraw-Hill, 1969.
62. Snedecor, G. and W. Cochran. *Statistical Methods*, 7th ed. Ames, Iowa: Iowa State University Press, 1980.
63. Springer, C., R. Herliky, R. Mall and R. Beggs. *Statistical Inference*. Homewood, Illinois: Richard D. Irwin, 1966.

64. Stephan, F. "Some Statistical Problems Involving Auditing and Inspection," *Proceedings of the American Statistical Association*, 1963.
65. Stevens, W. *Table of 105,000 Random Decimal Digits*. Washington, D.C.: Interstate Commerce Commission, 1949.
66. Stringer, K. "Practical Aspects of Statistical Sampling in Auditing," *Proceedings of Business and Economic Statistics Section*, American Statistical Association, 1963.
67. Sudakar, K. A Note On "Circular Systematic Sampling Design," *Sankhya, The Indian Journal of Statistics*, Series C, Vol. 40, Part 1, March 1978.
68. Taylor, R. "Error Analysis in Audit Tests," *Journal of Accountancy*, May 1974.
69. Teitlebaum, A. "Dollar-Unit Sampling in Auditing," Faculty of Management Working Paper, McGill University, and presented at the Annual Meeting, American Statistical Association, 1973.
70. Teitlebaum, A., D. Leslie and R. Anderson. "An Analysis of Recent Commentary on Dollar-Unit Sampling in Auditing," Faculty of Management Working Paper, McGill University, March 1975.
71. Teitlebaum, A. and C. Robinson. "The Real Risk in Audit Sampling," *Journal of Accounting Research*, Vol. 13 Supplement, 1975.
72. Teitlebaum, A., J. McCray and D. Leslie. "Approaches to Evaluating Dollar-Unit Samples," Faculty of Management Working Paper, McGill University, and presented at the Annual Meeting, American Accounting Association, 1978.
73. Teitlebaum, L. and M. Schwartz. "Practical Improvements in Audit Testing," *Internal Auditor*, September 1958.
74. Tippett, L. *Random Sampling Numbers*, Tracts for Computers No. 15. Cambridge: Cambridge University Press, 1927.
75. Tippett, L. "Statistical Methods in Textile Research, Uses of the Binomial and Poisson Distributions. A Snap-Reading Method of Making Time Studies of Machines and Operatives in Factor Surveys," *Journal of the Textile Institute Transactions*, February 1935.
76. Tracy, J. "Bayesian Statistical Methods in Auditing," *Accounting Review*, January 1969.
77. Trueblood, R. and R. Cyert. *Sampling Techniques in Accounting*. Englewood Cliffs, New Jersey: Prentice-Hall, 1957.
78. U.S. Air Force Auditor General. *Table of Probabilities for Use in Exploratory Sampling*, Washington, D.C.: Department of the Air Force, September 1959.

79. U.S. Air Force Auditor General. *Tables of Probabilities for Use in Stop-or-Go Sampling*, Washington, D.C.: Department of Air Force, August 1961.
80. Vance, L. and J. Neter. *Statistical Sampling for Auditors and Accountants*. New York: John Wiley & Sons, 1956.
81. Van Heerden, A. "Steekproeven als middel van accountantscontrole" (Statistical Sampling as a Means of Auditing), *Maandblad voor Accountancy* en Bedryfshuish aulkunde, 1961.
82. Wald, A. *Sequential Analysis*. John Wiley & Sons, New York, 1947.
83. Weinberg, G. and J. Schumaker. *Statistics, An Intuitive Approach*. Belmont, California: Wadsworth, 1962.

Appendix A: Glossary of Terms

Acceptance Sampling—A specialized sampling scheme developed for use with production line items such as industrial products. Acceptance sampling involves repeated sampling from larger lots or groups and based on criteria established in advance, a lot or group is either accepted or rejected. The proper application of acceptance sampling techniques and the establishment of a meaningful decision making rule requires accurate information about the proportions of a given characteristic (such as defects) in the universe. Acceptance or rejection of the entire lot or group depends upon the number of defective units found in the sample. This sampling technique is not recommended for use in substantive testings.

Aggregate—This is the total of a series of quantities in a universe. When using information from a sample, the aggregate for the universe from which the sample was selected may be estimated by multiplying the mean of the sample by the number of items in the universe.

Algorithm—A set of ordered procedures, steps or rules, usually applied to mathematical procedures, and expected to lead to the solution of a problem in a finite number of steps.

Arithmetic Mean—See mean.

Array—Arrangement of numerical values in sequence from lowest to highest, or vice versa.

Assurance—Audit synonym for confidence or reliability. The overall audit assurance is derived from inherent assurance (related to the nature of the items in question), assurance derived from internal control, and assurance from other auditing procedures, and analytical review and sampling assurance.

Attribute—A categorical descriptive characteristic of a class of items or observations expressed in percentages or proportions. For example, percentage of correct ledger entries, proportion of correct inventory counts, and so on.

Attributes Sampling—It is a specialized form of sampling for variables where values are either 0 or 1. An auditor may test a sampling frame either for qualitative feature or a quantitative element, or both. When testing primarily for qualitative characteristics, the process is commonly referred to as sampling for attributes. On the other hand, when the primary objective is to estimate an amount, such as dollars, average wage rate, or value of inventory, the procedure is called sampling for variables. The same sample may be used for both purposes; however, the estimation and evaluation procedures are somewhat different. Monetary-unit sampling is a modified form of sampling for attributes.

Audit Appraisal—This involves the subjective aspects of audit considerations of the sample results; whereas, statistical evaluation relates to statistical aspects. In audit appraisal, each finding or category of finding is considered separately for audit significance. Also, the impact of other sources of reliance are considered in conjunction with the sample results.

Audit Risk—The risk that material errors or irregularities, if they exist, will not be detected by the auditing procedures. It is the complement of assurance (confidence) from auditing procedures.

Audit Sampling—The application of auditing procedures to less than all of the units within an account balance of class of transactions for the purpose of estimating and evaluating some characteristic.

Audit Stratification—The separation or stratification, by the auditor based on judgment, experience and other information, transactions or segments he deems require more thorough testing or 100 percent review. For example, high value transactions, unusual charges, large credits, and so on. This procedure should not be confused with statistical stratification.

Audit Unit—An individual invoice, account balance, line-item, and so on, in the universe; the definition of sampling unit when audit-unit sampling is used.

Audit-Unit Sampling—A sampling procedure where the sampling units are defined as the individual items in a universe, such as invoices, with each individual audit unit being given a determinable non-zero chance of selection.

Audited Amount—The amount established by the auditor as the amount that should be in the records based on the results of the audit examination.

Average Sampling Interval (ASI)—For a monetary-unit sample, it is the quotient of the universe monetary value divided by the sample size. It is the average number of monetary units to be accumulated before reaching the next monetary-unit selection point.

Base precision (BP)—One of the components of the upper error limit in monetary-unit sampling. It is the upper error limit when no sample error is found. Basic precision can be expressed as a percentage, rate or in absolute monetary units, such as dollars.

Bayesian Statistics—A school of thought within statistics in which estimates or probabilities of events are based on subjective beliefs as modified by empirical data. In classical statistics, probability estimates are based solely on objective data. Bayesian statistics is considered more decision-oriented than classical statistics since the point of sufficient information for a decision is reached quicker. The Bayesian approach also explicitly considers the cost of obtaining additional information.

Bias—A systematic error, difference or variation in a set of data. This can result from inadequate information or preconceived notions. For example, the auditor who is not aware of the fact that certain categories of expenses are unallowable, will produce biased results concerning the extent of questionable expenses in a particular audit of expense vouchers. Also, a questionnaire could be biased if it allows only desired responses. Bias may originate from poor auditing and/or sampling procedures, or from deficiencies in performing the audit, or from an inherent characteristic of the estimating technique. The degree of bias related to an estimating technique is often so small as to be of no practical significance. For example, the use of ratios may not provide unbiased estimates; however, the bias is so small in comparison with the improved precision over the mean estimate.

Binomial Distribution—A discrete probability distribution used to find the probability of the occurrence of an event when the events can be classified into only two categories (yes or no, correct or incorrect, success or failure, etc.) and the probability of success remains constant from one trial to another. The binomial distribution is used when predicting the probability of the number of failures (or successes) in a series of units, when it is known that a certain percent of the units usually fail (succeed).

Block Test—A test in which the items are selected from periods of time or consecutive groupings, such as all expense vouchers in a particular month. Any inference concerning transactions in other months which were not tested would be judgmental and without statistical support.

Book Value or Amount—See Recorded Amount.

Characteristic—A particular feature of a universe item which may be either quantitative or qualitative. A quantitative characteristic is called a variable; a qualitative characteristic is called an attribute. The critical characteristic is the most significant parameter to be estimated among different parameters. The sample size is generally associated with the critical characteristic.

Class Interval—An interval setting limits of the values which are grouped together in a frequency table or distribution.

Cluster Sampling—The sampling process in which items or records are randomly selected in bunches, groups or clusters rather than individually. For example, randomly selecting a number of days in a year and then auditing all appropriate records in the selected days. Hence, each sampling unit is a cluster of elements or items. Generally, this is the least efficient type of statistical sampling.

Coefficient of Variation—A measure of relative disperson or variation computed by dividing the standard deviation by the corresponding mean and expressing the result as a percentage. The measure is useful in comparing the variation in two or more series of data when the means differ significantly or when the series are not expressed in the same units of measurement. Most often applied when appraisal is needed for a comparison in the degree of quality, uniformity of performance and stability of two or more processes, employees, machines, materials, and so on.

Compliance Testing—Audit testing where the objective is to detect, if they exist, critical compliance deviations, especially those associated with material monetary errors.

Computer Program—See Sampling Software.

Conditional Probability—The likelihood that an event will occur, given that another event has occurred or a specific condition exists.

Confidence Interval—A range of values of a sample statistic wherein the actual universe value is believed to lie a stated proportion of times if the same sampling procedures were applied many times to the sampling frame. The interval is computed on the basis of a known sample value, a desired precision and a specified level of confidence. The one-sided confidence interval measures the proportion of upper precision limits that exceeds the universe characteristic or the proportion of lower precision limits that fall below the universe characteristic.

Confidence (Assurance) Level—This indicates the probability that the "true" or actual value will be within the corresponding confidence, or precision, interval. For example, if a confidence interval based on a random sample were computed on the basis of a 95 percent confidence level, it would mean that in repeated samples of the same size from the same sampling frame in every 95 samples out of 100, the estimated confidence interval would be expected to include the actual universe value. The confidence level is the complement of sampling risk. The desired level of confidence depends upon judgments about the risks associated with a sample estimate and the availability of other sources of reliance.

Constant—A quantity or factor whose value remains fixed and does not vary from item to item or observation to observation. For example, each office (regardless of volume of activity) will be managed by two supervisors. Since two is the same for each office, two supervisors are referred to as a constant.

Convenience Test—The selection of test items without reason or randomness because they are most accessible.

Cumulative Frequency Distribution—A chart or table which shows the sum of the relative frequencies from the lowest class interval through each succeeding class interval.

Cumulative Monetary Amount (CMA) Sampling—A sampling procedure used by Deloitte, Haskins & Sells, which is based on selection with probability proportional to size. Its results are generally similar to those of dollar-unit sampling. This type of sampling (PPS) is described in the AICPA SAS No. 39, *Audit Sampling.*

Degrees of Freedom—The factor used in the evaluation of sample precision based on the type of estimator used. As the number of degrees of freedom is reduced the related variance and thus the standard deviation are increased. The degrees of freedom for the mean estimate is one less than the sample size. For the difference and ratio estimates, the degrees of freedom may be set as one less than the number of observed differences in the sample. For the regression estimate, the degrees of freedom may be set as two less than the number of differences in the sample. For stratified random sampling, the number of degrees of freedom is reduced from the corresponding unstratified sample by one less than the number of strata used.

Detection (Discovery) Sampling—A statistical sampling method which provides an indication of the probability of locating at least one irregularity or characteristic in question. It is not designed to estimate the error rate in the universe. Detection sampling table (Appendix C) indicates the probabilities that a simple random sample will contain at least one item with a designated characteristic if the sample is drawn from a universe which contains a specified number or proportion of items with this characteristic. Conversely, if a sample of a specified size reveals none of the characteristics, the table may be used to indicate the assurance (confidence) that no more than a specified number of items in the universe contains the characteristic. The sampling method is also referred to as discovery sampling and exploratory sampling.

Difference—The universe difference equals the total audited amount minus the total recorded amount. For an individual sampling unit, the difference equals the audited amount minus the recorded amount.

Difference Estimate—An estimating procedure which uses the amount of differences between sample audited values and book values to estimate the universe amount.

Dollar-unit Sampling—A method of sampling in which the sampling unit is defined as an individual dollar, and each given an equal chance of selection. Thus, the chance of selection of any audit unit (e.g., a voucher) is related to the number of dollar units it contains.

DUS-cell Selection—A method of selecting a dollar-unit sample in which the universe is divided into subdivisions called cells, each equal in size to the average sampling interval, with one dollar unit being selected randomly from each cell.

Efficiency—One estimator is said to be more efficient than another estimator if it produces a smaller variance. Thus efficiency is a way of measuring the relative merits of various estimators.

Error—A difference in either an amount or procedure from that considered acceptable under the circumstances. An apparent difference for which there is a satisfactory explanation would not be considered an error.

Error Analysis—An investigation of each observed compliance deviation or monetary difference to ascertain the nature, cause and audit significance.

Estimating Procedure or Estimator—A technique of projecting the results of a sample to estimate the universe value or characteristic.

Estimation Sampling—A form of statistical sampling in which an estimate is made of some characteristic in the universe on the basis of a sample selected from that universe. If a single figure is estimated for the universe, it is called a point estimate. If a confidence, or precision, interval is calculated, the result is called an interval estimate, which is a characteristic of estimation sampling.

Evaluation Matrix—The outline format of the audit findings derived from a tabulation of the sample results by characteristic and source. It is an excellent device for depicting the findings of the sample. After the sample data are tabulated in the matrix, clusters, outliers and trends are revealed.

Exploratory Sampling—See Detection Sampling.

Finite Correction Factor—Mathematical factor applied to the estimated standard error of estimate to adjust for sampling without replacement.

Flexible Sampling—This sampling strategy does not require a predetermined sample size. It begins with a minimum sample for a desired confidence, assuming no error is found. If errors are detected in the examination of the minimum sample, either stratification is extended, the sample is replicated or other auditing techniques are employed, depending upon the objective and the audit significance of the errors. Error analysis is an essential feature.

Frame—A listing of the sampling units; also, procedures that account for all the sampling units without the effort of actually listing them.

Frequency—The number of items or units in a statistical classification.

Frequency Distribution—A listing of the number of times values of a characteristic or variable have occurred or have been observed in different classes. It may be a table or chart or other kind of arrangement which shows the classes into which the data have been grouped and the corresponding frequencies in each class.

Hypergeometric Distribution—A discrete probability distribution used to find the probability of an occurrence when random samples are selected without replacement from a universe. Appendix C and Appendix D are based on the hypergeometric distribution.

Interim Evaluation—The statistical evaluation of the precision of a sample at any point during its examination, provided the randomly selected items at that point have had an opportunity to cover the entire frame of audit interest.

Interpenetrating Subsamples—A sampling method by which the frame is divided into separate zones of approximately the same size, and random selections are made from each zone, so that there are as many subsamples as there are zones. The comparison of results from individual subsamples provides an intuitive check. See Replicated Sampling.

Interval Estimate—An estimate of a universe characteristic or amount in terms of a range or interval of values. An interval estimate involves a two-sided confidence interval, bounded by an upper error limit and a lower error limit for a specified confidence level.

Interval Selection—See Systematic Selection.

Judgment Sampling—Any sample in which the items are selected subjectively and not by an appropriate random selection procedure.

Lower Precision Limit—For a two-sided confidence interval, it is the lower end of the confidence interval.

Materiality—The limit of accuracy to which it is practical and reasonable to measure financial statement data and to estimate the value of infractions.

Mean—The most commonly used index of central tendency of a set of data, which is computed by summing the values of the items and dividing this sum by the number of items. It is also called the arithmetic mean and the "average."

Mean-per-Unit Estimate—An estimating procedure which estimates the universe total value by multiplying the sample mean by the total number of items in the universe.

Median—An index of central tendency which divides an array of a set of data into two subsets of an equal number of items.

Minimum Sample Size—The smallest sample for a desired one-sided confidence level provided no error is found.

Mode—The value which occurs most frequently in a series of data. It is an index of central tendency in a frequency distribution. The modal value(s) corresponds to the highest point(s) on a frequency distribution. There may not be a single mode.

Model—A representation of the relationships which define a system, operation or a situation. A model, with its analytical features, may be a set of equations, a computer program or any other representation. Models permit the manipulation of relevant variables to ascertain how a process, operation or concept reacts under various circumstances.

Multistage Sampling—A sampling process involving several stages, in which units at each subsequent stage are subsampled from previously selected larger units. For example, in the first stage warehouses are selected; in the second stage file cabinets are selected within such selected warehouse; in the third stage file drawers are selected from each selected file cabinet; and so on.

Nonsampling Risk—The portion of the ultimate audit risk of not detecting a material error or gross infraction that exists because of limitations of the procedures used, the system being examined, the maturity, skill and care of the auditor, and the timing of the application of the procedures. Thus, it is the ultimate audit risk less the risk due to sampling.

Normal Distribution—It is a continuous probability distribution which is represented by a symmetrical bell-shaped curve. It approximates the sampling distribution for many statistical estimates. The normal distribution is determined by its mean and standard deviation. The standard normal distribution, which has a mean of 0 and a standard deviation of 1, is available in tables and is used to determined confidence coefficients.

Occurrence Rate—The percentage or proportion of the sampling units that possess a specified attribute or characteristic.

Optimum Allocation—A method for allocating a stratified sample among strata such that each stratum sample size is proportional to the product of the number of stratum universe items times the stratum standard deviation.

Parameter—A measure, quantitative or qualitative (such as a mean or an error rate), which is calculated directly from a universe or is estimated as applicable to the universe.

Point Estimate—A single, specific estimate for a universe characteristic or value. It is usually computed from sample results and is related to the type of estimator used, such as difference, ratio and regression.

Poisson Distribution—A discrete probability distribution used to approximate the binomial distribution when the number of items in the universe is large and the probability of occurrence of the event (error) is small. It is frequently used in dollar-unit sampling procedures.

Population—See Universe.

Posterior Confidence—In Bayesian statistics, it is the combination of subjective prior probabilities with subsequent statistical evidence.

Precision—A measure of closeness between a sample estimate and the corresponding unknown universe characteristic. It is computed by multiplying the standard error of the estimate by a factor corresponding to the desired confidence level.

Precision Interval—See Confidence Interval.

Prior Probabilities—The values of subjective probabilities prior to obtaining objective sampling evidence. The concept arises in Bayesian inference. Prior probabilities are relative in that when additional evidence is considered, the revised posterior probabilities may become the prior probabilities for subsequent evidence.

Probability—One commonly used definition defines probability as the ratio of the number of outcomes that would produce a specific event to the total number of possible outcomes; that is, the likelihood or chance of the occurrence of a specific event or outcome. It is usually expressed as a percentage or proportion from zero to unity. For example, in a universe of 10,000 vouchers containing 500 travel vouchers, 2,500 of another type and 7,000 of a third type, the probability that a randomly selected voucher is a travel voucher is .05 or 5 percent.

Probability Distribution—All possible values of a variable and the associated relative frequencies (probabilities) with which the values are expected to occur. There are two types of probability distributions: (i) discrete distributions, such as the binomial, Poisson and hypergeometric, in which the variables can assume only specific values (usually integers) and (ii) continuous distribution, such as the normal, t and F distributions, in which the variable can assume any value within a specified range or interval. When summed over all possible values of a variable, the associated probabilities equal unity.

Probability-Proportional-to-Size (PPS) Sampling—A generalization of the sampling procedures used in monetary-unit sampling, dollar-unit sampling and CMA sampling. It is a procedure for sampling for variables which uses attributes theory to express universe estimates in monetary amounts. With PPS sampling the probability of selecting an audit unit is proportional to its recorded value.

Probability Sample—See Statistical Sampling.

Probe Sample—This is a preliminary sample of an audit area used to obtain clues of the conditions as a basis for concentrating further audit effort through stratification and the use of other procedures.

Purposive Tests—Subjectively, or nonrandomly, selected test items chosen because there is evidence that they are illustrative of a condition or they concentrate effort on material or sensitive items.

Quartile—One of three values within an array of numbers positioned so that one-fourth of the items lie below the lower quartile, and half lie below the median quartile, and three-fourths lie below the upper quartile.

Random Digit Table—Table of digits appearing in no regular order which can be used to create random numbers.

Random Moment Sampling—See Work Sampling.

Random Numbers—A sequence of random digits. When the digits are grouped into numbers, no pattern in successive numbers can be detected and each number within the applicable range has an equal chance of occurring.

Random Sample—A sample obtained by a process of random selection. A simple random sample gives every sampling unit an equal chance of being chosen at each stage in the selection. Also, it is a sample drawn so that every combination of the same number of items (size) in the universe has an equal chance of being drawn. A random sample provides a mathematical means of estimating the precision.

Random Time Sampling—See Work Sampling.

Random Variable—A variable whose value is determined by chance and the possible values can be described by a probability distribution.

Range—The difference between the smallest and largest values in a set of values. It is the simplest index of the dispersion in a set of numbers.

Ratio Estimate—An estimating procedure which involves the ratio of two variables. In auditing, the point estimate of the universe value is estimated from the sample ratio of audited values to book values multiplied by the total recorded amount of the universe.

Recorded Amount—The amount appearing in the client's or auditee's records.

Regression Estimate—An estimating procedure which provides an estimate of the total value of a dependent variable by substituting the known value of the independent variable in a linear equation. The equation is derived from the sample values of pairs of audited and book values to estimate the universe audited amount.

Relative Frequency Distribution—A chart or table which shows the proportion or percentage (relative frequency) of items in each class interval.

Relative Precision—The ratio of the computed precision to the point estimate.

Replicated Sampling—A sampling procedure that selects from the entire universe two or more independent random samples of the same or approximately the same size. Each replicate is statistically evaluated and the resulting estimates are used to estimate the precision of the overall estimate based on all sample items. Replication provides a simple way of calculating estimates and their precisions, such as means, totals, proportions or ratios. Also, replicates provide data for comparative analyses.

Risk—Often a distinction is made between risk and uncertainty. Risk is measurable while uncertainty is not. In statistical sampling, risk is the complement of probability or assurance. For example, if one has 95 percent assurance that the universe value is within the computed confidence interval, then the risk is 5 percent that the value will not be within the interval. It is an uncertainty when it cannot be calculated mathematically in terms of probability. In other words, in risk situations, the probabilities associated with potential outcomes can be calculated. Risk is associated with situations of repeated events, each individually unpredictable but with the average outcome predictable in the long run.

Sample—A part of a universe selected by either judgment or random procedures for the purpose of investigating and/or estimating properties or characteristics of the universe. In auditing, a sample is related to test-checking.

Sample Design—All of the principal steps which are planned in advance of conducting the sample. Briefly, this involves defining the audit objective and related problems, identifying the sampling frame and critical characteristic, determining the selection method, deciding on extent of audit and statistical stratification, indicating minimum sample size and provisions for replication, selecting the estimation and evaluation procedures, indicating provisions for error analysis, and audit appraisal such as an evaluation matrix, providing for the integration of other evidence, making provisions for proper audit interpretation of all evidence such as use of graphic analysis, and presenting an opinion with supporting rationale.

Sample Mean—The sum of the sample values divided by the sample size. The sample mean is an unbiased estimate of the universe mean.

Sample Size—The number of sampling units selected for the sample. Although a number of factors influence the determination of sample size, major factors are the variability and the occurrence rate of the critical characteristic to be estimated, the precision desired and the effectiveness of the stratification.

Sample Standard Deviation—A measure of the dispersion of the sample items and used as an estimate of the universe standard deviation.

Sampling Distribution—It describes the probability of each possible estimate for samples of the same size from the same universe. The sampling distribution is dependent upon the universe distribution, the sample size, the method of selecting the sample, the nature and frequency of the critical characteristic, and the estimation and evaluation procedure.

Sampling Error—In statistics, it is a measure of the difference between the value derived from a statistical sample and the value of the same characteristic which would have been derived if the entire universe had been tested by the same individuals, using the same care and procedures, and during the same time period.

Sampling Fraction—Often denoted by the symbol "f", it represents the fraction or proportion of universe or sampling frame units included in the sample. It is derived by dividing the number of units in the sample (n) by the total number of units in the frame (N). The sampling fraction is used in computing the finite correction factor.

Sampling Frame—See Frame.

Sampling Plan—The method of selecting the sample.

Sampling Risk—The risk that the auditor's decision based on a sample may be different from the decision he would have made if the test had been applied in the same manner to the entire universe. For compliance testing, the sampling risk is the risk that the sample supports overreliance on internal accounting control. For substantive testing, sampling risk is the risk that the sample supports the conclusion that the records are not materially misstated when they are materially misstated.

Sampling Software—Computer programs designed to stratify, randomly select and evaluate statistical samples.

Sampling Unit—Any of the designated elements that comprise the universe of interest.

Simple Random Sample—A random sample of units selected with equal probability at each selection stage. See Random Sample.

Skewness—A lack of symmetry about the mean in a frequency distribution. Accounting universes are right skewed because there are many small amounts and few very large amounts. A skewness coefficient is a number which indicates the relative degree of skewness in a distribution. One index of skewness is obtained by subtracting the mode from the median and dividing the difference by the standard deviation of the distribution.

Standard Deviation—An index of the degree of dispersion among a set of values; an index of the tendency of individual values to vary from the mean value. The standard deviation of a set of values is computed by subtracting the mean value from each individual value, squaring each of these differences, summing the results, dividing this sum by the number of values, and then finding the square root of the result. The square of the standard deviation is the variance.

Standard Error of the Estimate—The standard deviation of the sampling distribution of a particular estimating procedure is called the standard error of the estimate.

Standard Normal Distribution—A normal distribution with a mean of zero and a standard deviation of one. This is accomplished by converting the estimate from original units into numbers of standard deviations from the mean. It is from this distribution that tables of areas under the normal curve are derived.

Statistic—A measure, quantity or value, such as a mean, total or proportion, which is calculated from a sample to estimate the corresponding parameter of the universe.

Statistical Efficiency—One statistical estimator is said to be more efficient than another if it requires a smaller sample size to achieve the same precision.

Statistical Inference—Using data obtained from a sample to make predictions about a larger set, the universe.

Statistical Sampling—Method of selecting a sample in such a way that each element of the universe has a known probability of being selected. This permits calculation of the sampling error. An advantage of statistical sampling over judgment sampling is the capability to mathematically calculate the precision.

Statistical Stratified Sampling—In auditing, statistical stratified sampling may differ from stratification for audit pruposes, wherein the stratified items or areas are often considered separately and not as a part of the sampling frame. In statistical stratified sampling, the entire universe of sampling units is divided into distinct subuniverses, called strata, and within each stratum a separate random sample is selected from all the sampling units in that stratum. From the sample obtained in each stratum, a separate stratum mean (or other statistic) is computed. These strata values are properly weighted and combined to form an overall estimate. The standard deviations are also computed separately within each stratum and then weighted and combined for an overall estimate.

Subjective Probability—A probability where the auditor, based on experience, relies on his judgment of the likelihood of various possible results or events.

Sufficient—A statistical estimator is sufficient when it makes use of all the information that can be obtained from a sample. As applied to accounting estimates, one usually has information about book values that may be used to improve estimates of audited amounts.

Systematic Error—A form of bias introduced with a degree of regularity. For example, a defective odometer might regularly measure miles traveled at 10 percent above actual mileage.

Systematic Sampling—A method of drawing a sample in which every nth item is drawn after at least two random starts; using 5 to 10 random starts is recommended.

Test-Check—The application of an auditing procedure to less than 100 percent of the audit units for the purpose of evaluating some characteristic, gaining an understanding of an operation, or clarifying an understanding of an internal control system.

Tainting—In monetary-unit sampling, dollar-unit sampling or PPS sampling, the ratio of the amount of the monetary error to the book amount in an audit unit is called tainting.

T-distribution—The approximate sampling distribution for the ratio of the sample mean to the sample standard deviation when sample is small. If the universe distribution is normal, it is the exact sampling distribution. It is useful in providing confidence coefficients for various sampling plans. See Degrees of Freedom.

Tolerable Error Rate—The maximum rate of deviation from a prescribed accounting control procedure that the auditor will tolerate without changing the nature, scope or timing of his substantive testing. This notion implies that the rate is not confined or clustered in a limited area.

Tolerable Monetary Error—In auditing, an estimate of the maximum monetary error that may exist in an account balance or class of transactions without causing material misstatements in financial documents.

Top (100%) Stratum—In dollar-unit sampling, a stratum of high value items, each exceeding the top stratum threshold, which will be examined 100 percent. It includes any other items to be examined 100 percent.

Top Stratum Threshold—The lower boundary of the stratum for which a 100 percent sample is selected. In dollar-unit sampling, it is usually set at a value equal to or less than the average sampling interval.

Ultimate Audit Risk—All the risks inherent in performing auditing procedures. It is a combination of the risk that material errors will occur in the accounting process from which the financial statements are developed and the risk that material errors which occur will not be detected by the auditor. The risk due to sampling is called sampling risk. All other risks are referred to as nonsampling risks.

Unbiased—A statistical estimator is said to be unbiased if its mean value taken over all possible samples equals the corresponding universe value. See Systematic Error.

Uncertainty—Risk and uncertainty are often used interchangeably; however, a distinction is sometimes made. An event may be risky if a probability distribution can be determined; it is uncertain if the probability of success or failure cannot be determined.

Unit—A single item or person or a group of items or persons regarded as an entity. Also one of the parts, or elements, into which a whole may be divided.

Universe—In auditing, the portion of items of an account or class of transactions of interest. The delimited universe excludes individual items that the auditor separates for special handling or detailed testing. See Audit Stratification.

Unrestricted Random Sample—See Simple Random Sample.

Upper Precision—A measure of how much the unknown universe characteristic may exceed the estimated amount at a specified one-sided confidence. It equals the standard error of the estimate multiplied by the corresponding one-sided confidence factor or coefficient.

Variable—A numerical characteristic which may vary from one observation to another. Since a variable can assume different values, it is represented by a symbol instead of a specific number.

Variables Sampling—Those sampling plans designed to estimate quantitative characteristics of a universe, such as dollar amounts. These plans utilize estimators which usually rely on large sample assumptions of normality.

Variance—An index of the degree of dispersion among a set of values. It is computed by subtracting the mean value from each individual value, squaring each of these differences, summing the results, and dividing this sum by the number of values. The square root of the variance is the standard deviation.

With Replacement—A sampling technique where any item in the universe may be included in the sample more than once, that is, as many times as its identification number is selected.

Without Replacement—A sampling technique where any item in the lected for the sample, cannot be selected again for that sample. Audit sampling is usually conducted without replacement.

Work Sampling—A method of random observations used to estimate idle time and the proportion of time spent on various work activities or operations.

Appendix B: Table of 105,000 Random Decimal Digits

Page 1

Line \ Col.	(1)	(2)	(3)	(4)	(5)	(6)	(7)	(8)	(9)	(10)	(11)	(12)	(13)	(14)
1	10480	15011	01536	02011	81647	91646	69179	14194	62590	36207	20969	99570	91291	90700
2	22368	46573	25595	85393	30995	89198	27982	53402	93965	34095	52666	19174	39615	99505
3	24130	48360	22527	97265	76393	64809	15179	24830	49340	32081	30680	19655	63348	58629
4	42167	93093	06243	61680	07856	16376	39440	53537	71341	57004	00849	74917	97758	16379
5	37570	39975	81837	16656	06121	91782	60468	81305	49684	60672	14110	06927	01263	54613
6	77921	06907	11008	42751	27756	53498	18602	70659	90655	15053	21916	81825	44394	42880
7	99562	72905	56420	69994	98872	31016	71194	18738	44013	48840	63213	21069	10634	12952
8	96301	91977	05463	07972	18876	20922	94595	56869	69014	60045	18425	84903	42508	32307
9	89579	14342	63661	10281	17453	18103	57740	84378	25331	12566	58678	44947	05585	56941
10	85475	36857	53342	53988	53060	59533	38867	62300	08158	17983	16439	11458	18593	64952
11	28918	69578	88231	33276	70997	79936	56865	05859	90106	31595	01547	85590	91610	78188
12	63553	40961	48235	03427	49626	69445	18663	72695	52180	20847	12234	90511	33703	90322
13	09429	93969	52636	92737	88974	33488	36320	17617	30015	08272	84115	27156	30613	74952
14	10365	61129	87529	85689	48237	52267	67689	93394	01511	26358	85104	20285	29975	89868
15	07119	97336	71048	08178	77233	13916	47564	81056	97735	85977	29372	74461	28551	90707
16	51085	12765	51821	51259	77452	16308	60756	92144	49442	53900	70960	63990	75601	40719
17	02368	21382	52404	60268	89368	19885	55322	44819	01188	65255	64835	44919	05944	55157
18	01011	54092	33362	94904	31273	04146	18594	29852	71585	85030	51132	01915	92747	64951
19	52162	53916	46369	58586	23216	14513	83149	98736	23495	64350	94738	17752	35156	35749
20	07056	97628	33787	09998	42698	06691	76988	13602	51851	46104	88916	19509	25625	58104
21	48663	91245	85828	14346	09172	30168	90229	04734	59193	22178	30421	61666	99904	32812
22	54164	58492	22421	74103	47070	25306	76468	26384	58151	06646	21524	15227	96909	44592
23	32639	32363	05597	24200	13363	38005	94342	28728	35806	06912	17012	64161	18296	22851
24	29334	27001	87637	87308	58731	00256	45834	15398	46557	41135	10367	07684	36188	18510
25	02488	33062	28834	07351	19731	92420	60952	61280	50001	67658	32586	86679	50720	94953
26	81525	72295	04839	96423	24878	82651	66566	14778	76797	14780	13300	87074	79666	95725
27	29676	20591	68086	26432	46901	20849	89768	81536	86645	12659	92259	57102	80428	25280
28	00742	57392	39064	66432	84673	40027	32832	61362	98947	96067	64760	64584	96096	98253
29	05366	04213	25669	26422	44407	44048	37937	63904	45766	66134	75470	66520	34693	90449
30	91921	26418	64117	94305	26766	25940	39972	22209	71500	64568	91402	42416	07844	69618
31	00582	04711	87917	77341	42206	35126	74087	99547	81817	42607	43808	76655	62028	76630
32	00725	69884	62797	56170	86324	88072	76222	36086	84637	93161	76038	65855	77919	88006
33	69011	65795	95876	55293	18988	27354	26575	08625	40801	59920	29841	80150	12777	48501
34	25976	57948	29888	88604	67917	48708	18912	82271	65424	69774	33611	54262	85963	03547
35	09763	83473	73577	12908	30883	18317	28290	35797	05998	41688	34952	37888	38917	88050
36	91567	42595	27958	30134	04024	86385	29880	99730	55536	84855	29080	09250	79656	73211
37	17955	56349	90999	49127	20044	59931	06115	20542	18059	02008	73708	83517	36103	42791
38	46503	18584	18845	49618	02304	51038	20655	58727	28168	15475	56942	53389	20562	87338
39	92157	89634	94824	78171	84610	82834	09922	25417	44137	48413	25555	21246	35509	20468
40	14577	62765	35605	81263	39667	47358	56873	56307	61607	49518	89686	20103	77490	18062
41	98427	07523	33362	64270	01638	92477	66969	98420	04880	45585	46565	04102	46880	45709
42	34914	63976	88720	82765	34476	17032	87589	40836	32427	70002	70663	88863	77775	69348
43	70060	28277	39475	46473	23219	53416	94970	25832	69975	94884	19661	72828	00102	66794
44	53976	54914	06990	67245	68350	82948	11398	42878	80287	88267	47363	46634	06541	97809
45	76072	29515	40980	07391	58745	25774	22987	80059	39911	96189	41151	14222	60697	59583
46	90725	52210	83974	29992	65831	38857	50490	83765	55657	14361	31720	57375	56228	41546
47	64364	67412	33339	31926	14883	24413	59744	92351	97473	89286	35931	04110	23726	51900
48	08962	00358	31662	25388	61642	34072	81249	35648	56891	69352	48373	45578	78547	81788
49	95012	68379	93526	70765	10592	04542	76463	54328	02349	17247	28865	14777	62730	92277
50	15664	10493	20492	38391	91132	21999	59516	81652	27195	48223	46751	22923	32261	85653

Source: Interstate Commerce Commission, 1949

Page 2

Line \ Col.	(1)	(2)	(3)	(4)	(5)	(6)	(7)	(8)	(9)	(10)	(11)	(12)	(13)	(14)
51	16408	81899	04153	53381	79401	21438	83035	92350	36693	31238	59649	91754	72772	02338
52	18629	81953	05520	91962	04739	13092	97662	24822	94730	06496	35090	04822	86774	98289
53	73115	35101	47498	87637	99016	71060	88824	71013	18735	20286	23153	72924	35165	43040
54	57491	16703	23167	49323	45021	33132	12544	41035	80780	45393	44812	12515	98931	91202
55	30405	83946	23792	14422	15059	45799	22716	19792	09983	74353	68668	30429	70735	25499
56	16631	35006	85900	98275	32388	52390	16815	69298	82732	38480	73817	32523	41961	44437
57	96773	20206	42559	78985	05300	22164	24369	54224	35083	19687	11052	91491	60383	19746
58	38935	64202	14349	82674	66523	44133	00697	35552	35970	19124	63318	29686	03387	59846
59	31624	76384	17403	53363	44167	64486	64758	75366	76554	31601	12614	33072	60332	92325
60	78919	19474	23632	27889	47914	02584	37680	20801	72152	39339	34806	08930	85001	87820
61	03931	33309	57047	74211	63445	17361	62825	39908	05607	91284	68833	25570	38818	46920
62	74426	33278	43972	10119	89917	15665	52872	73823	73144	88662	88970	74492	51805	99378
63	09066	00903	20795	95452	92648	45454	09552	88815	16553	51125	79375	97596	16296	66092
64	42238	12426	87025	14267	20979	04508	64535	31355	86064	29472	47689	05974	52468	16834
65	16153	08002	26504	41744	81959	65642	74240	56302	00033	67107	77510	70625	28725	34191
66	21457	40742	29820	96783	29400	21840	15035	34537	33310	06116	95240	15957	16578	06004
67	21581	57802	02050	89728	17937	37621	47075	42080	97403	48626	68995	43805	33386	21597
68	55612	78095	83197	33732	05810	24813	86902	60397	16489	03264	88525	42786	05269	92532
69	44657	66999	99324	51281	84463	60563	79312	93454	68876	25471	93911	25650	12682	73572
70	91340	84979	46949	81973	37949	61023	43997	15263	80644	43942	89203	71795	99533	50501
71	91227	21199	31935	27022	84067	05462	35216	14486	29891	68607	41867	14951	91696	85065
72	50001	38140	66321	19924	72163	09538	12151	06878	91903	18749	34405	56087	82790	70925
73	65390	05224	72958	28609	81406	39147	25549	48542	42627	45233	57202	94617	23772	07896
74	27504	96131	83944	41575	10573	08619	64482	73923	36152	05184	94142	25299	84387	34925
75	37169	94851	39117	89632	00959	16487	65536	49071	39782	17095	02330	74301	00275	48280
76	11508	70225	51111	38351	19444	66499	71945	05422	13442	78675	84081	66938	93654	59894
77	37449	30362	06694	54690	04052	53115	62757	95348	78662	11163	81651	50245	34971	52924
78	46515	70331	85922	38329	57015	15765	97161	17869	45349	61796	66345	81073	49106	79860
79	30986	81223	42416	58353	21532	30502	32305	86482	05174	07901	54339	58861	74818	46942
80	63798	64995	46583	09785	44160	78128	83991	42865	92520	83531	80377	35909	81250	54238
81	82486	84846	99254	67632	43218	50076	21361	64816	51202	88124	41870	52689	51275	83556
82	21885	32906	92431	09060	64297	51674	64126	62570	26123	05155	59194	52799	28225	85762
83	60336	98782	07408	53458	13564	59089	26445	29789	85205	41001	12535	12133	14645	23541
84	43937	46891	24010	25560	86355	33941	25786	54990	71899	15475	95434	98227	21824	19585
85	97656	63175	89303	16275	07100	92063	21942	18611	47348	20203	18534	03862	78095	50136
86	03299	01221	05418	38982	55758	92237	26759	86367	21216	98442	08303	56613	91511	75928
87	79626	06486	03574	17668	07785	76020	79924	25651	83325	88428	85076	72811	22717	50585
88	85636	68335	47539	03129	65651	11977	02510	26113	99447	68645	34327	15152	55230	93448
89	18039	14367	61337	06177	12143	46609	32989	74014	64708	00533	35398	58408	13261	47908
90	08362	15656	60627	36478	65648	16764	53412	09013	07832	41574	17639	82163	60859	75567
91	79556	29068	04142	16268	15387	12856	66227	38358	22478	73373	88732	09443	82558	05250
92	92608	82674	27072	32534	17075	27698	98204	63863	11951	34648	88022	56148	34925	57031
93	23982	25835	40055	67006	12293	02753	14827	23235	35071	99704	37543	11601	35503	85171
94	09915	96306	05908	97901	28395	14186	00821	80703	70426	75647	76310	88717	37890	40129
95	59037	33300	26695	62247	69927	76123	50842	43834	86654	70959	79725	93872	28117	19233
96	42488	78077	69882	61657	34136	79180	97526	43092	04098	73571	80799	76536	71255	64239
97	46764	86273	63003	93017	31204	36692	40202	35275	57306	55543	53203	18098	47625	88684
98	03237	45430	55417	63282	90816	17349	88298	90183	36600	78406	06216	95787	42579	90730
99	86591	81482	52667	61582	14972	90053	89534	76036	49199	43716	97548	04379	46370	28672
100	38534	01715	94964	87288	65680	43772	39560	12918	86537	62738	19636	51132	25739	56947

Page 3

Line \ Col.	(1)	(2)	(3)	(4)	(5)	(6)	(7)	(8)	(9)	(10)	(11)	(12)	(13)	(14)
101	13284	16834	74151	92027	24670	36665	00770	22878	02179	51602	07270	76517	97275	45960
102	21224	00370	30420	03883	94648	89428	41583	17564	27395	63904	41548	49197	82277	24120
103	99052	47887	81085	64933	66279	80432	65793	83287	34142	13241	30590	97760	35848	91983
104	00199	50993	98603	38452	87890	94624	69721	57484	67501	77638	44331	11257	71131	11059
105	60578	06483	28733	37867	07936	98710	98539	27186	31237	80612	44488	97819	70401	95419
106	91240	18312	17441	01929	18163	69201	31211	54288	39296	37318	65724	90401	79017	62077
107	97458	14229	12063	59611	32249	90466	33216	19358	02591	54263	88449	01912	07436	50813
108	35249	38646	34475	72417	60514	69257	12489	51924	86871	92446	36607	11458	30440	52639
109	38980	46600	11759	11900	46743	27860	77940	39298	97038	95145	32378	68038	89351	37005
110	10750	52745	38749	87365	58959	53731	89295	59062	39404	13198	59960	70408	29812	83126
111	36247	27850	73958	20673	37800	63835	71051	84724	52492	22342	78071	17456	96104	18327
112	70994	66986	99744	72438	01174	42159	11392	20724	54322	36923	70009	23233	65438	59685
113	99638	94702	11463	18148	81386	80431	90628	52506	02016	85151	88598	47821	00265	82525
114	72055	15774	43857	99805	10419	76939	25993	03544	21560	83471	43989	90770	22965	44247
115	24038	65541	85788	55835	38835	59399	13790	35112	01324	39520	76210	22467	83275	32286
116	74976	14631	35908	28221	39470	91548	12854	30166	09073	75887	36782	00268	97121	57676
117	35553	71628	70189	26436	63407	91178	90348	55359	80392	41012	36270	77786	89578	21059
118	35676	12797	51434	82976	42010	26344	92920	92155	58807	54644	58581	95331	78629	73344
119	74815	67523	72985	23183	02446	63594	98924	20633	58842	85961	07648	70164	34994	67662
120	45246	88048	65173	50989	91060	89894	36036	32819	68559	99221	49475	50558	34698	71800
121	76509	47069	86378	41797	11910	49672	88575	97966	32466	10083	54728	81972	58975	30761
122	19689	90332	04315	21358	97248	11188	39062	63312	52496	07349	79178	33692	57352	72862
123	42751	35318	97513	61537	54955	08159	00337	80778	27507	95478	21252	12746	37554	97775
124	11946	22681	45045	13964	57517	59419	58045	44067	58716	58840	45557	96345	33271	53464
125	96518	48688	20996	11090	48396	57177	83867	86464	14342	21545	46717	72364	86954	55580
126	35726	58643	76869	84622	39098	36083	72505	92265	23107	60278	05822	46760	44294	07672
127	39737	42750	48968	70536	84864	64952	38404	94317	65402	13589	01055	79044	19308	83623
128	97025	66492	56177	04049	80312	48028	26408	43591	75528	65341	49044	95495	81256	53214
129	62814	08075	09788	56350	76787	51591	54509	49295	85830	59860	30883	89660	96142	18354
130	25578	22950	15227	83291	41737	79599	96191	71845	86899	70694	24290	01551	80092	82118
131	68763	69576	88991	49662	46704	63362	56625	00481	73323	91427	15264	06969	57048	54149
132	17900	00813	64361	60725	88974	61005	99709	30666	26451	11528	44323	34778	60342	60388
133	71944	60227	63551	71109	05624	43836	58254	26160	32116	63403	35404	57146	10909	07346
134	54684	93691	85132	64399	29182	44324	14491	55226	78793	34107	30374	48429	51376	09559
135	25946	27623	11258	65204	52832	50880	22273	05554	99521	73791	85744	29276	70326	60251
136	01353	39318	44961	44972	91766	90262	56073	06606	51826	18893	83448	31915	97764	75091
137	99083	88191	27662	99113	57174	35571	99884	13951	71057	53961	61448	74909	07322	80960
138	52021	45406	37945	75234	24327	86978	22644	87779	23753	99926	63898	54886	18051	96314
139	78755	47744	43776	83098	03225	14281	83637	55984	13300	52212	58781	14905	46502	04472
140	25282	69106	59180	16257	22810	43609	12224	25643	89884	31149	85423	32581	34374	70873
141	11959	94202	02743	86847	79725	51811	12998	76844	05320	54236	53891	70226	38632	84776
142	11644	13792	98190	01424	30078	28197	55583	05197	47714	68440	22016	79204	06862	94451
143	06307	97912	68110	59812	95448	43244	31262	88880	13040	16458	43813	89416	42482	33939
144	76285	75714	89585	99296	52640	46518	55486	90754	88932	19937	57119	23251	55619	23679
145	55322	07598	39600	60866	63007	20007	66819	84164	61131	81429	60676	42807	78286	29015
146	78017	90928	90220	92503	83375	26986	74399	30885	88567	29169	72816	53357	15428	86932
147	44768	43342	20696	26331	43140	69744	82928	24988	94237	46138	77426	39039	55596	12655
148	25100	19336	14605	86603	51680	97678	24261	02464	86563	74812	60069	71674	15478	47642
149	83612	46623	62876	85197	07824	91392	58317	37726	84628	42221	10268	20692	15699	29167
150	41347	81666	82961	60413	71020	83658	02419	33322	66036	98712	46795	16308	28413	05417

Page 4

Line \ Col.	(1)	(2)	(3)	(4)	(5)	(6)	(7)	(8)	(9)	(10)	(11)	(12)	(13)	(14)
151	38128	51178	75096	13609	16110	73533	42564	59870	29399	67834	91055	89917	51096	89011
152	60950	00455	73254	96067	50717	13878	03216	78274	65863	37011	91283	33914	91303	49326
153	90524	17320	29832	96118	75792	25326	22940	24904	80523	38928	91374	55597	97567	38914
154	49897	18278	67160	39408	97056	43517	84426	59650	20247	19293	02019	14790	02852	05819
155	18494	99209	81060	19488	65596	59787	47939	91225	98768	43688	00438	05548	09443	82897
156	65373	72984	30171	37741	70203	94094	87261	30056	58124	70133	18936	02138	59372	09075
157	40653	12843	04213	70925	95360	55774	76439	61768	52817	81151	52188	31940	54273	49032
158	51638	22238	56344	44587	83231	50317	74541	07719	25472	41602	77318	15145	57515	07633
159	69742	99303	62578	83575	30337	07488	51941	84316	42067	49692	28616	29101	03013	73449
160	58012	74072	67488	74580	47992	69482	58624	17106	47538	13452	22620	24260	40155	74716
161	18348	19855	42887	08279	43206	47077	42637	45606	00011	20662	14642	49984	94509	56380
162	59614	09193	58064	29086	44385	45740	70752	05663	49081	26960	57454	99264	24142	74648
163	75688	28630	39210	52897	62748	72658	98059	67202	72789	01869	13496	14663	87645	89713
164	13941	77802	69101	70061	35460	34576	15412	81304	58757	35498	94830	75521	00603	97701
165	96656	86420	96475	86458	54463	96419	55417	41375	76886	19008	66877	35934	59801	00497
166	03363	82042	15942	14549	38324	87094	19069	67590	11087	68570	22591	65232	85915	91499
167	70366	08390	69155	25496	13240	57407	91407	49160	07379	34444	94567	66035	38918	65708
168	47870	36605	12927	16043	53257	93796	52721	73120	48025	76074	95605	67422	41646	14557
169	79504	77606	22761	30518	28373	73898	30550	76684	77366	32276	04690	61667	64798	66276
170	46967	74841	50923	15339	37755	98995	40162	89561	69199	42257	11647	47603	48779	97907
171	14558	50769	35444	59030	87516	48193	02945	00922	48189	04724	21263	20892	92955	90251
172	12440	25057	01138	38611	88135	68089	10954	10097	54243	06460	50856	65435	79377	53890
173	32293	29938	68653	10497	98919	46587	77701	99119	93165	67788	17638	23097	21468	36992
174	10640	21875	72462	77981	56550	55999	87310	69643	45124	00349	25748	00844	96831	30651
175	47615	23169	39571	56972	20628	21788	51736	33133	72696	32605	41569	76148	91544	21121
176	16948	11128	71624	78754	49084	96303	27830	45817	67867	18062	87453	17226	72904	71474
177	21258	61092	66634	70335	92448	17354	83432	49608	66520	06442	59664	20420	39201	69549
178	15072	48853	15178	30730	47481	48490	41436	25015	49932	20474	53821	51015	79841	32405
179	99154	57412	09858	65671	70655	71479	63520	31357	56968	06729	34465	70685	04184	25250
180	08759	61089	23706	32994	35426	36666	63988	98844	37533	08269	27081	45886	22835	78451
181	67323	57839	61114	62192	47547	58023	64630	34886	98777	75442	95592	06141	45096	73117
182	09255	13986	84834	20764	72206	89393	34548	93438	88730	61805	78955	18952	46436	58740
183	36304	74712	00374	10107	85061	69228	81969	92216	03568	39630	81869	52824	50937	27954
184	15884	67429	86612	47367	10242	44880	12060	44309	46629	55105	66793	93173	00480	13311
185	18745	32031	35303	08134	33925	03004	59929	95418	04917	57596	24878	61733	92834	64454
186	72934	40086	88292	65728	38300	42323	64068	98373	48971	09049	59943	36538	05976	82118
187	17626	02944	20910	57662	80181	38579	24580	90529	52303	50436	29401	57824	86039	81062
188	27117	61399	50967	41399	81636	16663	15634	79717	94696	59240	25543	97989	63306	90946
189	93995	18678	90012	63645	85701	85269	62263	68331	00389	72571	15210	20769	44686	96176
190	67392	89421	09623	80725	62620	84162	87368	29560	00519	84545	08004	24526	41252	14521
191	04910	12261	37566	80016	21245	69377	50420	85658	55263	68667	78770	04533	14513	18099
192	81453	20283	79929	59839	23875	13245	46808	74124	74703	35769	95588	21014	37078	39170
193	19480	75790	48539	83703	15537	48885	02861	86587	74539	65227	90799	58789	96257	02708
194	21456	13162	74608	81011	55512	07481	93551	72189	76261	91206	89941	15132	37738	59284
195	89406	20912	46189	76376	25538	87212	20748	12831	57166	35026	16817	79121	18929	40628
196	09866	07414	65977	16419	01101	69343	13305	94302	80703	57910	36933	57771	48546	03003
197	86541	24681	23421	13521	28000	94917	07423	57823	97234	63951	42876	46829	09781	58160
198	10414	96941	06205	72222	57167	83902	07460	69507	10600	08858	07685	44472	64220	27040
199	49942	06683	41479	58982	56288	42853	92196	20532	62045	78812	35895	51851	83534	10669
200	23995	68882	42291	23374	24299	27024	67460	94703	40937	16961	26053	78749	46704	21983

Page 5

Line \ Col.	(1)	(2)	(3)	(4)	(5)	(6)	(7)	(8)	(9)	(10)	(11)	(12)	(13)	(14)
201	78994	36244	02673	25475	84953	61793	50243	63423	69309	80308	49977	18075	43227	08266
202	04909	58485	70686	93930	34880	73059	06823	80257	44193	08337	47655	75932	29209	41954
203	46582	73570	33004	51795	86477	46736	60460	70345	37322	19987	67143	41129	89514	46892
204	29242	89792	88634	60285	07190	07795	27011	85941	01852	43096	31173	43730	48505	17958
205	68104	81339	97090	20601	78940	20228	22803	96070	10251	62711	66200	74330	13820	18966
206	17156	02182	82504	19880	93747	80910	78260	25136	62018	62919	73801	57195	83457	70597
207	50711	94789	07171	02103	99057	98775	37997	18325	88281	61091	97889	79977	04544	72963
208	39449	52409	75095	77720	39729	03205	09313	43545	43786	70443	41350	73369	42405	80516
209	75629	82729	76916	72657	58992	32756	01154	84890	04107	17469	69346	68651	97433	89491
210	01020	55151	36132	51971	32155	60735	64867	35424	25257	93844	39928	52519	34368	02114
211	08337	89989	24260	08618	66798	25889	52860	57375	52815	43539	18072	44270	27309	56535
212	76829	47229	19706	30094	69430	92399	98749	22081	52564	90431	35208	40323	87505	10227
213	39708	30641	21267	56501	95182	72442	21445	17276	90344	33199	02522	97883	09515	65930
214	89836	55817	56747	75195	06818	83043	47403	58266	52630	75573	91088	41118	27195	40650
215	85903	61370	66081	54076	67442	52964	23823	02718	28786	06121	29680	55295	67086	57574
216	71345	03422	01015	68025	19703	77313	04555	83425	46763	95315	23150	15116	18017	42730
217	61454	92263	14647	08473	34124	10740	40839	05620	62418	73374	92577	06755	21856	56272
218	80376	08909	30470	40200	46558	61742	11643	92121	22294	26648	69676	46198	00331	85186
219	45144	54373	05505	90074	24783	86299	20900	15144	26506	53770	76431	23861	71208	80694
220	12191	88527	58852	51175	11534	87218	04876	85584	78465	82182	03412	13217	14313	70593
221	62936	59120	73957	35969	21598	47287	39394	08778	38036	30140	89117	32054	44603	61849
222	31588	96798	43668	12611	01714	77266	55079	24690	84716	77732	35363	85525	17015	56344
223	20787	96048	84726	17512	39450	43618	30629	24356	05294	34236	65299	36922	46995	65765
224	45603	00745	84635	43079	52724	14262	05750	89373	79088	38088	65082	92504	80545	03090
225	31606	64782	34027	56734	09365	20008	93559	78384	99219	61747	96111	86965	33233	29812
226	10452	33074	76718	99556	16026	00013	78411	95107	10786	44886	44612	06830	27848	87597
227	37016	64633	67301	50949	91298	74968	73631	57397	08632	04762	59328	34926	07403	60916
228	66725	97865	25409	37498	00816	99268	14471	10232	19035	21695	07540	96447	20743	92472
229	07380	74438	82120	17890	40963	55757	13492	68294	87170	49468	40164	13374	23021	17006
230	71621	57688	58256	47702	74724	89419	08025	68519	95138	54788	32999	34374	05780	17506
231	03466	13263	23917	20417	11315	52805	33072	07723	87876	75258	22709	99869	11609	46666
232	12692	32931	97387	34822	53775	91674	76549	37635	91118	31062	89441	31839	88614	78168
233	52192	30941	44998	17833	94563	23062	95725	38463	03665	49189	46359	37401	73407	61817
234	56691	72529	66063	73570	86860	68125	40436	31303	79330	59083	34862	00540	21734	75535
235	74952	43041	58869	15677	78598	43520	97521	83248	52173	17636	77106	01044	22990	74874
236	18752	43693	32867	53017	22661	39610	03796	02622	78267	24503	73518	76545	99088	08369
237	61691	04944	43111	28325	82319	65589	66048	98498	46941	81427	44447	70357	18864	15525
238	49197	63948	38947	60207	70667	39843	60607	15328	09528	17277	84278	04463	12188	35359
239	19436	87291	71684	74859	76501	93456	95714	93518	10683	75617	78841	25315	74041	71554
240	39143	64893	14606	13543	09621	68301	69817	52140	03976	48795	60266	99592	68334	18790
241	82244	67549	76491	09761	74494	91307	64222	66592	67270	38593	18094	95095	08649	25047
242	55847	56155	42878	23708	97999	40131	52360	90390	73108	40475	80487	07787	35238	50990
243	94095	95970	67826	25991	37584	56966	68623	83454	49461	97707	12479	25041	40565	18313
244	11751	69469	25521	44097	07511	88976	30122	67542	54825	03274	02765	67162	40312	76127
245	69902	08995	27821	11758	64989	61902	32121	28165	21326	97375	44801	66977	08232	06807
246	21850	25352	25556	92161	23592	43294	10479	37879	21826	11453	29584	70067	09471	16319
247	75850	46992	25165	55906	62339	88958	91717	15756	78817	35541	01177	06869	10543	57652
248	29648	22086	42581	85677	20251	39641	65786	80689	49066	14456	91681	69371	18292	39377
249	82740	28443	42734	25518	82827	35825	90288	32911	79666	52959	01475	83321	24991	80102
250	36842	42092	52075	83926	42875	71500	69216	01350	92846	84792	87455	06842	22422	77379

Page 6

Line \ Col.	(1)	(2)	(3)	(4)	(5)	(6)	(7)	(8)	(9)	(10)	(11)	(12)	(13)	(14)
251	89429	26726	15563	94972	78739	04419	60523	31022	23728	37647	16476	11170	68376	56874
252	43427	25412	25587	21276	44426	17369	29010	45337	90845	98053	41447	14897	18753	68291
253	58575	81958	51846	02676	67781	95137	88430	78260	66962	31812	12759	06427	40337	50115
254	61888	71246	24246	23487	78639	92006	63846	92263	33212	26516	93662	72399	88244	33922
255	73891	47025	40937	71907	26827	98865	38882	25757	26662	91441	89357	87803	61521	80600
256	40938	73894	40854	15997	55293	95033	31736	75068	91314	75293	04895	39355	54837	57203
257	98053	43567	17292	86908	71364	06089	92394	73691	57883	09983	35643	79309	53449	95334
258	59774	29138	46993	39836	99596	59050	25419	04130	54632	17223	94604	22973	97731	99476
259	09765	07548	63043	59782	81449	13652	94420	74460	46707	94303	85523	95244	70995	10742
260	38991	64502	24770	29209	82909	66610	84418	66214	26001	78685	69117	72446	79783	22305
261	25622	27100	56128	62145	82388	45197	97609	83942	01120	71717	32858	58679	97165	02810
262	31864	74120	66231	82306	91784	33177	17681	18963	07216	49288	43185	62797	00735	27085
263	81171	75639	60863	49562	28846	81581	10249	23190	53440	32357	16472	99013	24328	93670
264	69874	52803	28544	51569	56090	44558	42095	92311	57915	13368	13719	15833	38744	56065
265	27848	51107	05761	02159	53911	01952	59273	32250	39647	89908	49075	23061	07795	95047
266	69407	69736	75375	31488	67528	84234	76462	13628	21886	13736	67478	45818	27867	93049
267	29418	03091	06364	13151	40663	43633	87954	69800	24773	62596	52476	60631	50503	94116
268	38222	31231	79415	44558	62490	26936	49682	16307	98535	44822	99574	58487	85020	68881
269	94720	83796	93251	03568	62484	29140	14152	37044	90398	92042	35099	31640	99753	44409
270	45275	16852	02284	41361	73733	61486	33189	08907	41189	08147	15472	33250	17361	79961
271	97260	09552	82626	42915	45847	87401	13339	53850	34931	00602	75307	99708	77863	04924
272	01990	65259	60684	78175	43825	45211	86287	78190	02431	66251	74970	50246	23975	80697
273	24633	42314	81192	50253	67516	59076	92006	65676	87343	89231	15760	73706	69426	01979
274	98071	52677	74920	74461	52266	26967	68284	31612	40335	28865	98949	64492	96905	29184
275	34101	79442	88403	48541	13010	16596	72001	38546	76305	22119	82668	84017	44111	40302
276	77186	93967	85918	66403	73837	73445	86663	15929	08237	05647	15785	70444	58670	95967
277	23114	05481	42335	51396	60823	22680	50459	05429	35227	92559	24136	13126	22099	52388
278	59988	49944	41038	99977	16348	41119	51548	19511	90148	65604	16147	63445	60585	10480
279	11852	42254	82304	05588	75165	20179	94198	25700	33473	59554	30974	69973	57629	38550
280	59992	87922	56299	01700	07003	97507	69260	53349	86947	27517	80159	01899	46890	53850
281	42116	86593	22828	41422	18176	03250	06079	85467	32052	56922	96804	51060	33157	83948
282	39663	61401	21471	42702	70588	53144	27087	05591	57759	51394	98873	45625	61069	78783
283	53542	72009	96296	68908	58657	87117	21483	28879	20480	57309	95552	09826	79928	17141
284	25996	76108	98476	36397	89457	19577	65877	04802	61938	25032	09190	74932	36985	82686
285	91106	26450	14461	50328	29084	32332	08635	25192	31337	20249	95073	93800	70022	99968
286	37133	88924	27845	13024	90687	23726	11212	30414	42185	49224	46560	80447	24334	74866
287	13982	25736	10087	16762	02564	27250	79316	83848	38684	80552	44402	85153	94526	41256
288	26663	36187	81688	25005	46677	75851	73938	73044	05132	61204	90384	90296	03182	36672
289	62572	08275	16313	24936	81690	53829	40412	01479	24241	58488	65341	93414	07135	43446
290	65925	95455	08383	24643	72962	08172	37824	87587	40698	34964	50166	74756	77033	41501
291	97978	74676	08942	48919	51592	71196	48534	16955	25759	95645	03148	10646	15660	86520
292	01914	42524	67820	47985	91773	10383	89514	07557	02084	16736	39198	69697	62485	61938
293	68565	44811	39238	70394	78555	33539	56310	40809	63204	14479	19635	97899	66947	58010
294	54370	31672	03893	32423	54092	69375	63308	00016	28407	98287	22874	57545	72695	01604
295	79954	89601	23881	46951	69084	33477	87968	15639	82409	34125	36864	52118	27102	87334
296	55479	01059	44229	56975	06785	80930	26443	44898	77561	51123	34495	31376	06238	15973
297	38114	70330	42157	86699	46212	74692	92603	91306	58558	57280	50639	20863	71670	81487
298	29766	83452	66202	02488	72704	97821	70614	53616	39050	30355	15340	97298	41795	35185
299	31771	70640	34779	41831	33456	53194	19602	74194	61154	51774	76822	73794	54182	45264
300	77582	87106	83577	99067	23835	48668	31503	54829	54783	13177	15307	86073	68915	88415

Page 7

Line \ Col.	(1)	(2)	(3)	(4)	(5)	(6)	(7)	(8)	(9)	(10)	(11)	(12)	(13)	(14)
301	64670	10396	82981	58320	71478	08143	48294	42631	45464	58092	14187	12271	98179	87812
302	25771	02205	73984	28436	88192	11470	11775	67385	66360	59884	93873	29948	66302	82227
303	27551	13537	54984	89406	88326	33993	92324	13249	35271	60400	70762	08343	76456	90068
304	91224	22417	44820	26189	57541	87558	45835	28461	54835	92411	44369	47512	49508	02841
305	75179	64320	71523	67868	38883	09674	27645	76240	47587	01677	38342	85598	12482	30749
306	64654	91085	65818	03313	39273	46384	66677	14148	87552	38383	67435	21072	63866	74644
307	98059	81123	67832	04102	66188	78200	67466	46043	65406	22834	08620	17509	51424	25187
308	38765	63585	18810	95805	11414	58096	00295	82626	42683	44518	12209	83245	53771	95469
309	01921	03564	71754	10213	80383	13473	94128	62199	59411	46782	62871	51149	87146	40129
310	16211	93671	27704	66778	96307	06732	63750	04191	40003	51653	54228	14916	05361	08884
311	70832	86076	61527	56123	48514	53935	86784	42351	67586	07432	61499	01773	97463	58615
312	22332	94265	67627	85815	00394	75271	98385	53697	56378	50592	77441	88505	89791	16331
313	81333	45965	64171	84367	15052	37965	03122	81914	69381	70034	92563	61804	58326	97895
314	39333	47453	66174	04546	10594	64271	61026	39471	55981	18628	67943	35599	37209	34061
315	29195	20825	50878	80273	26285	90070	79586	12449	77293	36577	59192	03658	90056	83145
316	74420	64037	06960	25109	08821	60143	34485	19257	29417	72713	72326	41572	41553	46946
317	22763	16508	24866	13177	07464	51730	65802	95718	28560	11332	74272	59189	53167	13133
318	72919	54618	40616	33287	51274	78491	53604	66742	97777	64468	98224	45485	17257	31561
319	92385	42402	15922	90033	21555	31647	22288	75692	20592	84620	58679	24587	83517	55327
320	85431	19857	97246	46118	71222	82744	67892	77155	10785	00344	19641	98279	18716	13895
321	40778	12451	14921	51464	45331	75822	46859	66829	35803	27645	76095	41535	25508	53066
322	88903	46592	60637	65231	08778	86813	47819	19218	46837	89671	77661	08518	85216	62664
323	29830	34899	85457	19548	83355	52479	77801	01596	48890	56104	68733	40830	58611	59181
324	22832	47422	08073	10107	46772	92299	42975	86376	27869	52954	07900	75918	51398	87598
325	75159	14809	11930	83531	51239	86298	72661	63015	98804	98491	99565	42801	71816	84000
326	99390	08217	56276	09263	82685	30451	25742	41105	74711	42007	02082	93025	86641	28952
327	68628	80897	08902	10867	91379	30068	84289	45020	92459	03831	08531	63496	98230	42884
328	92393	95901	41179	72129	72502	91097	09488	84896	37720	68104	73817	67626	16221	63527
329	53122	66033	38229	51879	29925	45574	53938	72801	64067	76328	28941	43645	37181	95329
330	43251	11941	86631	93264	53433	70281	55000	24550	74751	32855	25399	95743	85393	20261
331	16613	24901	34866	75002	55163	68308	20070	36953	39378	71191	84510	47599	93608	24379
332	12010	60852	92603	70393	17989	95755	14672	58786	41996	02893	94163	36156	54203	94138
333	85528	97879	27814	08219	02908	71582	31439	00360	72264	87245	65903	42298	28061	81889
334	32590	55079	33556	83169	92087	77939	53792	78795	58159	86394	41749	91623	26973	81474
335	92934	30650	16449	15805	61551	38689	59179	85485	18537	70496	98694	19796	76804	03673
336	80614	10150	09389	61892	79477	14522	40270	45744	29588	29717	39590	10223	43049	78775
337	62398	12034	90764	52872	22285	50592	42505	80560	38213	18917	10015	03887	62589	15851
338	02222	46811	05145	67916	15184	02636	59078	57773	21259	86090	56705	65566	04487	95954
339	08690	31785	61664	61382	24149	21471	23328	03093	31266	14840	30703	01640	07874	16630
340	61187	73897	66168	12885	73191	89432	65414	41886	75911	35708	43208	59193	04727	31037
341	12324	61149	85643	64999	63738	46671	25408	69313	54455	04917	35047	09951	72776	84697
342	47635	42279	98620	70677	52386	50904	97403	03931	42090	28179	98028	47728	45696	74176
343	70965	00390	08878	15373	70276	71889	86953	37931	23286	20508	40100	22486	37323	35429
344	58764	15262	96814	54548	00042	19721	78869	85937	36639	29135	12633	67225	69588	74178
345	07429	05609	31207	50254	68389	07714	92268	64698	32823	60122	46213	05646	54742	98304
346	15665	28659	54952	53217	76898	88931	25786	55912	85269	29212	84976	08888	94332	58528
347	64208	53232	99459	43605	04553	48451	68154	49436	49891	65524	65133	55163	76765	26006
348	17952	73276	52567	48489	64264	24220	55498	97548	98437	26033	39026	17377	43519	27425
349	60531	43217	39999	38615	97195	76928	87688	99010	90189	12522	00675	01995	82781	95130
350	76692	39999	43254	68110	88053	88727	14187	98623	84225	78440	67082	37425	40559	16838

Page 8

Line \ Col.	(1)	(2)	(3)	(4)	(5)	(6)	(7)	(8)	(9)	(10)	(11)	(12)	(13)	(14)
351	06433	80674	24520	18222	10610	05794	37515	48619	62866	33963	14045	79451	04934	45576
352	39298	47829	72648	37414	75755	04717	29899	78817	03509	78673	73181	29973	18664	04555
353	89884	59651	67533	68123	17730	95862	08034	19473	63971	37271	31445	49019	49405	46925
354	61512	32155	51906	61662	64130	16688	37275	51262	11569	08697	91120	64156	40365	74297
355	99653	47635	12506	88535	36553	23757	34209	55803	96275	26130	47949	14877	69594	83041
356	95913	11085	13772	76638	48423	25018	99041	77529	81360	18180	97421	55541	90275	18213
357	55864	44004	13122	44115	01601	50541	00147	77685	58788	33016	61173	93049	04694	43534
358	35334	82410	91601	40617	72876	33967	73830	15405	96554	88265	34537	38526	67924	40474
359	57729	88646	76487	11622	96297	24160	09903	14047	22917	60718	66487	46346	30949	03173
360	86648	89317	63677	70119	94739	25875	38829	68377	43918	77653	04127	69930	43283	35766
361	30574	06039	07967	32422	76791	39725	53711	93385	13421	67957	20384	58731	53396	59723
362	81307	13114	83580	79974	45929	65113	72268	09858	52104	32014	53115	03727	98624	84616
363	02410	96385	79007	54939	21410	86980	91772	93307	34116	49516	42148	57740	31198	70336
364	18969	87444	52233	62319	08598	09066	95288	04794	01534	92058	03157	91758	80611	45357
365	87863	80514	66860	62297	80198	19347	73234	86265	49096	97021	92882	61422	75890	86442
366	68397	10538	15438	62311	72844	60203	46412	65943	79232	45702	67055	39024	57383	44424
367	28529	45247	58729	10854	99058	18260	38765	90038	94209	04055	27393	61517	23002	96560
368	44285	09452	15867	70418	57012	72122	36634	97283	95943	78363	36498	40662	94188	18202
369	86299	22510	33571	23309	57040	29285	67870	21913	72958	75637	99936	58715	07943	23748
370	84842	05748	90894	61658	15001	94055	36308	41161	37341	81838	19389	80336	46346	91895
371	56970	10799	52098	04184	54967	72938	56834	23777	98392	31417	98547	92058	02277	50315
372	83125	85077	60490	44369	66130	72936	69848	59973	08144	61070	73094	27059	69181	55623
373	55503	21383	02464	26141	68779	66388	75248	82690	74099	77885	23813	10054	11900	44653
374	47019	06683	33203	29608	54553	25971	69573	83854	24715	48866	65745	31131	47636	45137
375	84828	61152	79526	29554	84580	37859	28504	61980	34997	41825	11623	07320	15003	56774
376	68921	31331	79827	05748	51276	57143	31986	99915	45821	97702	87125	44488	77613	56823
377	36458	28285	30424	98420	72925	40729	22337	48293	86847	43186	42951	37804	85129	28993
378	95752	96065	36847	87729	81679	59126	59437	33225	31280	41232	34750	91097	60752	69783
379	26768	02513	58454	56958	20575	76746	49878	06846	32828	24425	30249	78801	26977	92074
380	42613	72456	43636	58085	06766	60227	96414	32671	45587	79620	84831	38156	74211	82752
381	95457	12176	65482	25596	02678	54592	63607	82096	21913	75544	55228	89796	05694	91552
382	95276	67524	63564	95958	39750	64379	46059	51666	10433	10945	55306	78562	89630	41830
383	66954	53574	64776	92345	95110	59448	77249	54044	67942	24145	42294	27427	84875	37022
384	17457	44151	14113	62462	02798	54977	48349	66738	60184	75679	38180	17640	36242	99357
385	03704	23322	83214	59337	01695	60666	97410	55064	17427	89180	74018	44865	53197	74810
386	21538	16997	33210	60337	27976	70661	08250	69599	60264	84549	78007	88450	06488	72274
387	57178	16739	98310	70348	11317	71623	55510	64756	87759	92354	78694	63638	80939	98644
388	31048	40058	94953	55866	96283	46620	52087	80817	74533	68407	55862	32476	19326	95558
389	69799	83300	16498	80733	96422	58078	99643	39847	96884	84657	33697	39578	90197	80532
390	90595	65017	59231	17772	67831	33317	00520	90401	41700	95510	61166	33757	23279	85523
391	33570	34761	98939	78784	09977	29398	93896	78227	90110	81378	96659	37008	04050	04228
392	15340	82760	57477	13898	48431	72936	78160	87240	52716	87697	79433	16336	52862	69149
393	64079	07733	36512	56186	99098	48850	72527	08486	10951	26832	39763	02485	71688	90936
394	63491	84886	67118	62063	74958	20946	28147	39338	32169	03713	93510	61244	73774	01245
395	92003	76568	41034	28260	79708	00770	88643	21188	01850	69689	49486	49128	14660	14143
395	52360	46658	66511	04172	73085	11795	52594	13287	82531	04388	64693	11934	35061	68576
397	74622	12142	68355	65635	21828	39539	18988	53609	04001	19648	14083	49623	10840	31915
398	04157	50079	61343	64315	70836	82857	35335	87900	36194	31567	53506	34304	39910	79630
399	86003	60070	66241	32836	27573	11479	94114	81641	00496	36058	75899	46620	70024	88753
400	41268	80187	20351	09636	84668	42486	71303	19512	50277	71508	20116	79520	06269	74173

Page 9

Line \ Col.	(1)	(2)	(3)	(4)	(5)	(6)	(7)	(8)	(9)	(10)	(11)	(12)	(13)	(14)
401	05073	90103	85167	53900	19720	41488	57476	39458	16621	69774	47953	35039	39283	21573
402	93320	80269	56684	39192	53220	74539	26393	00787	94490	23386	38454	33466	32159	77439
403	18806	70257	96424	13606	14356	76599	25390	63236	04513	16358	30540	10551	32498	18685
404	22253	45923	29815	18578	23316	30896	64771	11220	86218	75956	22399	36234	61644	80682
405	93640	45982	40011	74142	29106	45729	43406	21457	04301	39651	76025	73819	11462	97385
406	47630	45980	76619	57138	57492	00030	77897	76236	64990	35985	57748	11606	72081	18359
407	01781	55061	07455	47083	71870	90597	10151	59606	96919	31174	99872	15843	99173	79512
408	69694	45054	33587	03664	95007	31567	25334	26433	75002	67607	33135	07076	82984	82675
409	51236	05052	26503	94661	29874	73492	88941	08488	09418	08173	63380	83067	68143	64983
410	89445	51039	73837	26720	38650	47322	68474	95047	20404	41577	46865	39849	78735	99192
411	40867	96834	02162	41517	88937	26099	56047	49164	35127	64916	75461	79160	14014	00445
412	92946	56944	93407	05010	54896	33173	30548	23667	43171	47849	40449	91072	91092	17613
413	75898	02275	90768	31902	52114	36634	46803	97970	92216	55398	75320	70475	82931	20172
414	22729	21695	90824	80500	09332	54667	46696	38166	02005	24615	85613	25948	75389	25765
415	28733	62663	23644	16416	47135	39137	62190	31032	58702	03805	67252	23712	92697	19071
416	51323	37770	42114	79742	59905	38480	25293	32993	36946	62701	51198	72941	52215	85257
417	69325	65551	49927	68073	56979	49454	79451	60753	70872	07422	06399	75240	80847	78231
418	11333	60801	36992	76128	27959	41306	93543	15926	99159	27102	98684	80175	98732	45405
419	86347	03703	36778	72501	95229	65735	14269	50220	77270	68604	05677	23347	43686	31584
420	73452	36179	82893	92262	43850	31888	71151	40682	49775	63628	45415	96270	31735	01509
421	75483	74009	73699	05870	36804	89338	73891	40740	98753	74566	74733	34777	05786	38294
422	73302	84917	75128	34085	86208	98399	79433	61960	01720	87458	24023	89971	09532	68155
423	42785	24350	05933	65282	12832	75382	29826	33197	81781	53542	63985	57022	22712	61343
424	40429	33209	68622	09308	38098	55947	12001	73526	23170	13721	37856	86502	74299	01346
425	92876	58271	99325	12301	72957	22690	62705	73892	01974	77759	92733	11331	08323	86196
426	32951	39844	99126	94838	48715	36586	42076	15283	19280	29166	24522	73131	83401	38920
427	09772	28139	48130	73301	35915	90923	19255	75242	84655	30163	75510	83315	98529	93805
428	78459	91322	50072	77941	65046	78363	21951	42319	46472	67617	34134	05905	61251	51040
429	14419	96517	99075	43664	81119	63487	95589	51785	07398	23245	10086	49097	46173	00507
430	97769	50967	24427	21011	92226	44380	23422	10654	43617	80504	90663	60751	79728	41132
431	09175	37545	39088	06879	21277	05153	81855	84043	35307	59465	75395	74758	09427	84460
432	52062	95519	54087	14072	50953	63477	64635	34552	75243	70222	75023	81454	70606	31861
433	70558	85169	01086	97202	10390	01819	88167	21851	87837	85287	69883	08289	74968	46947
434	22553	61317	08968	67521	16627	48855	97263	94242	93354	72446	28840	88195	82751	94352
435	95216	75263	60351	02643	00063	20824	67468	89441	84055	47035	29741	47972	61914	66864
436	49087	61399	47781	32173	96672	04528	15881	46764	20115	03226	79308	31970	49804	85150
437	24808	79068	70787	43106	97133	37236	77888	48451	20788	44648	70360	64965	57715	94826
438	89879	79942	43781	05069	80143	59176	47392	70372	26899	16228	71205	14564	97087	95690
439	61178	79295	58926	21977	28435	32631	23062	31822	70462	05965	22312	33013	74612	83733
440	37444	56047	23208	34710	12147	28568	58817	98807	56775	08129	08794	23646	92846	61706
441	99633	00363	16853	20789	87674	03938	36077	41012	08813	51168	78822	37353	61281	31172
442	87363	59239	42023	78056	51254	95644	90527	41398	74996	94977	22149	96616	54435	52469
443	23923	87269	85277	34727	78036	74471	12157	11655	25194	47557	26181	67825	80224	41490
444	45610	26370	13094	34500	36750	54517	85011	26567	01021	32485	58903	43529	24191	91832
445	44166	80095	08286	38126	48834	73423	13617	08853	16286	16023	77901	39118	14288	39385
446	81875	27486	53925	22330	37168	97954	11967	03309	97096	64221	11318	98720	01100	13651
447	79400	83852	52174	42577	18553	14023	69629	61913	41050	69689	57284	38160	57756	16762
448	42799	46647	36718	49704	17150	07935	62372	39933	20838	27652	54801	41067	08240	35163
449	09302	36408	64569	93033	95645	56791	14830	81699	45057	85796	63756	93944	60649	84847
450	88078	81456	17242	84590	93660	34619	51965	85618	36558	54410	68456	98504	83011	19393

Page 10

Line \ Col.	(1)	(2)	(3)	(4)	(5)	(6)	(7)	(8)	(9)	(10)	(11)	(12)	(13)	(14)
451	85018	23508	91507	76455	54941	72711	39406	94620	27963	96478	21559	19246	88097	44926
452	11904	73678	08272	62941	02349	71389	45605	60947	60775	73181	43264	56895	04232	59604
453	75344	98489	86268	73652	98210	44546	27174	27499	53523	63110	57106	20865	91683	80688
454	65566	65614	01443	07607	11826	91326	29664	01603	23156	89223	43429	95353	44662	59433
455	51872	72294	95432	53555	96810	17100	35066	00815	01552	06392	31437	70385	45863	75971
456	03805	37913	98633	81009	81060	33449	68055	83844	90942	74857	52419	68723	47830	63010
457	21055	78685	71250	10329	56135	80647	51404	06626	10042	93629	37609	57215	08409	81906
458	48977	36794	56054	59843	57361	65304	93858	56760	63348	24949	11859	29793	37457	59377
459	93077	72941	92779	23581	24548	56415	61927	64416	29934	00755	09418	14830	62887	92683
460	84533	26564	91583	83411	66504	02036	02922	63569	17906	38076	32135	19096	96970	75917
461	11338	12903	14514	27585	45068	05520	56321	22693	35089	07694	04252	23791	60249	83010
462	23853	68500	92274	87026	99717	01542	72990	43413	59744	44595	71326	91382	45114	20245
463	94096	74920	25822	98026	05394	61840	83089	09224	78530	33996	49965	04851	18280	14039
464	83160	88362	09350	98536	38155	42661	02363	67625	34683	95372	74733	63558	09665	22610
465	97425	47335	69709	01386	74319	04318	99387	86874	12549	38369	54952	91579	26023	81076
466	83951	11954	24317	20345	18134	90062	10761	54548	49505	52685	63903	13193	33905	66936
467	93085	35203	05740	03206	92012	42710	34650	73236	66167	21788	03581	40699	10396	81827
468	33762	83193	58045	89880	78101	44392	53767	15220	66319	72953	14071	59148	95154	72852
469	49665	85397	85137	30496	23469	42846	94810	16151	08029	50554	03891	38313	34016	18671
470	37541	82627	80051	72521	35342	56119	97190	43635	84249	61254	80993	55431	90793	62603
471	22145	85304	35348	82854	55846	18076	12415	30193	42776	85611	57635	51362	79907	77364
472	27153	08662	61078	52433	22184	33998	87436	37430	45246	11400	20986	43996	73122	88474
473	00301	49425	66682	25442	83668	66236	79655	88312	93047	12088	86937	70794	01041	74867
474	43815	43272	73778	63469	50083	70696	13558	98995	58159	04700	90443	13168	31553	67891
475	14689	86482	74157	46012	97765	27552	49617	51734	20849	70198	67906	00880	82899	66065
476	16680	55936	82453	19532	49988	13176	94219	88698	41755	56216	66852	17748	04963	54859
477	86938	60429	01137	86168	78257	86249	46134	51865	09836	73966	65711	41699	11732	17173
478	33944	29219	73161	46061	30946	22210	79302	40300	08852	27528	84648	79589	95295	72895
479	16045	67736	18608	18198	19468	76358	69203	02760	28625	70476	76410	32988	10194	94917
480	37044	52523	25627	63107	30806	80857	84383	78450	26245	91763	73117	33047	03577	62599
481	61471	45322	35340	35132	42163	69332	98851	50252	56911	62693	73817	98693	18728	94741
482	47422	21896	16785	66393	39249	51463	95963	07929	66728	47761	81472	44806	15592	71357
483	24133	39719	14484	58613	88717	29289	77360	09030	39605	87507	85446	51257	89555	75520
484	67253	67064	10748	16006	16767	57345	42285	56670	88445	85799	76200	21795	38894	58070
485	62382	76941	01635	35829	77516	98468	51686	48140	13583	94911	13318	64741	64336	95103
486	98011	16503	09201	03523	87192	66483	55649	36764	86132	12463	28385	94242	32063	45233
487	37366	24386	20654	85117	74078	64120	04643	14351	71381	28133	68269	65145	28152	39087
488	73587	83993	54176	05221	94119	20108	78101	81276	00835	63835	87174	42446	08882	27067
489	33583	68291	50547	96085	62180	27453	18567	55524	86068	00069	59254	24654	77371	26409
490	02878	33223	39199	49536	56199	05993	71201	78852	65889	32719	13788	23937	90740	16866
491	91498	41673	17195	33175	04994	09879	70337	11861	69038	51915	23510	32050	52058	24004
492	91127	19815	30219	55591	21725	43827	78862	67699	01009	07050	73324	06738	27510	33761
493	12997	55013	18662	81724	24305	37661	18956	50064	39500	17450	18030	63124	48061	59412
494	96098	13651	15393	69995	14762	69734	89150	93126	17700	94400	76075	08317	27324	72723
495	97627	17837	10472	18983	28387	99781	52977	01657	92602	41043	05686	15650	29970	95877
496	40064	47981	31484	76603	54088	91095	00010	13800	76690	75133	60486	28491	03845	11507
497	16239	68743	71374	55863	22678	91609	51514	98135	42870	48578	29036	69876	86563	61729
498	58354	24913	20435	30965	17453	65623	93058	08313	99293	00990	13595	77457	79969	11339
499	52567	65085	60220	84641	18273	49604	47418	90974	83965	62732	85161	54330	22406	86253
500	06236	29052	91392	07551	83532	68130	56970	33273	61993	88407	69399	17301	70975	99129

Page 11

Line \ Col.	(1)	(2)	(3)	(4)	(5)	(6)	(7)	(8)	(9)	(10)	(11)	(12)	(13)	(14)
501	88188	99345	94118	40373	50387	24802	81352	61640	56614	71506	75541	37818	88047	94144
502	05200	50533	59428	02797	16833	10038	18901	40743	99449	49825	44637	73724	42649	67052
503	82828	41316	92617	31346	89263	06589	07121	07151	23905	98435	50453	12983	04738	76421
504	71006	99318	19269	35233	79183	78538	06326	62715	28701	52809	56581	05925	85210	17745
505	05937	00875	32264	82808	00229	03868	71072	11619	44876	34508	07889	62424	54319	32842
506	06021	04370	93070	90737	05354	68427	25554	11165	00123	80338	03876	85648	24978	01687
507	54789	10960	44023	57857	56556	83993	70787	28193	65872	33723	00125	99818	85571	69509
508	90400	05707	29128	14859	84117	72206	53740	00464	51853	78852	83593	82926	48985	64355
509	51424	01651	99970	73521	82356	03297	36288	93531	69269	84798	78962	06336	95618	89718
510	79743	88757	43370	86536	07166	06401	14413	23643	21527	91903	91384	31444	54783	38760
511	77418	00322	98854	51507	00565	33066	65791	47857	32483	38493	52606	91078	13631	67863
512	17580	49302	16408	05678	75532	46218	74359	77556	82242	00134	70154	09027	79459	18730
513	15489	45559	88648	64330	42126	43145	81287	73884	69312	03395	06879	49662	40000	61598
514	56342	66773	16536	32600	73958	75993	84250	19254	06677	54192	53422	58200	74464	73949
515	20202	19216	23762	47856	04623	70728	86657	70801	53719	25214	65635	07565	49977	45525
516	84877	51708	69357	67914	55372	97225	52837	46723	00256	96221	26641	00309	36009	48392
517	01647	00311	44989	21900	96079	15793	13148	01433	78721	02647	25454	53913	97554	41578
518	45652	89311	45302	14539	32045	86727	40595	55953	93448	07805	53622	27330	18749	57867
519	79975	06153	08932	59185	71386	19070	87098	19392	13899	56096	83645	45871	35950	52272
520	49744	54713	37063	77467	15348	03383	96086	93296	12413	55774	97318	66402	11209	52495
521	40922	94903	29638	46870	14108	84391	87313	65969	43349	85142	25650	01896	48680	51236
522	53319	48020	77444	51447	07916	99506	83504	22290	63835	45589	04884	92760	70462	00538
523	76682	10559	85446	56236	85919	76388	59850	03262	60347	31077	07165	26588	31296	56112
524	48869	97229	69581	84581	71728	45150	16901	88717	62688	24828	89469	35483	76532	30256
525	95961	19279	38078	17473	43945	21562	90937	52140	73771	56084	08775	94820	78139	25987
526	16581	25945	94076	91281	92272	41233	58614	18912	58454	34011	85969	83621	92099	19131
527	78282	26332	44072	55104	16895	98311	56005	23331	21939	03463	53828	78930	30987	40988
528	43473	39179	53174	43498	72674	13087	54261	01844	45738	93150	13240	16694	59155	67589
529	06513	31352	09177	21367	64725	23784	18125	74873	83971	92678	96950	69821	41119	43312
530	48734	39737	03448	99009	98136	34562	30339	93143	07350	94289	76144	47238	08110	00037
531	54832	70111	48339	75270	11652	41697	43277	58089	70520	96997	71007	87803	52458	06637
532	55844	69515	22658	75438	83086	41325	04694	40359	28351	53492	73134	02370	72313	53039
533	42829	54398	93338	90705	00626	97752	93482	27726	51835	23966	50279	26329	25754	43530
534	81128	63461	10925	44382	73365	98875	77605	27351	49177	36914	50258	62361	38229	89608
535	62885	26354	10368	78026	00186	46783	02059	98892	98061	15330	31705	71923	29266	72716
536	19525	10375	27010	42791	49471	90607	98103	31752	04842	13693	84292	48485	76178	41716
537	26570	99202	73924	59888	01827	93314	63949	35394	12989	05867	11568	45056	16609	80470
538	04772	17749	01537	96036	02102	02622	06007	52239	61201	57415	35609	38761	19589	24238
539	49129	12491	62558	64323	44856	29045	76871	80449	81381	73642	48643	23848	48390	56829
540	19937	75104	57780	95871	94547	53541	77723	54114	90290	62627	65181	15687	81062	06729
541	52571	67962	72775	28480	87411	12075	45177	08796	99297	48807	88310	75454	45456	85394
542	54943	80723	81195	84069	28144	48106	04169	16575	62665	97861	71650	56981	61794	94285
543	16375	88048	29625	08111	92924	53335	09525	88290	17679	08945	25816	11848	95106	22031
544	38745	91458	30363	95005	55854	38628	13599	73065	40870	82576	37089	86738	16284	44725
545	09937	17776	86425	88916	80594	28347	08092	64255	55604	78635	13197	72213	95102	36723
546	30097	47192	27960	15937	42080	61048	14358	44508	72683	51088	55368	85587	27046	11198
547	02410	60124	62825	42947	74690	89730	16073	28184	30078	92578	83709	08044	76238	47599
548	44804	80165	19442	72194	76910	40274	93861	06568	92482	70037	66779	63312	00619	94053
549	37352	79142	51032	58844	03167	57351	51850	92810	35331	78995	44221	41532	51606	26430
550	60640	14199	48263	71533	94235	42431	44114	90993	41149	06159	39242	11163	14764	19246

Page 12

Line \ Col.	(1)	(2)	(3)	(4)	(5)	(6)	(7)	(8)	(9)	(10)	(11)	(12)	(13)	(14)
551	31630	67734	78201	94545	80152	62327	83165	31035	82295	11824	06765	29501	62849	50419
552	25101	98983	36993	40028	58036	14075	05980	57094	45527	18766	77741	12985	14112	65058
553	86207	09805	46240	70644	76012	37000	98321	97197	30645	56169	09363	44394	29087	96569
554	31611	47643	28795	48115	17223	63161	29677	69820	77189	20762	94296	94528	82984	71418
555	10649	89132	59781	12373	35999	30832	02508	93055	57173	79848	85439	18861	26742	54970
556	68210	16228	34601	40972	22887	89759	09095	00587	03998	13659	64179	98567	69313	84637
557	32367	69587	66162	44358	69844	73042	88091	07288	74971	47066	36987	53520	58309	58605
558	14684	42446	01751	37459	31945	03627	47690	97813	45272	42789	99315	26662	15833	37246
559	64260	04661	39957	01200	84800	27930	98937	76108	11043	29101	01767	78894	92922	66537
560	66035	77943	70861	32037	96699	56314	75755	68667	04730	15256	13987	52743	42306	87515
561	20966	71492	32323	11867	47523	24094	23334	78839	81588	67374	43885	24512	81956	75721
562	20498	68176	02027	22358	15907	13247	44784	82957	53009	73379	44093	58405	38515	85531
563	56320	79875	60634	17556	52153	63549	03661	78290	96447	04192	30187	63198	75932	02367
564	50559	90270	33571	86091	34749	56784	98486	06018	27447	00884	29564	51522	35571	69208
565	49366	90095	73459	12225	28483	35358	99941	63054	54358	80748	54049	85937	64718	21465
566	29022	19268	03003	96622	30239	20482	50028	16632	57708	78559	70241	47977	78645	48550
567	77212	76531	68842	37777	20085	38703	31753	18608	52524	08585	91711	63572	57007	11379
568	41121	90499	83459	71424	27596	74645	68790	66478	78885	53799	02026	24596	88692	55936
569	06613	95412	94751	60763	56611	73508	88479	86151	78563	31633	92321	23304	77153	86639
570	66430	95324	60108	42377	56350	67861	47478	11961	78516	95316	64393	52020	44994	88205
571	39380	09648	47285	62864	03421	34292	36084	39604	89838	03635	30064	72710	26327	65521
572	27595	63289	75149	03348	91237	28372	65330	00966	73904	17477	34953	08975	83142	48425
573	43525	45549	58819	48478	14007	11384	19576	68138	30774	51898	24711	72537	18360	95682
574	60024	79858	72015	01236	27444	47010	80938	15828	86484	92753	04322	27171	41828	79025
575	35914	73076	05158	40190	41294	72776	38528	22272	34709	34561	65554	53461	61776	03585
576	58253	32995	54370	34437	98365	17630	42196	49736	39619	16731	71798	38047	85559	56700
577	85887	54618	23532	73821	80904	05950	63441	86109	79900	14063	03152	39235	74289	42342
578	88988	60426	34636	40601	57718	93925	67763	67671	21739	87534	83385	91492	45796	04681
579	65381	17333	32358	49608	36893	43453	38580	27639	52832	01522	11108	59992	23168	04414
580	20214	88406	06098	07770	51679	64857	36998	95796	46745	36780	56791	34690	61634	85411
581	04970	56425	74303	94793	55055	44762	22800	02663	91182	13108	07408	50545	21312	11365
582	87556	31521	19669	16311	01767	67301	63016	01227	29273	79256	50368	88653	23329	86500
583	34615	85796	30299	56090	41453	82886	33077	99791	86553	64640	61529	88660	66941	15936
584	95769	07371	49569	42262	74097	44317	40304	05346	90342	73324	81555	82769	23559	55238
585	57338	58325	53918	08075	69395	08189	66668	11663	84852	93146	27182	34936	97267	63996
586	08882	30762	14602	54767	67683	57818	98756	49119	91258	71916	65948	24841	38607	39412
587	62644	32814	31337	06011	80623	36021	81846	71868	98242	72307	25917	56240	73499	45106
588	55358	48010	24440	51092	89263	82352	95072	68828	28001	48991	19201	90963	34192	63336
589	02254	43319	79888	86311	57615	69666	60651	49084	39681	66415	10201	53931	44245	42850
590	61023	48494	57279	47133	89534	37085	28495	39162	89121	52021	23143	14829	55792	84641
591	41900	90881	96080	35258	70734	59465	49227	01431	90694	96186	57811	54518	30108	01861
592	94500	19764	94752	73077	74726	86176	31623	14569	89225	09606	73438	95276	21237	36507
593	66375	34653	37125	84780	92759	09781	06912	96802	92502	97497	67702	49763	25950	49924
594	27141	20959	02318	57546	45467	06653	99212	60612	26046	53553	59787	70491	69632	46009
595	25670	88933	19316	57014	52797	83779	56495	33104	26858	07662	41253	97688	76883	29444
596	06590	67256	21117	69351	98168	81043	26682	00063	96223	30436	21987	15450	56574	45011
597	36429	99750	38004	64992	25021	45680	46501	92943	99165	20707	43410	53746	54716	17090
598	52011	93461	06334	23801	34422	13728	35016	80605	86628	21689	34082	26035	11928	41817
599	09443	77694	26882	15663	45983	29425	12085	96233	29036	91135	28288	27709	90674	09705
600	25862	81855	15254	28462	95680	42433	70311	09702	71615	72688	85922	58369	89734	08750

Page 13

Line \ Col.	(1)	(2)	(3)	(4)	(5)	(6)	(7)	(8)	(9)	(10)	(11)	(12)	(13)	(14)
601	60359	07603	81594	66235	48154	61257	27978	64695	63165	44593	08210	16863	09655	00855
602	34992	97880	79115	47587	76167	47086	57064	16730	74172	60317	83215	38133	06303	05466
603	04887	64208	71842	97885	32616	23280	11783	19852	64266	24446	14189	77419	30991	92130
604	09332	86232	88199	66094	72594	30100	23673	68705	66989	42666	81857	34651	36167	24221
605	42326	62962	06485	04978	96639	96214	91478	12408	21457	19862	99102	91426	10181	51762
606	49187	42836	17042	35179	31880	48444	89877	50915	37426	21556	25999	84256	82314	18813
607	09228	57404	42180	07949	98750	31506	78442	45809	12725	49774	11276	46371	81681	00623
608	69720	73477	91252	48009	81393	76401	48168	25967	33372	84414	21506	46131	46046	12354
609	82222	13787	98611	95257	34753	36674	44326	66070	61131	70620	42865	89251	54844	04013
610	30703	00513	54586	05623	43999	55387	67189	95058	91174	13121	27557	16512	77963	40635
611	86369	62151	70713	41166	79321	52215	94358	28962	35868	22796	87221	40014	68875	71420
612	83331	99035	68506	96734	91074	24356	03035	66926	32197	54944	76781	86722	11769	27368
613	43053	60600	98921	43720	77342	26186	64554	46226	64244	10703	49564	69737	32948	43060
614	57104	49148	18487	01775	71782	04679	64369	06208	71669	63046	10470	54194	96709	86502
615	33177	11409	13925	18130	54242	13460	39174	63528	22670	31810	04313	50669	20653	31779
616	05424	76714	05732	29415	01183	45054	13493	44006	61641	80304	96504	52181	05359	72203
617	92950	58665	41191	69259	50244	55322	75137	90193	31989	17381	43795	26981	15326	02303
618	54925	20502	71767	82737	64847	04496	35921	42670	08584	54090	52907	75331	09155	54187
619	41980	43710	55304	57526	29616	92314	85883	21584	55045	81997	62277	58884	01590	13532
620	83825	70977	67987	61545	92066	71215	93967	63071	69928	98917	05699	35957	04679	58769
621	84047	83627	37763	07081	33048	57895	42182	73279	08032	19165	01701	35656	03328	81785
622	12776	69127	67921	57611	85876	30744	40886	68396	79787	76434	71221	86769	15104	19062
623	81419	55440	69506	09115	45032	48343	78352	39075	31689	76469	64918	15149	88457	97144
624	59844	03603	96297	58028	93069	35674	38479	54639	54455	10300	73946	94827	53164	07458
625	18350	74940	07044	11210	53622	00779	36027	51496	01694	57895	84570	18271	54461	42210
626	79960	18784	13376	03415	84450	78874	22050	19730	92598	54291	60658	73188	03446	49864
627	45420	24157	16374	22384	56892	84941	97157	99656	33978	81436	10955	98991	10456	35727
628	13945	09559	68152	56960	39453	51654	10617	55628	47933	85161	52998	75414	59552	03546
629	91206	33871	60730	96821	95808	29763	39678	73104	43398	38181	44314	58343	28884	94613
630	24847	08724	81499	72905	95102	63004	22223	19808	90777	54986	97234	18458	22889	83960
631	94303	08209	27804	49372	66392	50578	02966	90907	33164	83044	97985	78526	00983	29271
632	22732	95331	60954	93333	71142	38827	48222	21779	35598	95957	58844	82319	19780	08330
633	82809	24004	65983	01091	70431	91145	88207	52216	94846	75303	85105	89486	08182	56504
634	62700	79965	09610	97213	48579	43574	37652	12447	80233	42473	94585	84840	99926	74778
635	89870	73755	48525	32765	50818	71468	37876	28334	07762	16180	45346	78324	20422	85784
636	81493	24124	67928	12735	41249	24180	54740	44290	58903	38681	04066	69393	84595	42173
637	43630	32189	08532	43055	08080	84208	06295	07813	24068	67549	43081	78581	02095	03471
638	60234	18992	13283	96334	39746	07272	25295	07871	34201	49620	52178	07290	89767	63890
639	00107	21861	60367	48999	71634	34053	28265	02064	06290	10620	17941	81086	51759	57028
640	09657	36088	05976	88267	62683	57675	85265	10856	06525	37911	52332	55752	25054	30436
641	93948	38350	63464	08008	96607	73505	75513	91238	11042	40972	62837	30260	84002	99947
642	42746	29761	72298	48186	88584	90141	72879	54531	99127	60063	82374	76895	63812	94877
643	12939	04181	27698	48297	20574	30169	45545	04462	91067	43847	62739	31141	30385	30098
644	71032	55283	94804	00202	12254	22920	73225	51484	73943	08431	45681	32663	67097	15644
645	09188	78876	95736	70659	32725	23024	04656	48102	15904	19019	09882	87431	16879	61253
646	79236	54729	47052	49717	22312	06735	58347	04402	03838	97049	35378	38579	24489	86899
647	41337	52635	48056	43317	11599	26382	41305	04589	92877	52732	53130	45275	30183	15962
648	73732	99966	30485	45994	30195	40239	52751	64124	67778	60982	12167	63134	10730	11350
649	92113	55625	03726	76886	64237	33300	48004	37440	76329	80441	74766	70630	97855	88039
650	63797	22667	74860	99731	06975	63055	72287	81976	43983	97018	25589	96618	93350	67143

Page 14

Line \ Col.	(1)	(2)	(3)	(4)	(5)	(6)	(7)	(8)	(9)	(10)	(11)	(12)	(13)	(14)
651	98707	41348	74106	43550	16638	72858	32375	50634	80903	43867	54215	65017	70776	09596
652	65496	87176	71726	09722	85667	37498	03926	72585	15193	45615	67073	65110	42404	28419
653	17617	18337	83583	73510	52998	02570	37334	82042	07261	97361	93327	81104	17979	15192
654	85006	32658	83348	38599	77549	70275	85147	08030	20980	86223	58024	56948	10793	81658
655	77279	66357	38044	75041	06698	16798	83023	84686	53539	86491	70402	31779	09591	22398
656	84133	32224	55350	73251	64120	86718	61061	66146	36275	48583	74783	85824	87328	39207
657	86535	01806	18470	94806	18228	71262	39195	99049	75632	67645	26117	89606	83976	53307
658	58459	99005	64939	57060	94609	06739	40489	08146	74014	24803	50699	05588	73658	24352
659	38783	61870	02744	23773	74163	30029	64988	67851	68279	53341	88122	03631	97916	27003
660	49454	76832	17745	75922	40087	61566	38620	92988	84531	99883	40785	53456	03864	20593
661	71219	82115	51031	28586	91328	59635	15746	09832	53937	07006	00272	00914	31619	73222
662	57770	63699	49430	82846	13285	69178	66391	44979	69923	02920	98234	68478	52861	04824
663	94539	93056	65825	27167	69783	48976	07466	76421	71439	98358	21989	77098	86413	62724
664	94992	45377	75471	45547	34348	61707	77657	29541	67815	56777	91185	00390	17361	18119
665	60717	63093	60684	08910	30296	12345	22178	71824	11973	61293	25937	61542	76603	48032
666	28040	08603	43675	64665	40567	76211	18715	62133	75336	98118	88984	44341	11202	34471
667	69841	63836	53186	49970	05485	65171	40135	67168	42593	37922	55146	40354	32803	63430
668	80931	69365	75077	19929	67852	63937	67390	94560	76131	58351	72302	68973	72689	28187
669	94000	24074	95072	32602	89373	78009	38009	60982	27122	29283	08881	91436	80644	09688
670	70609	17467	26861	51035	44534	49774	88833	15577	80076	68625	42230	94087	97450	13569
671	24016	76531	77282	21820	61170	37198	16941	14009	97386	61256	11687	70360	18405	85586
672	36008	76472	60324	45386	67266	98846	17763	98235	53433	01295	74947	71106	33976	55405
673	65769	02131	95850	61875	52572	82826	09571	61125	81326	72388	40377	93419	38458	13301
674	71033	43026	99601	18102	33654	29885	48692	09691	95837	77697	69896	47214	85897	59723
675	25152	11755	33527	05149	71696	63492	32871	48386	37657	82062	02643	34598	36128	42411
676	81251	78313	53908	66912	84868	51859	15959	31119	92964	94054	57594	61310	42452	76262
677	37401	18809	83908	01638	72548	80521	02945	29410	37736	96301	32848	16332	63069	35975
678	01383	28893	13771	99613	39531	20895	92322	01032	66104	86756	84812	10692	73677	58045
679	03921	79304	91058	03175	33155	58344	13242	02686	57998	84750	88695	99642	86560	79808
680	19882	28056	10093	45334	89019	23914	12549	69492	09859	46366	22224	63308	95897	20408
681	55339	29655	37282	26504	70059	91780	59955	01729	92886	28616	52061	47284	02989	20385
682	01725	18314	40197	00964	44112	60484	07486	10883	23106	18254	23711	26458	09878	12981
683	45073	49109	77778	46092	30928	16111	15372	76016	15402	98293	24495	79618	69649	19768
684	60861	39467	17392	05446	26083	65352	24501	20502	13346	11151	37472	05548	35893	47964
685	88775	27006	69664	22246	77064	40312	14388	70861	47198	24261	69913	07368	78156	07858
686	20497	65297	17965	97094	26451	47473	07292	03191	47612	36599	67827	74007	48783	12329
687	76692	21880	39663	77289	36681	47935	76260	96157	44019	73403	83985	84210	58091	88097
688	00528	63679	57762	88094	26402	25925	10598	09606	41596	90916	25311	91501	63286	87423
689	57718	09889	94003	43899	32989	87707	46639	53512	25509	92751	60744	24239	13608	90176
690	85652	55278	23342	30372	47987	19936	89623	38221	45073	67291	57082	13194	39749	75753
691	16543	93477	41421	34710	59498	55069	70504	08544	80118	77512	88070	03943	84969	02116
692	50315	17150	14006	10453	96007	41512	36027	60557	57141	50554	80556	47441	56506	62857
693	73876	27245	79551	56241	53843	66714	24050	26354	14915	62391	75724	57468	34621	22325
694	26749	07215	99485	51013	50210	69934	73613	95543	94599	13955	29586	62507	90618	52247
695	89438	39457	58265	75693	72570	17885	07259	10273	25229	52788	55762	14772	50200	55909
696	24388	81801	22581	26331	47945	11717	42187	06031	02488	09199	74782	65757	27989	81532
697	19688	39199	20531	96205	50855	14725	38656	79167	44771	43966	73485	79632	77181	84817
698	07943	07018	03516	05747	94259	73583	43619	99509	02102	08279	57372	53487	10885	95017
699	90391	97135	97952	83722	35578	82905	41761	66670	43482	34931	94438	93341	60927	31368
700	61564	65108	76451	71430	08671	84975	36182	70787	79442	29461	01209	20022	93085	28312

Page 15

Line \ Col.	(1)	(2)	(3)	(4)	(5)	(6)	(7)	(8)	(9)	(10)	(11)	(12)	(13)	(14)
701	79681	63467	02907	86515	71330	04490	47372	68791	27576	02044	03784	94581	60105	75131
702	30305	20743	10302	71391	18138	23412	11858	47818	22324	52031	10600	49892	34101	71430
703	32763	33847	58250	64362	87550	94978	22888	78355	34651	41604	51892	89533	81610	72641
704	59166	21978	40556	13084	31782	00518	18621	60508	93095	74017	17416	76900	25261	63227
705	55843	94845	30006	51045	17428	50657	68237	02969	30500	43569	28051	22505	07159	78162
706	33537	82468	52422	32155	54419	61661	64835	06496	16377	92607	86248	89492	49306	95414
707	13533	29605	31430	07663	95274	11484	42579	15718	54485	08857	93691	02973	26687	90437
708	24626	54219	12284	06890	05239	42846	24773	15025	15161	51340	54739	99433	54328	47800
709	48002	32024	17230	37523	47488	31080	29352	61444	04011	56275	19259	72475	53451	43397
710	03742	00004	98249	12256	94253	95378	88918	98167	46646	19727	24181	83358	28999	11769
711	17749	89193	37944	53702	49918	65397	72597	63520	77429	68355	21003	00657	02157	68031
712	34837	36219	22048	97047	68804	09633	28689	80484	89331	77577	30376	10021	78267	78049
713	99451	37922	90191	39229	07564	41077	91554	46657	74652	84677	49671	94805	82406	99797
714	74045	00036	53137	15250	19646	20451	46677	53620	74712	17246	96626	28587	33618	46845
715	98998	98774	98159	00032	97323	81490	21552	35001	10913	48910	91005	62408	83253	19770
716	61513	02266	36871	85993	23028	67082	93486	45110	86288	34493	65710	04268	04955	49074
717	67056	19960	53863	63917	68283	31123	17443	32019	19695	85622	46808	03535	91566	36785
718	83036	04625	93284	14368	10979	95800	72182	77004	07320	79516	00915	53209	00884	65464
719	71901	25497	76987	74388	41605	39295	75622	41203	87987	09672	81312	08728	49867	95245
720	46484	77860	02062	92917	70275	40593	93265	92722	39193	47099	39046	85989	24607	72287
721	18312	05137	64361	86541	17794	32313	52847	08862	36752	32624	11035	92500	35016	18519
722	63093	94089	17729	19607	19340	19022	50080	21998	49864	07107	64287	41647	75264	09230
723	38109	69439	62094	49578	37728	17809	11563	10073	17299	69238	88068	04754	51698	11641
724	41421	22003	36770	32741	10325	30892	14112	34880	92387	45169	96668	89183	10099	44382
725	92320	12828	57972	83551	63054	95028	50857	40315	04962	36431	54964	33961	89397	70359
726	42226	72413	67949	96906	17848	21446	35722	10376	84226	16403	14642	23253	07162	57664
727	01094	08525	21349	41981	55232	76652	00857	77173	63362	64936	96601	95816	14729	35398
728	75760	51119	37218	16828	89127	42801	01084	78402	28359	41533	83339	69672	64909	55192
729	62568	56665	42394	67135	03069	93275	11662	23607	00878	53800	14840	28975	51693	97620
730	77151	67677	85258	46925	92504	87860	85299	45952	76388	72989	12170	03449	28315	95994
731	71920	39074	15464	36753	86550	24330	17873	20798	15221	80763	69974	95552	66857	85097
732	99411	04216	66076	90718	67214	03688	71088	92479	27623	97466	27560	63689	35799	28078
733	05654	88507	03119	93043	06951	35126	26154	26820	68861	37807	17485	58902	92005	57597
734	65937	81013	09884	97787	85851	00011	66801	02686	73801	19522	67200	80477	73121	88815
735	87649	70531	88258	21822	97418	67341	54342	80836	46142	04718	08348	75316	53030	90533
736	57827	19642	95661	23788	19164	78112	07304	32337	12845	12588	70054	12267	96360	43459
737	78911	81376	22392	42570	33512	17996	29406	47329	66928	89312	16994	87633	71914	52038
738	91302	54963	94112	60597	31843	40120	00386	35486	27379	02873	69868	03803	29671	74165
739	82950	87509	65708	14385	86299	23769	87183	66267	13819	58266	16843	06546	00621	60286
740	21888	66504	85577	67163	46317	92073	68325	46664	03841	58572	46048	04635	78350	53233
741	73799	60026	87226	26744	12037	98558	66640	32882	11415	64686	78236	05844	76840	80802
742	48237	10339	99550	86134	30229	39131	76976	56296	89453	79556	48059	70552	45852	06593
743	24293	69496	20243	17738	55798	96178	86022	42073	16407	53031	48671	82154	04054	90741
744	18748	01580	73315	84924	81621	67021	99065	43590	82522	71919	25097	49536	03140	63610
745	94470	36824	89203	23689	37016	18462	59404	27230	78689	88837	41119	95462	94394	13374
746	87639	11791	63380	25952	20838	13638	32782	23841	16936	91384	20472	81876	85484	35003
747	65676	78482	33343	65797	56005	15782	27311	64066	28230	36207	06446	09976	09463	56698
748	94357	62236	54083	37960	43467	79372	26370	08273	18180	84100	30757	33315	28583	53434
749	06595	83512	74524	10051	97759	64738	10370	07874	28301	08201	07624	63508	82486	82993
750	34033	69035	18588	88893	83679	27789	22340	12208	00381	06023	77844	28666	24220	19220

Page 16

Line \ Col.	(1)	(2)	(3)	(4)	(5)	(6)	(7)	(8)	(9)	(10)	(11)	(12)	(13)	(14)
751	03548	52011	53722	62927	01693	90948	75340	16660	00939	77148	52778	20615	97851	58353
752	05066	74263	11659	84658	52063	42299	94340	20391	00080	43359	44831	06891	86588	88565
753	19814	89956	36256	19896	56654	69424	84446	95294	00919	60267	34349	64353	92469	01606
754	44928	11206	21377	35086	62233	93761	56000	42066	35898	48944	38852	33177	16239	76624
755	52158	08923	94812	04443	32028	96465	94430	42834	22836	88818	44467	18329	30144	31536
756	53211	58616	05135	52204	51079	11341	20366	86248	35160	31485	92436	69726	12722	52722
757	27334	43976	14685	38119	29486	57290	34141	18058	91299	58001	17944	00296	13949	94904
758	66166	06709	69495	42150	38018	43875	44534	48712	98089	76889	55780	09038	95667	57007
759	01837	16750	96491	33095	30383	48804	21228	10863	08350	25610	80866	00115	27666	94607
760	38295	51093	63495	13203	93562	94132	54810	11410	73776	55752	36138	52297	31528	77790
761	72510	80830	39948	94133	00780	14167	83322	32747	97291	66126	64996	21354	64615	64234
762	16354	00336	06494	30078	46134	62486	70923	17400	28797	83599	78655	67488	09715	69232
763	45168	70700	04592	35281	47737	28881	57221	79935	99756	61564	36936	55668	87148	95939
764	68137	13619	84666	78104	83546	72551	40848	52138	36343	93975	09556	58888	08185	80961
765	31848	95753	76858	89517	91138	62356	66405	98171	66239	49761	88800	29069	30175	31267
766	17216	37292	05495	50885	98994	32966	83496	54614	75214	34959	15500	32107	81638	46696
767	73211	42922	57386	95490	56100	08977	78499	53540	41748	09805	67110	53329	92776	32312
768	29217	56753	90171	87554	57421	35839	60961	87828	84777	19688	42464	35569	84904	13130
769	66780	34571	71684	28798	47123	25232	67995	39562	35855	61645	64448	65170	46792	77081
770	55780	48081	93674	70837	92534	65892	32422	02887	03170	05372	48068	65758	62376	94055
771	05788	64237	58140	63279	60170	17229	69038	02500	21071	26294	95005	47815	87991	93358
772	21911	51065	51525	65122	52608	52836	86157	19943	63173	34921	11943	36931	74722	06193
773	32645	38861	25181	18042	31903	46525	96290	24323	68269	45841	73918	13720	52336	48416
774	72304	15382	01151	63162	23656	69649	54580	33479	62899	33716	53668	86735	36746	21939
775	27637	04122	86132	22538	98976	90718	81840	33461	28526	96231	28082	89710	16038	26648
776	66305	85906	87925	50081	37585	63674	65144	75814	87596	04642	07590	82276	59336	84262
777	53470	40332	81044	00558	50403	48029	07719	83340	20258	49737	45499	72241	90103	96141
778	01355	77096	64828	02445	94511	09503	12283	15264	32845	55594	03668	05664	67999	06001
779	66954	04728	86153	10933	86557	10877	97049	40124	87071	03864	29526	42888	09057	82966
780	53734	15628	08080	24011	04187	65722	82863	59616	60664	76761	84417	72560	77869	42603
781	23114	08743	07186	23825	04298	72839	43963	52136	60923	88568	55458	87335	81939	67361
782	16803	91335	64192	13631	20332	74852	67813	68131	56695	93253	71650	81735	13302	30319
783	57233	99034	38028	32038	81270	77809	99813	67091	56997	62005	95516	34011	52035	79500
784	83203	59665	72314	67942	01380	24467	27813	74783	56696	84101	93852	12016	20869	57009
785	79299	97340	04568	65386	04876	31514	78813	77887	66697	51048	28395	62626	42911	99130
786	99442	63865	14360	83898	98873	17471	01813	81541	66913	52874	10563	59450	16885	16009
787	77773	19621	81557	84629	18808	89056	48194	65899	86244	74431	24232	33299	58332	18314
788	18581	12572	15185	57989	87644	83902	49946	76235	92633	10232	84241	66467	86018	92272
789	80978	13327	19682	53353	96223	04775	05590	55237	81695	31870	48893	94058	88943	77040
790	96884	95522	04791	93463	20316	84054	68980	56907	77188	54803	67622	57968	78532	66688
791	61113	82511	44196	73740	16111	73200	65980	66320	77327	14715	24789	08145	28077	32303
792	01272	75657	28365	17431	93603	32457	57980	66157	87797	55887	57788	29337	73409	35682
793	82357	77578	75628	93073	38281	72103	39980	71150	87282	01585	25850	46138	21777	97555
794	20434	70899	44243	19741	59954	73617	11661	81574	05903	25909	38638	32196	41345	22152
795	38580	10399	44894	25476	04984	64543	14141	97911	89418	97980	49571	60483	94714	51753
796	20211	12980	45861	82537	87534	39406	03701	94381	23777	61546	81541	69603	61768	40328
797	98954	03124	09433	19894	01380	18962	40701	05185	23280	04945	49801	98051	77848	62320
798	99749	02140	46641	56354	78746	89410	35701	00250	23508	51246	09460	02313	71279	74223
799	47167	24716	84417	40097	46608	11667	40701	06155	23784	68961	34289	04531	12752	08516
800	65812	77947	27864	98144	01818	28214	83701	10257	33945	08795	30976	14017	59161	31388

Page 17

Line \ Col.	(1)	(2)	(3)	(4)	(5)	(6)	(7)	(8)	(9)	(10)	(11)	(12)	(13)	(14)
801	33993	51249	78123	16507	57399	77922	38198	63494	00278	30782	33119	64943	17239	69020
802	39041	05779	74278	75301	01779	60768	22023	07510	67883	55288	67391	54188	31913	29733
803	56011	26839	38501	03321	43259	73148	43615	49093	91641	77179	50837	48734	85187	41210
804	07397	95853	45764	43803	76659	57736	44801	45623	23714	69657	87971	24757	94493	78723
805	74998	53337	13860	89430	95825	65893	96572	73975	19577	87947	23962	78235	64839	73456
806	59572	95893	69765	43597	90570	60909	06478	77692	30911	08272	81887	57749	02952	51524
807	74645	13940	28640	00127	04261	17650	34050	78788	57948	36189	88382	72324	59253	30258
808	42765	23855	38451	11462	32671	62126	23800	02691	57034	34532	19711	71567	90495	55980
809	66561	56130	30356	54034	53996	98874	78001	29707	91938	72016	16429	69726	41990	33673
810	50670	13172	31460	20224	34293	59458	24410	01366	68885	22798	52873	18370	15577	63271
811	53971	08701	38356	36149	10891	05178	55653	31553	20037	39346	28591	13505	04446	92130
812	47177	03085	37432	94053	87057	61859	97943	81113	62161	11369	54419	58886	89956	12857
813	41494	89270	48063	12253	00383	96010	41457	54657	46881	75255	29242	07537	53186	95083
814	07409	32874	03514	84943	74421	86708	34267	66071	62262	99391	61245	95839	75203	93984
815	03097	12212	43093	46224	14431	15065	18267	60039	62089	38572	70988	17279	05469	28591
816	34722	88896	59205	18004	96431	41366	50982	92400	59369	43605	26404	04176	05106	08366
817	48117	83879	52509	29339	87735	97499	42848	81449	80024	81312	59469	91169	70851	90165
818	14628	89161	66972	19180	40852	91738	23920	75518	32041	13411	61334	52386	33582	72143
819	61512	79376	88184	29415	50716	93393	96220	82277	64510	43374	09107	28813	41848	08813
820	99954	55656	01946	57035	64418	29700	99242	42586	11583	82768	44966	39192	82144	05810
821	61455	28229	82511	11622	60786	18442	36508	98936	19050	57242	33045	54278	21720	87812
822	10398	50239	70191	37585	98373	04651	67804	84062	27380	75486	63171	24529	60070	66939
823	59075	81492	40669	16391	12148	38538	73873	68596	25538	83646	61066	45210	24182	18687
824	91497	76797	82557	55301	61570	69577	23301	31921	09862	73089	69329	41916	41165	34503
825	74619	62316	80041	53053	81252	32739	65201	92165	93792	30912	59105	76944	70998	00317
826	12536	80792	44581	12616	49740	86946	41819	85104	25705	92481	95287	61769	29390	05764
827	10246	49556	07610	59950	34387	70013	64460	96719	43056	24268	23303	19863	43644	76986
828	92506	24397	19145	24185	24479	70118	42708	54311	95989	08402	77608	98356	47034	01635
829	65745	27223	22831	39446	65808	95534	03348	11435	24166	62726	99878	59302	81164	08010
830	01707	04494	48168	58480	74983	63091	81027	72579	67249	48089	34219	71727	86665	94975
831	66959	80109	88908	38759	80716	36340	30082	43295	37551	18531	43903	94975	31049	19033
832	79278	02746	50718	90196	28394	82035	03255	39574	41483	12450	32494	65192	54772	97431
833	11343	22312	41379	22297	71703	78729	65082	57759	79579	41516	46248	37348	34631	88164
834	40415	10553	65932	34938	43977	39262	95828	98617	27401	50226	17322	44024	23133	57899
835	72774	25480	30264	08291	93796	22281	51434	66771	20118	00502	07738	31841	90200	46348
836	75886	86543	47020	14493	38363	64238	16322	45503	90723	35607	43715	85751	15888	80645
837	64628	20234	07967	46676	42907	60909	73293	38588	31035	12226	37746	45008	43271	32015
838	45905	77701	96976	70056	80502	68650	24469	15574	40018	90057	96540	47174	03943	37553
839	77691	00408	64191	11006	39212	26862	99863	58155	66052	96864	61790	11064	49308	94510
840	39172	12824	43379	57590	45307	72206	53283	75882	93481	44830	06300	45456	49567	51673
841	67120	01558	99762	79752	17139	52265	97997	66806	55559	62043	51324	32423	88325	99634
842	88264	85390	92841	63811	64423	50910	38189	88183	56685	22910	58250	70491	71111	37202
843	78097	59495	45090	74592	47474	56157	88287	47032	66341	38328	70538	91105	12056	36125
844	41888	69798	82296	09312	04150	07616	34572	83202	58691	27354	37015	11278	49697	65667
845	46618	07254	28714	18244	53214	39560	68753	16825	48639	38228	95166	53649	05071	26894
846	29213	42101	25089	11881	77558	72738	57234	28458	74313	29665	97366	94714	48704	07033
847	38601	25735	04726	36544	67842	93937	68745	62979	97780	28293	75881	08362	71546	17993
848	92207	10011	64210	77096	00011	79218	52123	29841	76145	82364	55774	15462	44555	26844
849	30610	13236	33241	68731	30955	40587	45206	11949	28295	12666	98479	82498	49195	46254
850	74544	72806	62226	65685	37996	00377	59917	91100	07993	15046	51303	19515	25055	56386

Page 18

Line \ Col.	(1)	(2)	(3)	(4)	(5)	(6)	(7)	(8)	(9)	(10)	(11)	(12)	(13)	(14)
851	76385	05431	82252	79850	31192	86315	75612	59985	76421	39300	64976	27951	17855	02220
852	08059	15958	10514	86124	29817	19044	03555	80725	67857	31395	68780	16560	79952	41739
853	30636	03463	50326	69684	38422	59826	47858	90601	50834	88109	43882	15687	06212	19886
854	23794	51463	67574	48953	73512	46239	10953	04628	60650	35048	34705	90502	31011	81004
855	01117	60216	29314	65537	84029	00741	40851	96344	13861	43421	57107	60813	06877	52161
856	29527	19577	01414	35290	70174	37619	80223	62206	22928	63414	03940	02188	20345	13183
857	64236	24229	17970	92022	64164	17873	41189	28240	60697	26495	87634	75899	09741	84939
858	92331	30325	61918	71623	38040	51375	91127	93903	83715	93244	04366	57679	70829	90088
859	93454	37190	23790	40058	03758	01774	90696	81674	53791	15559	42798	46892	57960	06575
860	17101	42181	45798	68745	24190	16539	32330	21732	65547	94356	38651	35108	16327	17886
861	30742	93358	95730	52535	34404	76057	21325	87526	93020	94861	83865	61393	89645	00773
862	02472	01280	67106	47893	93551	76697	56598	67982	77376	33312	58893	69370	59118	95277
863	80718	72187	67178	77179	06212	37409	46788	68930	21672	88783	59304	82369	19410	93050
864	85406	73687	02116	57637	94701	46754	54019	96344	93343	47764	57490	21321	29075	40086
865	00563	67156	88141	13491	92592	35746	72117	37593	72780	78271	75915	85972	58615	71755
866	89190	58965	55215	24337	58807	36123	09235	95541	96979	03336	34380	66288	98659	46572
867	01438	81590	83758	45361	76209	65081	34785	68423	04408	73827	78494	02765	46174	83192
868	79127	53282	50510	80129	23960	78423	31988	78571	79458	95043	23997	97528	21631	63898
869	33952	92823	32840	94420	51193	69652	04332	81675	64644	68673	33718	02256	34414	87710
870	57146	14126	71734	56942	83371	31526	13444	11912	03152	66411	42853	08437	35667	26251
871	33158	61761	73207	01764	81696	55137	41834	81860	81310	14711	36599	78042	62086	41752
872	63615	69083	00118	47991	99521	88655	94451	67445	99377	75528	40794	30140	82298	85868
873	89010	46915	70186	55657	76955	25430	91951	56473	34225	68103	25353	89595	04715	20102
874	32547	43398	30909	62599	53105	27460	56734	41954	47696	82113	38508	88941	49983	36899
875	61992	28258	27359	61002	16882	44018	85376	66756	14395	66865	67036	78374	43612	44134
876	78326	74541	22198	48380	45919	76160	10974	03127	58980	18350	22089	54977	94019	84739
877	35493	53008	78622	38329	27611	12327	52541	00861	62380	65890	79729	99710	64836	43706
878	19130	59917	28850	76593	02389	80759	18481	02724	57578	35705	89265	25033	13767	95888
879	00317	05769	03497	42174	32653	23663	29569	36342	85908	29572	60063	41170	59957	14755
880	84122	36454	70776	17000	83017	07027	98058	41274	22476	27436	30798	62287	21235	00249
881	76320	32120	91585	39640	23470	86000	68204	23980	17625	53197	35128	76385	02848	61680
882	09234	36233	94404	42812	39210	25967	12232	38195	16649	96739	64610	96067	89561	15772
883	16206	70598	95378	70573	42636	53862	81334	65439	28858	07619	59608	61460	00581	43226
884	04071	51662	67884	73911	08708	66287	89261	73451	81146	77733	70162	42449	44755	56401
885	97545	87732	83795	38027	90239	80044	64677	47912	82144	85918	93508	05816	57549	74831
886	53253	56120	42720	25660	36921	30891	42042	80370	97880	62507	01218	19202	22323	81363
887	66817	18439	53188	35155	24309	88284	74644	25454	19606	61460	52684	36568	68108	45653
888	28077	26409	11443	22200	23129	32407	52401	78416	63693	35633	77724	86835	89829	81383
889	18889	00291	13701	12401	26466	67700	55805	63818	16067	95185	97241	66126	16774	39342
890	10598	64974	66296	33329	30560	73380	94905	04959	80213	14228	97242	94826	64276	37466
891	18656	81152	45498	14400	92435	67664	86229	74358	76537	87066	42293	91743	49462	42808
892	79044	10440	25777	05486	65659	22183	82080	04351	19530	49941	29181	34667	28910	70927
893	74042	20365	42672	34850	60670	56980	88333	75288	64996	26913	62379	55068	91239	85752
894	87249	06640	09090	03242	68467	85678	23411	53443	19526	03205	29261	36061	34325	03761
895	82839	52537	00518	60559	43669	44297	75071	17146	35492	60718	38106	06409	75657	66013
896	07749	62249	01611	43795	17129	44447	95197	25088	22150	54487	11578	77560	26460	55002
897	37171	34598	87234	28324	85927	23465	80833	62872	40826	10066	64888	33605	24848	30681
898	55432	45030	85336	49128	40487	63959	25879	60415	26744	38584	51543	17333	47300	13834
899	43658	35437	83506	11209	24770	87123	21494	85056	56630	75919	26005	90077	50380	09261
900	59536	79475	04874	50831	16996	04750	02246	08846	82410	50997	45824	55547	08168	16679

Page 19

Line \ Col.	(1)	(2)	(3)	(4)	(5)	(6)	(7)	(8)	(9)	(10)	(11)	(12)	(13)	(14)
901	77583	63193	14378	66314	92154	31173	75223	92947	81041	91385	89091	83989	34982	66565
902	41435	82033	40363	74800	79198	03991	01635	43666	02630	46039	53273	86262	91450	79883
903	17163	74517	11281	83105	16146	27577	33565	39621	27321	06387	51838	37591	54290	54527
904	47909	53647	99896	13016	53708	18549	75241	22312	53261	80185	69888	12646	64464	98505
905	88024	64045	70954	56434	73880	50310	29601	52204	93295	05735	15486	39857	83584	59587
906	51540	26148	65528	66246	74154	47254	65546	75309	90415	17346	20559	29568	30124	11974
907	29122	04072	85663	33914	63587	20151	36720	25827	13447	17082	74817	92523	81519	65724
908	34873	53537	85834	11394	44898	92403	01979	22785	94354	33271	44579	40756	70108	30514
909	26132	07856	14223	02824	66094	88093	35794	69015	67498	97087	25936	52239	07472	74585
910	26892	74387	97124	32851	83465	51300	58546	12579	43451	48661	79351	50231	11824	77723
911	73485	32215	29894	87362	47818	52494	27684	83573	91237	03585	13266	63905	94600	43823
912	45678	13831	93961	93558	34050	42936	00860	66594	36783	56939	47057	55257	16555	65642
913	71687	79185	46096	22818	19051	39123	08793	18798	79528	05621	12131	88365	51154	46175
914	11833	28572	61695	36928	50851	98599	76466	73848	76675	68072	50077	27709	60695	55306
915	90529	59895	55565	57955	75623	55721	43773	98277	38842	05069	31590	31482	38040	71541
916	31697	92763	05134	45963	15725	39771	54430	76890	52975	52037	06071	14710	84160	05126
917	47212	92010	40601	23364	72705	15989	02965	78394	53956	48542	35568	09897	20287	54478
918	56686	61285	36139	18867	12877	89109	48963	58636	18825	47664	74509	53250	61778	43202
919	50669	06740	74638	62054	53083	22790	82404	85132	17634	35664	26876	54191	61412	08062
920	16821	64195	64269	62235	41829	93374	74377	17728	84164	32762	18275	91425	96275	97711
921	30608	94750	38092	42230	85443	34743	00348	36016	74853	00499	03387	07739	96416	28578
922	83211	45796	33370	31291	76074	40065	50006	74244	44355	96978	79518	51941	26627	39503
923	51774	53905	66945	65882	70352	38447	50881	25003	88424	98825	47209	16332	48104	50130
924	89095	90943	76279	78323	37210	77141	50300	95786	00463	82949	55660	93650	41154	43309
925	54047	41638	45626	98549	20648	89365	10934	10799	37862	71684	98288	94592	70961	51028
926	81549	98773	68979	28308	53686	57192	65423	47225	27490	78045	57512	11431	50502	11301
927	72335	08233	56777	99798	57340	54203	12334	16800	91872	82213	79086	33896	31707	48739
928	02540	14032	90858	26020	28174	91563	65006	14401	58189	32658	13812	72382	19169	24151
929	09226	10658	79614	84200	53035	16839	83058	55540	32716	03834	05299	58331	90479	34808
930	18050	79976	70461	19890	08214	77015	86657	29965	21658	72049	74959	95217	71503	74950
931	44378	42627	16636	69923	89122	92792	95380	20397	06649	02752	46279	99681	05154	56400
932	92101	13475	15373	53642	77422	26662	58069	86649	38783	30050	20087	26681	05859	41218
933	27926	69613	88880	95537	08518	66627	61173	27188	55705	29607	24361	83286	00052	41181
934	06900	75222	25385	63569	03483	90973	86572	53485	23353	18345	24553	43494	84651	48483
935	08717	31244	56529	97336	17309	32629	34184	47240	07490	86150	78423	82309	01762	33463
936	79966	02658	18011	59422	86272	69447	45731	19344	49685	95236	60922	09665	50176	98674
937	85668	03060	61274	14020	16594	72642	59596	70634	58195	78404	80798	60696	00686	75274
938	29530	85435	54205	01781	88806	28369	69095	26494	77711	62699	92045	74777	97537	69768
939	69284	91923	11744	13051	69350	55403	75969	64260	42303	54467	61374	22426	70625	96732
940	41503	48684	36452	06825	80218	76395	85109	67786	71072	04047	38520	31872	00118	84181
941	47063	45686	78929	82808	67899	16936	94990	90513	44712	51443	78685	61446	11529	64531
942	90503	44995	53753	88362	99077	14510	95018	53155	71938	99420	15408	44461	27981	90368
943	08367	68828	82856	56380	26969	16987	20563	11277	42307	32736	99532	65361	64300	76678
944	05207	94071	56234	12584	21319	90050	20768	27895	83856	69778	39159	68796	17739	30795
945	15445	53683	92875	02502	48973	00564	70805	92775	34093	00317	34349	79168	87667	98240
946	99603	19507	68314	12524	57700	39240	96194	11150	67555	95582	47540	91542	35023	35235
947	82736	02198	84682	18512	42010	69026	15920	14305	46477	35022	30006	35516	48513	09493
948	04925	66489	08883	69176	95712	25057	80789	87303	24848	46375	92999	26747	77435	49766
949	48692	63679	55426	75873	21338	78242	14907	30788	79299	34428	88050	39956	35155	55514
950	23766	44510	30176	02784	65111	89090	99467	74257	98806	10498	63125	99067	24489	32313

Page 20

Line \ Col.	(1)	(2)	(3)	(4)	(5)	(6)	(7)	(8)	(9)	(10)	(11)	(12)	(13)	(14)
951	17222	24847	14225	43238	39943	20269	52593	99929	69993	66806	16035	74175	04974	75631
952	51117	86027	85128	33442	59152	67511	36916	93946	25402	53934	83623	18774	21413	37540
953	98122	85707	47835	60583	81100	24151	58693	46274	83783	68899	42100	14798	30325	01988
954	07079	03751	09486	74453	95050	28861	19001	56932	51925	60421	43291	13854	87524	26618
955	91827	97079	74080	85168	67841	92785	85584	97685	18369	02716	04330	45215	67343	98098
956	63924	46330	97808	84837	61032	98609	67177	96231	44689	05224	72408	37347	30369	45386
957	27527	61690	11334	66616	74361	26430	06353	29841	87500	03627	42526	26924	64139	45052
958	62640	79415	84174	76251	22673	44026	59218	78743	90058	54715	88307	88689	51172	47812
959	24485	14761	49583	21948	88300	95215	58519	81521	21773	21559	07775	96603	12156	58338
960	31163	12555	57763	73074	54090	95819	02558	92149	15685	15315	84184	89401	53301	70641
961	75829	90951	85627	98027	69242	44145	74974	58554	71020	36095	41778	51657	99796	93408
962	50265	07807	43138	62890	82977	38769	11754	81428	56568	40473	43510	66127	10806	57208
963	71200	80801	81289	44767	02901	61530	67617	50078	88909	40539	31521	68964	47991	59963
964	88107	27970	99234	22295	36871	14761	74494	02770	94610	89209	52688	92319	56341	04119
965	61197	02791	14924	93662	36173	00145	81019	63834	09891	16660	20541	48016	48620	66704
966	83405	26615	87854	03164	98061	64144	81948	49844	07816	04927	68917	11008	96965	74719
967	02276	39832	64484	17204	96556	46324	12068	68780	30396	01443	36666	76408	18082	19438
968	13020	23251	09546	06687	39893	65402	96369	71504	04242	80485	66151	39141	52873	13346
969	50766	20504	47573	87991	62684	10921	29624	44384	82938	30828	72266	20455	32794	18715
970	74813	19126	38598	77307	59250	62431	25603	49204	72792	87154	34140	45891	44767	55879
971	34150	68179	87177	64247	40979	62952	88828	81184	83310	49826	78268	84810	34755	76708
972	95035	52598	27655	90383	76468	08022	90814	78356	95077	65538	77477	12978	72000	34428
973	76846	66076	01650	77548	03009	48350	61844	17850	10898	75323	23270	74928	29066	50899
974	24146	49880	75124	54340	74297	31307	87627	14922	55793	78307	61406	73263	29927	21784
975	27427	83596	07262	58388	43212	81674	09745	74206	19310	11767	88334	52762	50280	77990
976	62277	00386	94965	57568	04965	52742	85423	87258	25805	97116	03662	40715	26670	51038
977	26218	09494	96963	28005	15942	13443	96330	50222	50856	27308	78353	83501	28171	69667
978	95828	98777	58855	21092	97414	28481	39244	80530	45700	78547	36536	86420	71140	25607
979	06052	38502	47874	51676	13316	43804	85364	25972	77144	32777	81828	16275	99252	28637
980	79103	19996	36294	42929	95081	45359	82772	10679	34236	27171	90309	42230	97146	91977
981	69040	75170	31286	93852	19394	43743	79837	38978	99397	91989	25596	24940	91964	97849
982	02688	26065	29875	54689	12247	67229	69268	72790	59095	68115	52432	03486	47945	00001
983	59669	62317	56242	90656	43992	64306	01038	22952	49953	72014	05599	06801	92765	39212
984	03409	78084	23818	21489	87944	11732	69494	98213	41356	65383	25792	96224	57038	19936
985	05188	62591	48883	62080	49460	32772	09094	13547	94514	38094	93755	39313	56688	30338
986	91857	24665	14441	83430	90899	23068	39436	65648	90795	04028	44670	88737	54796	80249
987	78853	88643	78621	02377	02253	06181	15150	41361	65890	86631	08289	15996	31647	89466
988	57803	03002	43966	52273	88993	47814	41363	97015	32977	09224	77829	77519	42395	21926
989	20414	62347	23572	02107	86764	95612	22638	02471	49752	86400	21017	92299	36523	28074
990	76748	64133	70898	44429	28121	56265	53054	85704	25631	56706	54899	25569	71394	48333
991	98506	47155	93726	32281	64505	56978	48469	71323	70581	72004	67535	79545	88388	46777
992	54242	81976	67665	70890	07185	20131	73487	53010	31677	34437	88622	03580	41811	40951
993	81990	00775	05836	34512	64536	69060	23223	57511	61764	85828	21119	66098	83996	70640
994	44250	04275	16365	24813	43750	62646	97455	01174	23358	71401	28580	92308	62973	72123
995	27942	19415	30541	10707	42525	84144	17237	38447	76318	03660	59989	76447	37567	66990
996	64300	38758	56657	85280	28224	84314	58118	97734	35701	06593	03690	50308	97271	97034
997	14653	08589	09404	61557	16276	63464	79223	12993	51056	57233	75099	31082	01867	25094
998	95087	67947	29851	06124	04054	73563	67625	44199	40748	52598	76137	26030	34138	76445
999	08774	32723	44960	29042	94665	36955	46376	14300	13280	88986	81039	34211	38168	67630
1000	37039	97547	64673	31546	99314	66854	97855	99965	26373	61701	24760	52014	96547	65403

Page 21

Line \ Col.	(1)	(2)	(3)	(4)	(5)	(6)	(7)	(8)	(9)	(10)	(11)	(12)	(13)	(14)
1001	25145	84834	23009	51584	66754	77785	52357	25532	46346	39806	70361	41836	33098	03790
1002	98433	54725	18864	65866	76918	78825	58210	76835	72248	43023	85884	08785	41152	11830
1003	97965	68548	81545	82933	93545	85959	63282	61454	27337	10560	76390	36976	62785	77529
1004	78049	67830	14624	17563	25697	07734	48243	94318	85992	15716	48891	74601	90030	82376
1005	50203	25658	91478	08509	23308	48130	65047	77873	02912	52675	62971	40993	79767	19178
1006	40059	67825	18934	64998	49807	71126	77818	56893	06714	25147	69782	56399	55494	29547
1007	84350	67241	54031	34535	04093	35062	58168	14205	87257	59443	29927	30667	99040	61060
1008	30954	51637	91500	48722	60988	60029	60873	37423	62206	78386	10008	03027	69882	99663
1009	86723	36464	98305	08009	00866	29255	18514	49158	93679	84854	27953	73627	45315	94563
1010	50188	22554	86160	92250	14021	63859	16237	72296	86715	17588	24327	69905	43592	52283
1011	50014	00463	13906	35936	71761	95755	87002	71667	21899	50605	44017	48019	01348	33230
1012	66023	21428	14742	94674	23308	58533	26507	11208	85354	34288	33031	06081	45402	54274
1013	04458	61862	63119	09541	01715	87901	91260	03079	17350	91991	46955	16951	42276	18627
1014	57510	36314	30458	09712	37714	95482	30507	68475	49167	54279	89734	24807	23802	60439
1015	43373	58939	95848	28288	60341	52174	11879	18115	91925	93506	10331	52809	06032	98707
1016	40704	48823	65963	39359	12717	56201	22811	24863	62130	86396	12606	40352	44181	56860
1017	07318	44623	00843	33299	59872	86774	06926	12672	23096	62738	48877	98889	77375	72317
1018	94550	23299	45557	07923	75126	00808	01312	46689	56900	74653	65614	94333	44775	42054
1019	34348	81191	21027	77087	10909	03676	97723	34469	08852	51465	38544	25247	00105	14825
1020	92277	57115	50789	68111	75305	58289	39751	45760	76922	62810	47392	47765	69009	41520
1021	61500	12763	64433	02268	57905	72347	49498	21871	80128	71588	34936	93724	47479	30366
1022	78938	71312	99705	71546	42274	23915	38906	18779	96970	81430	77457	90793	53509	79535
1023	64287	93218	35793	43671	64055	88729	11168	60260	97668	29687	78958	71480	75638	32812
1024	35314	29631	06937	54546	04470	75463	77112	77126	46942	35873	21104	70323	07091	25408
1025	96864	11554	70445	24841	04779	76774	96129	73594	67259	08958	45986	91442	28529	91358
1026	96093	58302	52236	64756	50273	61566	61962	93280	54750	91572	22825	35238	34955	77692
1027	16623	17849	96701	94971	94758	08845	32280	59823	97638	63913	81772	20605	13104	86762
1028	50848	93982	66451	32143	05441	10399	17775	74169	04324	00199	97321	71104	70286	44675
1029	48006	58200	98367	66577	68583	21108	41361	20732	39538	95746	91363	37416	08833	72639
1030	56640	27890	28825	96509	21363	53657	60119	75385	83475	94949	94360	64076	24305	66042
1031	01554	40592	26557	79189	81099	03951	60438	94263	21708	82714	33487	15713	24149	90641
1032	63799	48415	28087	17196	42784	58377	02026	79893	28219	95848	29099	68424	97854	87967
1033	77654	05943	43283	63181	72300	50360	38411	08823	83158	85734	18917	92697	20102	13956
1034	51058	95654	72648	60349	59563	59021	18616	93244	67666	36344	60506	10937	69837	92511
1035	14982	68789	41999	84797	17743	18775	61437	55258	36401	21686	80416	77429	99664	16803
1036	22305	50900	77277	32383	41723	35868	61829	75281	73850	09094	10748	87176	35537	42489
1037	51375	62050	18718	88109	36767	27440	73438	87774	65688	13701	38111	44531	10728	25248
1038	14590	70151	63862	02246	33710	46456	50223	35742	46423	27373	95388	90238	81178	95956
1039	60312	56009	52580	65722	46072	26273	32049	89647	81900	49167	21910	98790	66208	92721
1040	48513	68172	46711	56744	58250	95301	20401	96568	22600	59000	44669	97303	25080	61986
1041	19052	03511	71346	50933	10108	19747	84718	35431	27652	36130	67396	98295	58395	54493
1042	11246	92882	40354	14700	66716	93470	48587	12514	20234	20769	86575	53071	84995	46531
1043	34268	97508	69711	07165	69844	07591	64588	58016	26532	17957	59119	70048	14130	72848
1044	22931	31857	95595	48438	78029	07045	04453	46783	97341	34526	02731	34808	65028	61054
1045	11476	16553	10297	25643	06192	72478	75968	23807	70533	73805	64118	60682	17286	45369
1046	57005	88551	02117	87827	00151	39027	66060	38915	55240	10473	52382	74523	13452	18918
1047	76056	48776	41451	27779	86241	63562	98013	21109	57835	56161	76084	85338	65491	85719
1048	11961	67214	64606	51982	87588	67464	54389	52910	37447	71173	85880	06581	73833	62987
1049	28009	20616	02614	15759	65618	92266	13545	39578	36021	09262	71240	16627	28570	44003
1050	09194	83950	36645	00734	74652	87328	53507	71924	47057	83163	16248	76396	17886	46311

Page 22

Line \ Col.	(1)	(2)	(3)	(4)	(5)	(6)	(7)	(8)	(9)	(10)	(11)	(12)	(13)	(14)
1051	02946	96520	81881	56247	17623	47441	27821	91845	15714	47300	41511	44365	99898	56027
1052	85697	62000	87957	07258	45054	58410	92081	97624	03586	82294	64114	94726	31820	74067
1053	26734	68426	52067	23123	73700	58730	06111	64486	82045	25637	90031	43964	39009	79954
1054	47829	32353	95941	72169	58374	03905	06865	95353	23101	41360	76055	74426	59237	70699
1055	76603	99339	40571	41186	04981	17531	97372	39558	69068	07420	84874	90220	46055	42857
1056	47526	26522	11045	83565	45639	02485	43905	01823	11997	65770	04894	13660	18373	20740
1057	70100	85732	19741	92951	98832	38188	24080	24519	62748	54356	00260	10051	82738	90022
1058	86819	50200	50889	06493	66638	03619	90906	95370	43420	00980	29684	66876	70294	47371
1059	41614	30074	23403	03656	77580	87772	86877	57085	88032	58290	15614	62605	45459	69371
1060	17930	26194	53836	53692	67125	98175	00912	11246	60393	34510	50509	62330	70152	20836
1061	24649	31845	25736	75231	83808	98997	71829	99430	99629	44443	24165	50988	92208	49492
1062	79899	34061	54308	59358	56462	58166	97302	86828	08415	18111	94115	47281	41826	38774
1063	76801	49594	81002	30397	52728	15101	72070	33706	83438	02756	46441	21655	30827	35016
1064	62567	08480	61873	63162	44873	35302	04511	38088	96236	03668	94183	03773	63502	40017
1065	49723	15275	09399	11211	67352	41526	23497	75440	46317	38123	75503	45577	45525	38725
1066	42658	70183	89417	57676	35370	14915	16569	54945	35358	73388	28224	21814	02558	40824
1067	65080	35569	79392	14937	06081	74957	87787	68849	78085	63685	57537	39437	41895	16115
1068	02906	38119	72407	71427	58478	99297	43519	62410	12242	90647	87380	53942	08294	91798
1069	75153	86376	63852	60557	21211	77299	74967	99038	57612	97072	51406	96836	26372	52131
1070	14192	49525	78844	13664	98964	64425	33536	15079	00091	87240	54274	86634	86702	23050
1071	32059	11548	86264	74406	81496	23996	56872	71401	59284	42455	71966	48720	99905	13769
1072	81716	80301	96704	57214	71361	41989	92589	69788	20721	82236	18597	40883	61703	57145
1073	43315	50483	02950	09611	36341	20326	37489	34626	81690	52379	75898	20851	36483	90898
1074	27510	10769	09921	46721	34183	22856	18724	60422	94578	44051	01563	48610	67004	21667
1075	81782	04769	36716	82519	98272	13969	12429	03093	29469	41177	17115	79207	64792	38623
1076	19975	48346	91029	78902	75689	70722	88553	83300	24330	95346	45931	28180	98118	66069
1077	98356	76855	18769	52843	64204	95212	31320	03783	26906	24265	17840	50625	30699	53192
1078	29708	17814	31556	68610	16574	42305	56300	84227	59147	18934	36923	83101	14609	61318
1079	88014	27583	78167	25057	93552	74363	30951	41367	81193	86710	51226	27067	33267	60753
1080	94491	19238	17396	10592	48907	79840	34607	62668	29680	14786	17902	83847	35528	34860
1081	56957	05072	53948	07850	42569	82391	20435	79306	91119	02988	58792	03362	84272	31787
1082	50915	31924	80621	17495	81618	15125	48087	01250	59915	47784	06495	30072	17575	59835
1083	49631	93771	80200	84622	31413	33756	15218	81976	79409	94895	23730	66665	06133	52448
1084	99683	58162	45516	39761	77600	15175	67415	88801	77378	91513	89678	07672	57062	91729
1085	86017	20264	94618	85979	42009	78616	45210	73186	03869	02881	23638	41792	11325	91653
1086	77339	64605	82583	85011	02955	84348	46436	77911	50191	19372	16950	39034	15032	84081
1087	61714	57933	37342	26000	93611	93346	71212	24405	18846	87224	59101	45552	54241	36523
1088	15232	48027	15832	62924	11509	95853	02747	61889	08567	76588	07718	32505	37265	26390
1089	41447	34275	10779	83515	63899	30932	90572	98971	66287	38099	12635	87644	33457	43804
1090	23244	43524	16382	36340	73581	76780	03842	64009	95845	60495	72208	80379	99160	55477
1091	53460	83542	25224	70378	49604	14809	12317	78062	38320	21331	50681	10089	52276	93264
1092	53442	16897	61578	05032	81825	76822	87170	77235	43138	65496	50955	38017	12609	16724
1093	55548	19096	04130	23104	60534	44842	16954	99466	37192	28518	63412	54607	62493	60940
1094	18185	69329	02340	63111	41788	74409	76177	55519	16443	60181	32185	54956	17203	62876
1095	02372	45690	38595	23121	73818	74454	02371	94693	72065	88784	28349	89472	24577	29429
1096	51715	35492	61371	87132	81585	55439	98095	55578	32831	61333	70640	95003	76601	18201
1097	24717	16786	42786	86985	21858	39489	39251	70450	32283	74130	21070	45390	65047	47577
1098	78022	32604	87259	93708	99438	68184	62119	20229	90328	25634	90890	65331	25416	67162
1099	35995	08275	62405	43313	03249	74135	43003	63132	00852	40147	53963	13009	85861	81690
1100	89192	86922	31908	42703	59638	31226	89860	45191	13098	00618	76390	85918	54584	59212

Page 23

Line \ Col.	(1)	(2)	(3)	(4)	(5)	(6)	(7)	(8)	(9)	(10)	(11)	(12)	(13)	(14)
1101	84192	90150	02904	26835	17174	42301	66230	51875	12019	90446	44891	53640	49491	40974
1102	21791	24764	53674	30093	45134	24073	24316	59727	95062	88513	63804	44547	19920	37719
1103	63501	05040	71881	17759	91881	69614	11891	27729	60387	80028	56757	90152	85907	27714
1104	07149	69285	55481	21889	67061	06631	27033	80489	97300	74647	95818	13557	92510	22063
1105	59443	98962	74778	96920	65620	36794	21088	06901	66167	48737	32171	41508	64576	30029
1106	39099	58021	28486	43052	99201	44400	12566	40060	00110	92074	99887	52372	31852	04306
1107	78176	58913	22638	69769	21102	72292	79494	53854	21234	84810	79995	74837	77566	54222
1108	11851	09065	98033	02752	58232	56504	22127	40028	20458	54283	37743	03336	89554	49006
1109	37515	25668	55785	66463	52758	67588	89891	47358	60805	96396	96460	70424	24511	69709
1110	45324	00016	46818	04873	75360	87519	65700	83430	99497	68213	98839	10962	61685	14741
1111	39338	62200	12200	22056	51447	92050	36908	42699	03714	77249	72501	85634	60027	90534
1112	48107	22333	37919	63359	60491	12582	23078	39200	81789	94129	06981	17417	36781	88738
1113	34582	52425	66816	98008	48509	23046	64998	75587	39526	73181	69189	73080	90698	22391
1114	11726	87723	17350	66656	25153	29295	30539	68093	19700	72068	46802	15591	10475	59195
1115	62681	84947	71441	44095	06799	51965	96630	34517	86340	70145	32256	99884	21945	24901
1116	19024	78592	25380	36560	49782	92033	98834	37738	21520	97739	66229	01420	45984	61386
1117	03788	52694	45452	93740	39643	73505	90883	10053	90470	24877	02811	03948	92737	21250
1118	65679	13628	89003	91555	21634	65067	85299	89657	64650	31564	74350	80779	73446	32001
1119	88772	32763	90526	17112	30425	24155	68202	53980	54149	02083	39615	17901	05289	75823
1120	63199	69880	72978	20867	57997	58471	23808	18518	39160	39924	56656	62181	29719	15680
1121	57013	40327	22838	70866	60522	03192	37722	52364	46511	78976	93796	29389	77418	00645
1122	81991	17216	21981	56841	77089	90232	98280	62900	04347	79116	98308	90018	68662	84028
1123	95637	56711	21483	23419	57836	94890	75997	95806	69840	96440	38152	54219	95388	69676
1124	37603	31906	85392	61483	94920	28886	87488	62857	26680	81452	19190	13618	86274	91991
1125	14316	23028	93147	58643	26761	34390	04023	04460	36456	36566	78618	99437	86280	47753
1126	42133	15695	19836	69538	92244	86830	08195	57578	23807	64850	23740	22393	84405	46730
1127	80104	03086	43208	65429	61701	99787	33860	86452	32956	55928	88298	50320	11042	33439
1128	56644	71608	78722	00126	37206	18748	72277	86832	94521	40759	25945	42118	60792	71332
1129	22240	27792	22943	51028	15020	12828	80359	39004	84963	95302	51079	16914	47121	00530
1130	36691	35957	30564	38452	20611	32423	01639	33168	43487	49113	73030	22466	26159	72248
1131	80883	38042	40889	86235	29617	02462	13285	26584	97124	01278	01371	82210	37741	07007
1132	86028	90782	03808	87781	13877	66987	62561	19821	92836	83691	09632	66384	95619	48847
1133	72796	20010	73633	78747	43470	61915	89760	68086	64966	47841	76044	29599	81088	31699
1134	01295	76818	47221	52383	17760	23087	57890	42237	86058	29897	32146	69086	01147	48562
1135	38336	16121	60770	27481	61366	80454	59127	46917	96207	28423	69102	42569	96585	01468
1136	60159	45585	84667	76241	25040	64123	62982	01259	16079	73837	30734	07512	89405	24384
1137	13620	69565	19833	25094	71527	27540	31297	49742	86668	81255	94208	72963	85699	65130
1138	61129	82150	46133	85056	87098	74354	15694	29175	34091	05041	83851	76151	01422	87630
1139	10683	69313	42006	68280	07110	79227	68750	77140	04429	01015	64858	92715	38553	31158
1140	90057	80871	40180	59371	15844	50034	38356	92986	65617	83627	87124	23160	98540	01507
1141	90903	02994	19299	37195	96116	59300	05753	78285	66608	94618	07285	24214	83835	46617
1142	76759	88718	84511	41002	12443	90015	37661	45306	70387	87002	82771	52847	96796	51101
1143	42876	06965	87207	01521	72643	54376	51819	67447	15548	81039	21389	18600	90951	76959
1144	34053	46074	00909	93541	17568	78648	21245	51523	96825	58814	06923	89805	08217	36825
1145	02707	56000	48515	92480	31701	28374	57525	94047	14995	32368	67478	93903	28352	18313
1146	26946	26519	81177	17594	55005	80011	76723	49927	74005	74816	31431	51417	90446	34492
1147	08609	40279	37024	17744	24046	86794	89036	38806	98386	72575	75105	13100	01522	54198
1148	23692	96268	38476	24281	76576	66333	12180	45407	00564	41920	10188	34969	18223	84251
1149	47164	96557	14200	88122	98488	47335	24463	66676	90028	42198	78152	37817	99951	01722
1150	16382	15528	53067	84067	79343	42662	56819	40187	40001	48553	35741	03862	26449	98922

Page 24

Line \ Col.	(1)	(2)	(3)	(4)	(5)	(6)	(7)	(8)	(9)	(10)	(11)	(12)	(13)	(14)
1151	01654	50375	23941	44848	79154	30193	15271	93296	12981	14087	67236	29163	18369	53614
1152	73750	68343	40727	81203	91727	06463	12248	57567	88750	57207	74771	63286	52041	37502
1153	64163	22132	22896	14305	45642	71580	21558	66457	87216	14927	02015	01218	11727	19528
1154	88445	85544	23627	79176	32502	34568	78777	35179	90882	48719	56575	54346	32740	80844
1155	94180	71108	19121	11958	33308	75930	24865	17426	46053	16954	09323	46726	52842	80029
1156	11433	12220	36719	35435	77727	78493	94580	70091	88108	68433	93301	41968	22860	06589
1157	61838	68801	49856	21739	95701	11180	73936	17628	12311	68680	73549	42393	32358	14055
1158	59908	68103	36855	19127	56637	06383	83182	76927	03134	60092	04961	22440	28532	75148
1159	41018	69556	06402	03436	89803	95604	55735	66978	70815	20734	24410	38543	81884	30318
1160	85095	01581	92299	06166	71078	31823	64316	95567	42508	63549	67909	19847	75922	42309
1161	59705	78103	66740	41743	69177	59277	47629	63874	95777	27606	55163	66908	77090	73600
1162	75094	55208	77905	20705	32408	16630	01242	63119	24272	40966	59365	45037	94748	45130
1163	78425	68672	79455	94334	23292	90422	75540	15843	03536	78929	27952	51325	38999	71222
1164	89088	86918	20787	05691	97309	96673	16389	51437	67463	62091	04531	68337	14487	03803
1165	83345	95889	39333	86027	24680	51909	97230	58136	00901	36504	79092	24556	64013	55752
1166	62896	00342	66647	57096	84913	67895	18804	17691	79985	02249	05018	93491	90608	78421
1167	96498	38270	80532	54307	07885	38892	50990	96766	10391	47494	61462	26236	50785	70081
1168	30974	47335	04918	42974	19294	72581	77377	04652	32773	77247	07213	68838	84135	88987
1169	57901	06163	99162	53285	27341	02507	41858	08436	95117	59801	64130	02595	05587	07478
1170	88494	80633	47785	53996	57058	04222	54488	12019	19961	98141	35105	55638	06283	21760
1171	34883	00045	89682	86664	92195	42593	56488	35402	03458	41818	29466	74119	02268	45858
1172	24373	46438	28935	63903	14722	10715	58795	42800	72900	58210	69550	27267	08158	59599
1173	16828	79262	23678	05509	23733	95318	77730	87614	75047	86064	83464	25290	76619	95392
1174	33723	27646	92335	87136	88062	21506	01750	71326	79990	72900	72778	42614	61003	76848
1175	01542	75066	73921	97188	31250	41996	31680	41783	58423	77216	06815	12302	03074	86293
1176	00100	12787	74100	95536	42359	01761	28842	71562	50387	85776	89723	95928	48924	40650
1177	82697	03389	19303	21646	22532	81701	03425	28914	68714	07516	21639	51991	86496	31881
1178	28137	17549	22698	72955	59849	02370	02784	13711	16466	51514	85699	30894	56668	95708
1179	02248	21570	33796	83789	72981	96423	68791	91684	63140	27870	37517	54420	33404	15825
1180	56175	82515	23348	42207	87644	57353	90349	16448	19161	99292	42009	31469	78108	59662
1181	80020	21622	67659	07878	17586	65524	20162	04712	31781	90129	72060	41170	23012	05969
1182	20271	23094	48372	77621	32889	19595	66500	28064	72134	85573	13428	55943	09667	75600
1183	38734	98044	02658	90698	72563	15076	23780	52815	24558	63962	95247	66493	79512	06464
1184	48183	24263	49297	32923	94406	63865	44336	27224	15045	59875	89505	20869	82810	55930
1185	48163	34158	03177	51696	57795	31725	14403	29856	64364	74978	12011	03976	55525	21112
1186	45658	15024	66664	18730	40671	92727	68626	81631	82377	33816	95294	34344	17329	58211
1187	71128	15524	55666	14763	13729	51708	54104	81331	03302	62068	55783	19096	59544	63944
1188	19041	42899	49464	93965	14960	88896	72784	82054	54161	77417	15678	87019	56787	91129
1189	32672	67506	93040	94527	31556	80163	80203	90928	95719	97206	69445	26889	57665	54865
1190	15823	48310	04391	15521	79255	69253	60254	01653	79162	36475	30155	35981	45176	75237
1191	82810	18981	62581	31642	42693	78972	60322	90462	04911	81323	80654	01210	18122	21060
1192	74772	80840	05816	29023	67410	12916	87933	78840	35477	28422	06146	02961	58363	15967
1193	52931	38199	85632	23761	99084	48028	07184	41635	90549	85312	04208	21329	92628	28400
1194	95395	87644	09722	99251	97129	70847	91864	08549	43675	39935	29510	57670	10593	45735
1195	76695	33451	57139	90612	11918	90871	60965	23555	29024	09147	34486	53759	75675	19516
1196	83560	50374	04410	57272	36705	51302	93147	29479	09307	32397	16313	31683	74323	03129
1197	28355	62002	85994	35807	84610	14186	51153	78998	40817	17031	97968	26048	86716	93709
1198	84684	54861	41330	66808	65231	14168	45193	27156	63562	37368	57576	03175	05539	96923
1199	21135	92001	43896	55887	35319	03793	60344	95970	86742	16043	87224	01591	77173	26690
1200	24236	01536	43897	41294	45551	46877	58631	82654	51804	67284	36564	70762	40619	86658

Page 25

Line \ Col.	(1)	(2)	(3)	(4)	(5)	(6)	(7)	(8)	(9)	(10)	(11)	(12)	(13)	(14)
1201	73496	09882	08999	21396	92878	92896	56836	62778	19251	20530	14361	02859	56813	92189
1202	93024	46250	20788	59964	81609	87694	40186	77855	91912	90065	25480	90939	62906	77020
1203	00974	03860	69142	00782	53189	44463	38322	68645	04153	73446	72483	96092	19099	27844
1204	85980	89568	86897	28819	93015	60804	45389	22679	62921	82470	41981	56496	43516	53703
1205	21926	84853	40364	14457	53939	61653	71979	81588	76284	98952	93527	00755	40680	16398
1206	96433	60242	22116	90230	87423	30668	35055	11813	81689	26586	80483	56892	72824	36970
1207	75535	03894	97949	64908	05890	70809	62170	04536	22023	23225	32445	48256	60669	35184
1208	15524	33007	92091	29915	57545	46078	03871	20516	23655	25994	45375	33270	01238	81788
1209	24189	85050	98663	38560	52553	81991	75859	52012	84186	14314	02003	12032	34204	30000
1210	83846	38182	07792	45967	41951	49621	82062	95925	48928	31720	92157	85869	51119	04031
1211	46739	32153	32959	10736	18932	13905	88740	19460	37435	43506	71957	72845	61462	76830
1212	41769	93463	66194	41318	13775	85724	36153	61195	49203	30323	62780	50706	07190	36282
1213	51182	90259	87138	92471	60295	10007	83366	18389	51091	48781	96234	72536	13147	93207
1214	47466	94179	64871	72693	41484	88010	67198	38679	56009	38993	19716	75096	35421	78977
1215	07835	74209	01787	22698	83567	81719	66085	75038	77919	56063	24978	68642	40122	44761
1216	56347	54336	56090	23776	75111	25829	40173	09882	22120	51197	37782	33145	60469	86554
1217	06888	32638	76714	83789	06638	09669	17406	40741	81363	93117	50682	40594	74399	33035
1218	35128	85695	59201	85555	41254	00703	73159	42562	58915	25518	10960	33930	27294	90639
1219	39612	93090	62253	15598	25911	65423	48300	94823	64886	92193	30423	35724	50001	70181
1220	78265	13257	29335	28987	55262	11962	21679	49924	29840	58689	81127	26594	22253	00935
1221	59329	90504	14855	84017	46761	29773	39746	37306	65238	60160	55801	48715	07263	07887
1222	09419	16715	89918	73266	36691	81481	40067	08081	14640	82156	24494	93889	93302	81493
1223	70316	73177	86655	51109	23593	43166	22575	19703	68559	74737	04236	49146	05485	66292
1224	42520	64431	56085	94988	65818	42961	92714	18925	58377	65806	60920	28335	08675	76171
1225	77795	37676	38164	39446	70684	39200	60688	02706	96807	55638	22967	06037	24814	36489
1226	96827	83921	94896	83396	27299	39388	45625	87659	42207	87885	23017	39734	60416	85488
1227	30509	37575	34107	61451	57192	28198	40088	01627	45649	08101	87770	53605	47531	54647
1228	48141	52622	68769	88507	15873	77512	43892	06278	68781	51205	69464	86145	61598	31280
1229	76775	13782	08331	26338	75128	26329	00011	82457	05784	18745	83927	50635	71277	38591
1230	80039	80106	61471	85010	57652	86208	51828	36666	67629	93189	82485	47927	17545	45710
1231	50113	18798	40463	27292	63786	89526	00611	38869	57250	46947	03270	49364	77561	51844
1232	70366	77355	78738	22944	28371	13278	75571	67914	11870	91882	72954	77650	91559	26554
1233	06521	04544	32907	33270	11277	03409	49440	15467	86398	11106	25506	18874	76940	02591
1234	44674	13886	70818	44288	98331	86481	28733	92684	50536	81567	22092	98683	67957	69799
1235	28082	15034	60185	14396	71652	17739	20900	54226	16312	91740	64103	04591	11191	97244
1236	74421	99841	80044	34782	16680	04551	63025	15146	24583	49797	60371	78626	03758	99361
1237	54667	63363	58192	26084	31078	11497	40343	65865	57238	74354	66207	57910	14211	25688
1238	17153	99397	13760	73178	37357	45190	06190	17062	21561	89720	37380	98109	90904	54969
1239	71721	67421	24805	50388	77723	26202	58475	75317	05427	79345	11986	07102	54013	40134
1240	78150	57466	19071	89828	07223	96310	80678	99686	00442	60341	81608	66515	98409	23116
1241	22575	85710	42407	09281	07074	02259	47903	43970	59180	18448	47771	82050	49732	66456
1242	51757	06816	15999	06704	12411	46373	08176	51644	16545	41250	80827	02713	45634	35497
1243	76793	84715	95520	96722	44927	14891	42580	05714	56252	75024	66387	90298	18461	58897
1244	64706	43607	95687	84579	65005	50318	98010	43800	47285	86501	22830	42132	51268	61228
1245	01619	24617	23748	43240	20646	66685	06769	14617	25884	06363	15587	68373	09650	53757
1246	66396	79480	02417	07971	42258	66401	89016	00576	56554	92936	43549	70432	96456	13052
1247	88247	92465	86970	27058	75111	35954	04908	39278	35048	22782	57676	02777	85138	13520
1248	34573	11739	68894	31311	68741	06781	16665	56810	54709	71875	86136	52822	03607	16646
1249	16426	72091	47835	13637	03540	50764	33141	61687	44597	55932	70676	38083	66787	93793
1250	12577	67820	43932	31786	09878	30754	35826	55219	49221	51126	11788	45411	53835	37723

Page 26

Line \ Col.	(1)	(2)	(3)	(4)	(5)	(6)	(7)	(8)	(9)	(10)	(11)	(12)	(13)	(14)
1251	69449	85916	03305	22339	40350	49711	45687	45329	24811	60540	43029	33811	08655	76553
1252	63881	61836	27585	66165	30016	44324	52080	61392	53547	07529	19329	60928	77147	79514
1253	88956	58983	35978	69628	35418	27869	00330	57312	39710	95596	41840	54045	46055	47783
1254	86027	57571	51844	97271	13490	05891	10381	21075	49707	16163	24928	72412	55791	37027
1255	65484	08659	79586	22646	00116	48807	24045	30468	25964	94816	25391	25628	77531	40804
1256	71995	61936	01207	83302	48590	99785	67234	86537	19243	19807	09504	70473	00090	99520
1257	00850	34022	65122	18191	10762	27345	31013	97627	07453	47702	37245	69370	61961	19905
1258	53680	89928	28841	15902	39075	80965	92066	90991	88070	83977	11311	30138	43493	06360
1259	55407	83873	51660	17483	95370	39259	84818	91371	81137	91256	71984	83527	94086	27779
1260	04987	87420	44247	02316	28463	68237	54671	82964	62580	78699	95062	27476	40466	47810
1261	63601	34750	89718	27983	33003	71523	25577	87683	44167	86715	83986	21460	62964	98014
1262	06022	73871	60775	96719	56248	39782	41168	24996	68906	94024	53417	75097	47998	72263
1263	55827	46978	02154	68985	45032	67401	04250	57385	20548	61567	89919	64690	44920	52051
1264	81793	87918	04365	13926	92315	99819	78247	76837	23149	52692	35562	93829	32027	79359
1265	80327	52129	14380	51652	36547	72102	50933	63154	15615	24613	31607	02546	27760	37113
1266	96604	46637	95906	20743	63780	74217	52251	71786	11087	68626	67083	73924	02870	93292
1267	55404	84898	08063	89735	55805	23658	29708	95836	22118	54763	65647	61295	41510	87822
1268	82280	18547	56385	97424	09154	23655	27047	91188	16529	87563	37573	09953	31184	59565
1269	37273	98782	97942	02581	42888	86591	41277	09557	85653	07982	53078	67678	44742	91929
1270	93916	22832	41840	80845	86732	38658	59807	92159	77308	80451	77710	98961	42769	79959
1271	91981	50674	00549	34128	91791	22038	19425	12454	96737	70866	63815	52298	12923	34504
1272	21469	00551	57201	62980	94587	82590	69092	87076	35047	16038	45835	03996	36051	78422
1273	48054	05959	42107	91681	12828	63540	64947	30075	62562	91902	32250	23449	77427	63056
1274	37028	44622	00003	43252	20801	21471	57108	09676	23441	49907	10860	36185	88165	53308
1275	10886	26404	74644	35326	59263	74350	44764	52972	61260	82303	60964	67652	78663	78897
1276	32114	73552	42712	90340	69113	15264	75283	71691	52501	39935	88235	13054	28302	80067
1277	39455	56072	78837	30889	77420	50665	66459	95502	79620	67491	22148	28665	14136	48381
1278	46816	21642	83735	29882	77347	05473	99982	26803	40956	06154	14388	30948	50668	09995
1279	93602	81212	40112	54182	41541	45483	28368	75654	57203	35645	10373	26386	32425	75690
1280	46298	56343	51571	06440	69964	62475	10141	87013	13304	02829	67839	02555	56749	24332
1281	89785	01137	59872	76654	59377	83786	52320	95958	99286	71778	87509	97834	68203	48432
1282	02551	24264	46696	36551	00607	91847	18293	53507	70951	04387	16061	06037	81343	72442
1283	12062	31292	07140	26560	14319	67176	14371	08219	48673	49326	42044	79912	81552	82594
1284	62099	74454	55839	65296	08071	26920	44302	15893	19019	41157	99645	78865	88914	21298
1285	28885	03268	14621	39884	26254	70133	03426	81597	53643	50161	41289	56119	36809	55000
1286	25201	31100	37809	54713	38221	74943	08927	36800	71246	22876	87044	60237	24663	64854
1287	15345	30805	29745	29458	50886	28616	73329	36538	80556	88777	42274	07024	33002	88991
1288	94577	14992	20745	45458	24084	93912	80562	74999	38324	43368	38866	66852	59048	78052
1289	25019	59941	60506	34956	57998	22550	81977	00813	74228	44453	92715	20627	81079	16026
1890	25351	56613	71994	27085	51520	93368	80568	68099	61123	51693	12202	70826	12858	67448
1291	50168	78838	91534	91717	36226	94416	24862	56716	35914	97583	32181	31349	23563	28028
1292	10869	70559	50634	57934	13690	54523	47036	74469	32565	60813	52573	41336	38667	51538
1293	63240	92582	42630	27213	07265	84138	97151	92011	81631	41962	28891	02431	42162	33967
1294	02621	06819	57734	42310	84837	38393	86348	28392	35434	50708	68887	59358	38391	09192
1295	15630	47503	12280	14240	33624	02234	90596	27274	84794	71578	04380	09055	74134	19932
1296	60414	59965	39416	29289	02401	21799	67753	48692	74170	25539	19140	55590	09538	69247
1297	85971	32373	19968	83504	94775	96495	55894	49629	92983	45823	42005	51128	40306	29257
1298	44925	45553	84250	28008	79125	37347	95355	98121	09597	58239	58694	55868	48200	11960
1299	45318	02208	17185	45131	37076	29921	67122	82739	73643	09402	50818	05274	13097	15318
1300	05098	40947	51260	92874	88221	16349	26289	57662	73294	93178	89288	06691	30731	97937

Page 27

Line \ Col.	(1)	(2)	(3)	(4)	(5)	(6)	(7)	(8)	(9)	(10)	(11)	(12)	(13)	(14)
1301	60366	78175	36293	76568	20636	49285	57768	68121	57201	01468	83031	66571	65129	94746
1302	17685	29660	98173	66736	69463	64668	51852	64196	88426	41842	13234	58631	01918	92237
1303	69965	78091	34664	81398	98458	19161	64210	63318	65013	45690	43974	66428	62244	22645
1304	13980	30753	09792	78154	26957	58175	91133	24664	15360	84004	62275	34009	82494	31361
1305	50268	09400	12618	03982	21793	66856	70162	07023	60933	93633	84343	50064	98209	68144
1306	37644	76570	52902	20848	18880	08452	55613	87058	05385	27034	15553	25714	85848	53664
1307	98780	79716	13950	35230	75151	56297	17556	33788	96348	71888	83684	33670	28768	74459
1308	14689	29968	04027	93099	25937	41948	78953	97916	36720	44967	03773	60400	87404	53861
1309	86898	89352	39497	32679	30083	01788	69579	97735	04320	09731	78767	05538	74401	60757
1310	37786	95630	15065	79149	78000	34641	90106	98703	96611	65082	37074	57526	71735	79913
1311	40405	31263	31468	66829	00613	20895	95135	37009	78410	74277	20199	09895	15077	60679
1312	89263	27679	44380	38310	13449	23891	46659	68246	25640	89674	31216	85820	53406	91206
1313	34865	25498	08316	11709	29980	79482	37776	12721	75768	27905	58014	38006	76745	93526
1314	10398	03254	95238	32342	99630	58887	23242	80720	86315	11141	07131	37848	68740	29146
1315	42382	67381	10759	33671	17625	32189	54575	57719	66426	45404	60468	88992	44805	01573
1316	08626	06528	78355	19779	60631	51403	94602	74319	34524	35961	41725	36570	67274	32763
1317	21190	62839	02901	47222	49014	01821	75171	34851	85520	49930	60830	01893	84615	83944
1318	26470	60864	60869	08931	69519	29584	12516	48733	16685	71865	56051	02172	05543	34667
1319	83960	44750	06525	71014	63725	01136	73102	75134	84744	00849	63140	30735	80256	66665
1320	60337	63586	45043	62781	25125	50656	11616	98062	85841	39323	49450	96131	86588	63363
1321	05405	48703	02975	92158	47045	45189	05458	16307	15880	72868	27060	93055	62377	68795
1322	33641	16073	36045	88475	36407	59251	23184	88492	22752	09899	84236	69575	05768	12512
1323	00435	46551	93038	27998	92168	04207	68063	82144	07909	64024	03583	02382	71875	26927
1324	62714	33646	71617	63976	01408	79439	11055	25935	27612	11181	99776	51240	07948	64174
1325	98811	03407	72032	69541	07709	85693	16591	26400	71174	45612	94903	59006	13869	62263
1326	44922	18477	82134	04268	42176	15397	49255	02055	08351	54276	33548	85723	96211	26502
1327	52023	15635	93237	96613	88740	43972	86000	27832	95896	27775	75715	78972	33849	71024
1328	53099	00222	08473	49333	66949	29693	65965	29851	71850	73937	00015	10617	22622	11171
1329	09681	40588	50684	24608	72574	81616	11724	23970	64884	05469	57419	14006	15723	89671
1330	86821	47710	17986	25064	63821	36301	49724	74320	38142	24864	20343	11702	08667	63514
1331	49084	82735	92900	80895	72431	99575	32669	08419	93009	60511	71995	64804	30923	30843
1332	47365	93121	90857	09806	89949	40510	36868	55389	58381	48433	26485	12804	42354	13480
1333	80475	03736	01954	58128	28287	08428	02771	94054	37769	31318	57619	41138	97533	01109
1334	90523	86240	06286	42082	54219	36983	22628	01476	46245	31344	77554	14390	30349	55904
1335	53474	83807	72336	87792	67077	56997	10432	33116	29976	45767	45953	00209	03218	40518
1336	53946	30368	28053	80412	02319	25354	25270	17059	07977	37641	79744	35193	20527	33004
1337	60801	89324	10494	06655	22834	67048	77373	46614	46567	73255	39149	33013	99682	55149
1338	15851	81289	99027	30178	64405	72936	03962	60509	13441	60798	52212	96297	69214	44348
1339	40512	67121	29336	96219	67383	64131	53616	33844	64039	76791	79266	58877	70616	47383
1340	38358	33301	57562	94624	60067	97146	21007	19851	55793	23088	99735	86817	38092	04621
1341	32529	00483	40638	64142	72208	51840	63013	11081	25210	60880	64136	31142	95715	25346
1342	16635	77774	73418	22678	56533	93821	39801	74037	78090	91236	43409	73638	44102	34784
1343	48220	43224	43890	05927	37025	60027	52196	91676	27637	23819	63097	13604	71435	16343
1344	05876	67302	63235	79069	35633	46093	01093	88949	21671	27063	72352	42853	07430	54263
1345	97129	55374	34685	72377	52025	98663	20311	04447	15333	03433	72159	44391	07910	44562
1346	41362	95497	55789	23179	65683	39353	49076	77776	42859	16137	00076	92647	83921	94799
1347	34532	90651	26632	39041	76740	24742	93466	41525	69868	89958	80514	23944	72703	66774
1348	12267	74978	74187	18103	34081	32015	47005	28974	27357	77084	27826	18687	39867	29756
1349	82863	24016	86446	62886	78103	78754	41941	48033	61949	37145	58920	64252	38732	88034
1350	17551	05912	55287	59322	34223	91493	38028	31388	50198	07503	75783	55385	19836	74730

Page 28

Line \ Col.	(1)	(2)	(3)	(4)	(5)	(6)	(7)	(8)	(9)	(10)	(11)	(12)	(13)	(14)
1351	47404	65603	65995	73759	05133	91015	39114	51408	45012	73196	51428	07489	49680	79004
1352	88185	47637	31280	44881	37547	27031	48582	48734	78568	39599	65434	35336	74548	06485
1353	43083	14876	99082	36151	16431	75355	39829	25167	14161	24305	11024	25514	13343	43079
1354	43228	64318	08651	50931	99446	01394	17394	85799	63497	40704	74826	05191	81759	81697
1355	72569	70591	45597	24910	36042	37737	94530	63225	81004	33156	57454	73741	88150	25305
1356	85723	35065	35010	49108	44254	25921	16876	39802	28917	66883	76432	69179	55582	39472
1357	07674	45281	79114	80423	35234	66899	16321	95496	02988	48603	58631	10332	07985	79635
1358	24371	85107	29303	19773	23482	49465	80345	34923	84981	24954	36370	30459	90920	33314
1359	92986	30371	87633	11174	54733	59745	15042	05502	19662	17636	61805	93327	54918	26770
1360	09284	26947	32468	82071	21313	35405	75779	46887	13686	72738	95167	56717	52030	71791
1361	50705	10000	92249	21784	39572	61912	67176	60971	58167	23516	06361	79851	76775	41078
1362	01399	98840	90412	97879	18097	68033	87243	59952	53176	72820	49419	40521	77891	32268
1363	48481	19537	51454	35710	23489	31569	10847	23933	74062	52064	79003	77919	66075	72351
1364	56274	21660	71567	37946	98883	73493	14879	44880	36508	56520	48816	96411	50527	19730
1365	23945	22771	09995	51119	57954	40956	67845	35605	76096	99893	12530	93785	14101	00603
1366	32786	25583	76137	49518	23040	27109	29328	74653	76314	95895	40772	69520	92256	55830
1367	96812	26093	22093	69157	43615	22575	38932	77496	37941	90722	84368	22184	04540	14421
1368	33127	58380	03808	02086	09202	45870	88112	71603	40009	86928	54008	13860	17795	40916
1369	27066	53942	24841	63350	06488	30075	30571	02046	07758	83615	97418	54342	38405	26364
1370	21033	13901	11525	27306	22546	41864	03552	67124	61483	85280	25259	89909	24761	62468
1371	02085	11182	57294	50965	57514	68744	87780	12626	82379	63410	13115	56403	24420	32802
1372	33431	16508	09725	01239	17748	55122	23390	29603	40112	32069	23531	80846	31417	13947
1373	31078	66351	90560	88924	95920	66727	09408	08843	25851	77238	19721	54516	79644	90522
1374	76030	79940	96569	98360	93988	56792	14457	14623	06338	09295	76658	13503	73871	29465
1375	18949	79885	82606	70510	59616	16345	21673	94644	26295	02866	93168	03292	12232	59680
1376	60488	20808	66381	35016	53865	18780	51609	76650	40831	37975	14026	09732	83563	96492
1377	96274	48730	10237	06490	22744	05231	08209	15355	95829	40167	55418	44540	64345	60253
1378	53209	88700	50956	65080	40880	04606	14695	40349	98037	17915	52744	57992	84675	99450
1379	09373	00553	88906	55737	57287	08439	06860	46565	89966	38774	76843	16712	80066	15296
1380	56535	86612	38352	37568	03479	70740	51283	85020	96367	69401	99787	03710	04474	85716
1382	72389	31159	98419	64819	09421	36740	43675	98593	93570	61623	95157	43561	55848	09846
1381	22087	37901	71096	89958	72519	05638	12662	52211	39454	20733	25225	33830	85277	53931
1383	22926	55101	69474	29510	80339	89195	51613	43231	06725	73779	27004	42162	87396	76752
1384	71225	96695	24021	15689	33804	99155	33889	78498	20368	55654	71722	57324	53446	14299
1385	41560	50531	18195	39965	27944	05386	37932	84917	97466	63178	51526	68959	71204	07977
1386	01862	37716	77614	72491	90338	62710	45689	89097	11153	38980	06320	24859	12757	63390
1387	52828	27393	44973	99166	61276	65732	16465	58624	77169	90194	71324	34551	34332	00340
1388	02368	60151	23378	91706	99658	28720	32798	18600	03073	23120	57235	95640	91723	18770
1389	90441	25732	45765	92022	08568	97671	02219	29566	78427	36549	47486	74697	34962	88696
1390	39284	99528	89455	49412	64125	91446	77972	60481	21634	83578	83050	56922	90020	64218
1391	49921	53976	98275	07921	67076	92533	09361	57168	59234	65805	03716	79186	15950	66582
1392	84174	93059	87298	31346	60845	98802	61457	22201	92528	44774	96269	70604	90520	62150
1393	96668	00377	70352	47613	96093	05801	22332	34252	67613	65937	27145	37959	00210	98261
1394	85387	39534	07436	01098	79865	35162	69056	62750	38043	13392	75744	89126	12976	23229
1395	53941	76161	35482	15450	16144	28354	04027	44422	28626	34870	57613	69425	61457	51492
1396	66418	63867	63981	28747	80882	80649	46441	59876	74028	86239	97207	33100	30963	43933
1397	48032	75602	58982	59365	91656	93365	36514	37065	51895	55746	80804	28470	12433	82419
1398	71399	11223	19608	90510	94364	18779	24214	55673	59983	39875	97610	59171	20207	64998
1399	64966	52295	44551	56501	14594	18795	83690	66628	43553	16608	62249	83161	58241	44630
1400	05364	58782	80069	02567	12667	16137	02457	95925	49430	37932	91675	56396	59065	87900

Page 29

Line \ Col.	(1)	(2)	(3)	(4)	(5)	(6)	(7)	(8)	(9)	(10)	(11)	(12)	(13)	(14)
1401	59844	58723	00069	27149	37807	36802	07016	17174	43461	49223	88994	35492	63846	08310
1402	12262	18803	94209	54054	05420	53401	88902	70088	74501	30083	98813	28964	65996	51411
1403	65215	18830	55843	35296	26268	27101	06971	49455	38249	87252	97477	93142	93713	11423
1404	72447	80493	18477	39502	10723	95205	74636	58282	43756	76555	65080	92290	01212	57613
1405	20203	37288	88676	31283	17594	40044	05289	19754	97974	12055	86892	26816	09427	62018
1406	42979	67093	75476	25312	00227	42334	96117	63572	81234	70337	76197	07579	05213	34084
1407	35173	36173	77500	32921	36632	99822	98524	17900	34985	71520	40720	31440	81039	90211
1408	07330	36615	20972	84490	38899	01743	45385	44457	67442	73146	60977	76262	08791	37165
1409	48596	90718	92884	31200	92818	12775	55860	86902	88241	55608	78793	99396	42732	12818
1410	77495	30824	50323	44441	37427	14990	15222	94033	11416	19086	88814	00623	23411	09612
1411	40259	67695	33361	55230	55307	49130	75436	27240	41914	77587	56275	91830	32096	84928
1412	58246	21582	46177	71390	61062	21690	37129	58241	71122	59036	15551	79279	82071	64794
1413	61101	13033	31768	67129	86532	64024	69389	28702	61633	21678	69862	50690	92855	05422
1414	04765	31821	44593	46422	51428	58416	91270	55106	70638	59222	25845	33319	77335	29238
1415	18824	06096	70524	95836	24943	25359	78956	98953	88610	76011	88314	76641	47188	63556
1416	47405	97469	90408	49420	45421	08755	98228	34796	88746	66820	26942	20890	93616	12668
1417	43258	95534	87336	44598	21812	47574	37461	71134	51379	02560	59287	18163	32017	34079
1418	78054	38053	76625	40189	85260	15936	19764	06393	52717	26632	03845	84789	57660	06534
1419	98009	48697	99745	57271	27845	89662	97060	34924	74216	40330	62163	84100	03510	87018
1420	89539	20545	01572	80160	99113	60898	15979	21744	31802	00506	70014	34224	24303	60213
1421	50971	68414	12749	98328	01385	03650	28584	46795	61672	18108	53097	98729	74866	73116
1422	08985	78498	12409	90670	70068	44249	86142	58168	49415	81671	40515	47715	14092	65155
1423	56387	98581	50840	43657	79899	79340	56881	39843	73894	04555	17874	01733	77115	24638
1424	95291	50890	88317	18046	23068	10820	87401	91948	77129	11243	71657	03462	75057	05700
1425	48269	12957	89398	73619	77078	38993	52967	80016	64932	44532	18121	93764	49174	82724
1426	30129	45255	61883	86928	40119	10822	81133	28040	49345	19326	81373	61492	45306	82317
1427	60384	92751	24437	92878	41118	45281	95549	34677	50057	11832	95817	72517	29479	09776
1428	90697	21527	45677	01301	32776	90315	24528	24802	44516	96779	92902	60511	33634	52737
1429	69348	59490	41924	15472	63892	04948	54391	86001	21344	19270	71069	18425	29397	24754
1430	31541	95745	00621	81689	34455	81882	85116	57733	55992	50017	69765	09465	52352	30175
1431	07365	84787	80609	25283	43880	74610	57662	59244	24173	03354	79635	76874	22007	15860
1432	32863	12770	71142	62515	68375	82877	56590	32717	83166	75115	57813	88814	40352	80827
1433	41459	65791	87317	31296	21168	07674	03736	96448	44439	32621	21884	41834	09015	86305
1434	70246	16564	57312	07083	91043	80663	33807	91347	09334	14579	22283	25043	51437	21623
1435	81769	34679	87801	46900	19251	48485	98608	68556	71292	63616	23231	96106	74690	98186
1436	58710	07935	41379	88812	98238	51991	27332	16278	22901	83504	91681	99962	98007	67523
1437	93137	62174	33395	37878	94351	38641	42854	62566	31140	73489	51051	44116	71495	49005
1438	92028	33566	76951	38293	59501	26781	35827	91719	09878	42461	06902	65924	35159	77071
1439	77643	97236	96376	43223	94208	43244	21477	64603	07954	96465	86425	04285	59298	94413
1440	48743	03178	22000	50815	29181	80767	27226	27128	35260	56008	03467	79998	23686	51235
1441	94061	44835	98120	78478	95414	08809	49294	85821	42752	70197	41772	01040	37368	53650
1442	43017	28338	32308	95011	44996	58065	14245	38794	96652	13214	79685	49385	08327	55004
1443	99631	98431	76974	63349	87461	39090	18052	66650	96938	68938	96831	33274	12612	62788
1444	88846	58974	93606	11833	28431	42696	49063	37026	04726	54248	44787	20724	91378	08937
1445	62332	76895	13101	41226	11923	94970	16685	23620	11918	86257	11694	43880	24531	27272
1446	26084	74818	49139	99047	02534	87700	27332	08393	39222	23943	91980	85104	07614	16278
1447	85785	03962	98057	67638	92405	44683	82124	07060	87337	09739	60577	36052	40571	56199
1448	08537	80929	66352	83924	21331	79310	93710	60398	74570	91901	94755	55198	11729	49801
1449	61985	45155	39295	67906	49595	08100	64610	68547	43692	80002	63518	31691	05598	66601
1450	92834	25766	03415	86498	18061	78501	20573	85001	77480	41572	36330	71213	68582	32234

Page 30

Line \ Col	(1)	(2)	(3)	(4)	(5)	(6)	(7)	(8)	(9)	(10)	(11)	(12)	(13)	(14)
1451	12066	40826	41486	23853	91969	03168	30629	50804	67753	08584	19882	23713	81435	90352
1452	57442	56762	76110	24363	72447	34084	80879	10938	91391	25386	36962	51084	90753	23549
1453	72078	39251	50886	46430	53994	56049	36030	71580	10161	70116	16556	95669	62817	05335
1454	63675	45810	44199	84436	13081	31652	36509	10949	05446	30913	60149	66850	89498	07155
1455	60595	74791	23267	37039	84875	27192	93298	46426	39072	98984	17404	28804	99841	61663
1456	45204	01095	86043	30604	28361	63342	48430	41094	64356	01440	86840	33363	12019	49913
1457	39994	25083	60371	97716	52655	58508	50026	50928	20567	17379	96141	06955	69389	92222
1458	06909	70912	30714	90653	83792	38309	42841	39378	16075	50103	15226	71790	42099	51455
1459	54780	92942	41783	68934	98006	71249	98263	59085	19490	59014	78394	92519	32837	33379
1460	34355	05525	52357	80287	89998	78576	11541	62443	17505	02287	58373	28620	70154	40433
1461	44229	68964	76398	97736	29356	19720	30957	40275	55062	07333	94037	64528	45791	95969
1462	79210	48613	03454	95486	70690	82321	21989	52789	88459	40665	98123	90591	43018	52535
1463	79746	45302	92695	80348	90693	32522	98359	82507	29568	74058	82667	05911	64996	69916
1464	41917	70277	95306	28321	40298	63054	80470	38497	41582	40297	10731	21011	76868	76451
1465	81682	67920	71356	90892	73481	32239	75242	22339	27678	21353	18881	51285	26395	76095
1466	03662	33387	37376	28181	10523	94005	06120	42216	15987	81374	19623	03216	33089	39776
1467	44950	55990	23779	93368	46063	58980	84151	09106	62520	65630	41521	27747	23516	97984
1468	19099	87872	79283	53751	80925	24221	74310	16851	21678	65600	78803	32680	75685	43132
1469	60213	68931	09542	71109	12758	25464	82318	08841	21851	19698	11087	71475	20908	85700
1470	58554	88248	81547	61101	13373	31787	17293	86404	54460	03941	62491	64924	13200	71641
1471	75826	10975	99991	75722	67780	84197	71507	75758	76840	92233	05862	88639	58977	82808
1472	35226	47738	87250	02131	19885	15470	44646	09361	73397	07561	90046	75443	83917	06696
1473	90137	47396	34336	75022	66586	04426	31440	41319	98881	80893	05531	91483	39232	69625
1474	61030	68038	88654	05986	85869	80044	46235	87084	59602	47088	13474	52449	98129	36933
1475	29523	08283	69819	14429	43131	80487	84369	03735	62873	39409	02302	52572	03324	91672
1476	05835	36457	15516	59987	43545	84159	39716	59021	20280	81563	62981	69666	74467	35031
1477	89456	73108	57063	46675	89930	85054	75680	04749	24886	48813	81276	89273	51686	84876
1478	44830	07509	23981	71725	10318	27204	23296	30808	11774	64813	41995	94873	75090	92732
1479	39864	47322	31251	82524	45643	31038	20455	52442	07923	84812	55808	92871	45642	27748
1480	94135	03380	81698	48694	52326	75025	50761	49392	01977	73439	78111	27989	68394	42038
1481	69240	52865	81175	30481	17890	64355	96345	39127	40278	98317	98973	92922	35319	59978
1482	79725	53403	89131	02804	26309	78157	29432	63674	41654	94507	32291	12131	48062	02443
1483	71198	93772	90685	29615	42387	72649	38219	30609	12534	23021	27494	87883	57993	82685
1484	09108	41317	42548	80675	32823	12150	66274	86189	66343	34571	75186	90986	08158	77741
1485	16637	65387	79852	38250	44973	45612	44938	93563	84733	82490	26738	95624	30208	40971
1486	63847	13687	45157	64816	50626	31236	34127	66409	15890	00972	48832	40660	16763	99106
1487	33924	92613	47288	49738	37532	82272	44805	75734	29660	59459	97949	35695	40757	84117
1488	72094	93704	50741	90556	72074	27652	85568	57142	58994	17869	85244	19737	96043	32528
1489	10274	87910	51552	22331	60297	22908	96064	11109	49773	00279	19375	84259	20675	89986
1490	45853	62407	77906	62399	95568	32481	10186	89765	63761	33384	05725	42448	97040	20117
1491	90163	21938	53874	51545	72222	78362	48573	18920	00208	95789	69375	93781	77688	47444
1492	61992	66865	35404	71123	23733	88910	45229	98444	94558	76975	06041	68354	82879	92208
1493	13829	31033	62790	91935	56721	21673	52965	76889	49018	24644	18625	61261	38054	52462
1494	09775	54186	21927	51460	26755	22921	50617	29806	37217	66079	73265	49426	86030	26047
1495	09592	17859	56164	45701	20939	92689	69421	75699	41219	29399	60373	56207	79282	01324
1496	42639	84884	69790	90784	34940	18891	69850	88330	51508	90991	53445	91947	67709	98816
1497	21655	81382	78529	20469	22066	96281	97864	17825	51975	23291	68971	41836	79125	96935
1498	80111	69552	54281	44357	98385	21425	72915	35041	27883	43938	64536	07481	51673	54470
1499	09761	22879	57602	28894	07030	91155	95164	52878	01407	63108	81643	95870	00561	52117
1500	47808	83766	69073	93127	76351	14177	05713	33110	29841	40650	83094	93585	20405	91235

Appendix C: Table of Probabilities for Use in Detection or Exploratory Sampling

Page 1 PROBABILITY, IN PER CENT, OF FINDING AT LEAST ONE ERROR IF TOTAL NO. OF ERRORS IN UNIVERSE IS AS INDICATED

SAMPLE SIZE	TOTAL ERRORS IN UNIVERSE SIZE OF 200.																		
	1	2	3	4	5	10	15	20	25	30	40	50	75	100	200	300	500	1000	2000
5.	2.5	4.9	7.4	9.7	12.0	22.8	32.6	41.3	49.1	56.0	67.6	76.7	90.8	97.0	100.0	-	-	-	-
10.	5.0	9.8	14.3	18.7	22.8	40.9	55.0	66.0	74.6	81.1	89.9	94.8	99.2	99.9	100.0	-	-	-	-
15.	7.5	14.5	21.0	27.0	32.6	55.0	70.3	80.6	87.5	92.1	96.9	98.9	99.9	100.0	100.0	-	-	-	-
20.	10.0	19.0	27.2	34.6	41.3	66.0	80.6	89.1	94.0	96.8	99.1	99.8	100.0	100.0	100.0	-	-	-	-
25.	12.5	23.5	33.2	41.6	49.1	74.6	87.5	94.0	97.2	98.7	99.8	100.0	100.0	100.0	100.0	-	-	-	-
30.	15.0	27.8	38.8	48.1	56.0	81.1	92.1	96.8	98.7	99.5	99.9	100.0	100.0	100.0	100.0	-	-	-	-
35.	17.5	32.0	44.0	54.0	62.2	86.1	95.0	98.3	99.4	99.8	100.0	100.0	100.0	100.0	100.0	-	-	-	-
40.	20.0	36.1	49.0	59.4	67.6	89.9	96.9	99.1	99.8	99.9	100.0	100.0	100.0	100.0	100.0	-	-	-	-
45.	22.5	40.0	53.7	64.2	72.5	92.7	98.1	99.5	99.9	100.0	100.0	100.0	100.0	100.0	100.0	-	-	-	-
50.	25.0	43.8	58.0	68.7	76.7	94.8	98.9	99.8	100.0	100.0	100.0	100.0	100.0	100.0	100.0	-	-	-	-
55.	27.5	47.5	62.1	72.7	80.4	96.3	99.3	99.9	100.0	100.0	100.0	100.0	100.0	100.0	100.0	-	-	-	-
60.	30.0	51.1	65.9	76.3	83.6	97.4	99.6	99.9	100.0	100.0	100.0	100.0	100.0	100.0	100.0	-	-	-	-
65.	32.5	54.5	69.5	79.5	86.3	98.2	99.8	100.0	100.0	100.0	100.0	100.0	100.0	100.0	100.0	-	-	-	-
70.	35.0	57.9	72.8	82.4	88.7	98.8	99.9	100.0	100.0	100.0	100.0	100.0	100.0	100.0	100.0	-	-	-	-
75.	37.5	61.1	75.8	85.0	90.8	99.2	99.9	100.0	100.0	100.0	100.0	100.0	100.0	100.0	100.0	-	-	-	-
80.	40.0	64.1	78.6	87.3	92.5	99.5	100.0	100.0	100.0	100.0	100.0	100.0	100.0	100.0	100.0	-	-	-	-
85.	42.5	67.1	81.2	89.3	93.9	99.7	100.0	100.0	100.0	100.0	100.0	100.0	100.0	100.0	100.0	-	-	-	-
90.	45.0	69.9	83.6	91.1	95.2	99.8	100.0	100.0	100.0	100.0	100.0	100.0	100.0	100.0	100.0	-	-	-	-
95.	47.5	72.6	85.7	92.6	96.2	99.9	100.0	100.0	100.0	100.0	100.0	100.0	100.0	100.0	100.0	-	-	-	-
100.	50.0	75.1	87.7	93.9	97.0	99.9	100.0	100.0	100.0	100.0	100.0	100.0	100.0	100.0	100.0	-	-	-	-
125.	62.5	86.1	94.9	98.1	99.3	100.0	100.0	100.0	100.0	100.0	100.0	100.0	100.0	100.0	100.0	-	-	-	-
150.	75.0	93.8	98.5	99.6	99.9	100.0	100.0	100.0	100.0	100.0	100.0	100.0	100.0	100.0	100.0	-	-	-	-
175.	87.5	98.5	99.8	100.0	100.0	100.0	100.0	100.0	100.0	100.0	100.0	100.0	100.0	100.0	100.0	-	-	-	-
200.	100.0	100.0	100.0	100.0	100.0	100.0	100.0	100.0	100.0	100.0	100.0	100.0	100.0	100.0	100.0	-	-	-	-
225.	-	-	-	-	-	-	-	-	-	-	-	-	-	-	-	-	-	-	-
250.	-	-	-	-	-	-	-	-	-	-	-	-	-	-	-	-	-	-	-
275.	-	-	-	-	-	-	-	-	-	-	-	-	-	-	-	-	-	-	-
300.	-	-	-	-	-	-	-	-	-	-	-	-	-	-	-	-	-	-	-
325.	-	-	-	-	-	-	-	-	-	-	-	-	-	-	-	-	-	-	-
350.	-	-	-	-	-	-	-	-	-	-	-	-	-	-	-	-	-	-	-
375.	-	-	-	-	-	-	-	-	-	-	-	-	-	-	-	-	-	-	-
400.	-	-	-	-	-	-	-	-	-	-	-	-	-	-	-	-	-	-	-
425.	-	-	-	-	-	-	-	-	-	-	-	-	-	-	-	-	-	-	-
450.	-	-	-	-	-	-	-	-	-	-	-	-	-	-	-	-	-	-	-
475.	-	-	-	-	-	-	-	-	-	-	-	-	-	-	-	-	-	-	-
500.	-	-	-	-	-	-	-	-	-	-	-	-	-	-	-	-	-	-	-
550.	-	-	-	-	-	-	-	-	-	-	-	-	-	-	-	-	-	-	-
600.	-	-	-	-	-	-	-	-	-	-	-	-	-	-	-	-	-	-	-
650.	-	-	-	-	-	-	-	-	-	-	-	-	-	-	-	-	-	-	-
700.	-	-	-	-	-	-	-	-	-	-	-	-	-	-	-	-	-	-	-

Source: USAF Auditor General Table, 1959

Page 2

PROBABILITY, IN PER CENT, OF FINDING AT LEAST ONE ERROR IF TOTAL NO. OF ERRORS IN UNIVERSE IS AS INDICATED

TOTAL ERRORS IN UNIVERSE SIZE OF 500.

SAMPLE SIZE	1	2	3	4	5	10	15	20	25	30	40	50	75	100	200	300	500	1000	2000
5.	1.0	2.0	3.0	4.0	4.9	9.6	14.2	18.5	22.7	26.7	34.2	41.1	55.8	67.4	92.3	99.0	100.0	-	-
10.	2.0	4.0	5.9	7.8	9.6	18.4	26.5	33.8	40.4	46.5	56.9	65.5	80.6	89.5	99.4	100.0	100.0	-	-
15.	3.0	5.9	8.7	11.5	14.2	26.5	37.1	46.3	54.2	61.0	71.9	79.9	91.6	96.7	100.0	100.0	100.0	-	-
20.	4.0	7.8	11.5	15.1	18.5	33.8	46.3	56.5	64.9	71.7	81.8	88.4	96.4	99.0	100.0	100.0	100.0	-	-
25.	5.0	9.8	14.3	18.6	22.7	40.4	54.2	64.9	73.2	79.5	88.2	93.3	98.5	99.7	100.0	100.0	100.0	-	-
30.	6.0	11.7	17.0	22.0	26.7	46.5	61.0	71.7	79.5	85.3	92.4	96.2	99.4	99.9	100.0	100.0	100.0	-	-
35.	7.0	13.5	19.6	25.3	30.5	51.9	66.9	77.3	84.4	89.4	95.2	97.8	99.7	100.0	100.0	100.0	100.0	-	-
40.	8.0	15.4	22.2	28.4	34.2	56.9	71.9	81.8	88.2	92.4	96.9	98.8	99.9	100.0	100.0	100.0	100.0	-	-
45.	9.0	17.2	24.7	31.5	37.7	61.4	76.2	85.4	91.1	94.6	98.0	99.3	100.0	100.0	100.0	100.0	100.0	-	-
50.	10.0	19.0	27.1	34.5	41.1	65.5	79.9	88.4	93.3	96.2	98.8	99.6	100.0	100.0	100.0	100.0	100.0	-	-
55.	11.0	20.8	29.6	37.4	44.3	69.2	83.0	90.7	95.0	97.3	99.2	99.8	100.0	100.0	100.0	100.0	100.0	-	-
60.	12.0	22.6	31.9	40.1	47.4	72.5	85.7	92.6	96.2	98.1	99.5	99.9	100.0	100.0	100.0	100.0	100.0	-	-
65.	13.0	24.3	34.2	42.8	50.3	75.5	88.0	94.2	97.2	98.7	99.7	99.9	100.0	100.0	100.0	100.0	100.0	-	-
70.	14.0	26.1	36.5	45.4	53.1	78.2	89.9	95.4	97.9	99.1	99.8	100.0	100.0	100.0	100.0	100.0	100.0	-	-
75.	15.0	27.8	38.7	47.9	55.8	80.6	91.6	96.4	98.5	99.4	99.9	100.0	100.0	100.0	100.0	100.0	100.0	-	-
80.	16.0	29.5	40.8	50.3	58.3	82.8	93.0	97.2	98.9	99.6	99.9	100.0	100.0	100.0	100.0	100.0	100.0	-	-
85.	17.0	31.1	42.9	52.7	60.8	84.8	94.2	97.8	99.2	99.7	100.0	100.0	100.0	100.0	100.0	100.0	100.0	-	-
90.	18.0	32.8	44.9	54.9	63.1	86.5	95.1	98.3	99.4	99.8	100.0	100.0	100.0	100.0	100.0	100.0	100.0	-	-
95.	19.0	34.4	46.9	57.1	65.3	88.1	96.0	98.7	99.6	99.9	100.0	100.0	100.0	100.0	100.0	100.0	100.0	-	-
100.	20.0	36.0	48.9	59.2	67.4	89.5	96.7	99.0	99.7	99.9	100.0	100.0	100.0	100.0	100.0	100.0	100.0	-	-
125.	25.0	43.8	57.9	68.5	76.4	94.5	98.8	99.7	99.9	100.0	100.0	100.0	100.0	100.0	100.0	100.0	100.0	-	-
150.	30.0	51.0	65.8	76.1	83.3	97.3	99.6	99.9	100.0	100.0	100.0	100.0	100.0	100.0	100.0	100.0	100.0	-	-
175.	35.0	57.8	72.6	82.3	88.5	98.7	99.9	100.0	100.0	100.0	100.0	100.0	100.0	100.0	100.0	100.0	100.0	-	-
200.	40.0	64.0	78.5	87.1	92.3	99.4	100.0	100.0	100.0	100.0	100.0	100.0	100.0	100.0	100.0	100.0	100.0	-	-
225.	45.0	69.8	83.4	90.9	95.0	99.8	100.0	100.0	100.0	100.0	100.0	100.0	100.0	100.0	100.0	100.0	100.0	-	-
250.	50.0	75.1	87.6	93.8	96.9	99.9	100.0	100.0	100.0	100.0	100.0	100.0	100.0	100.0	100.0	100.0	100.0	-	-
275.	55.0	79.8	91.0	96.0	98.2	100.0	100.0	100.0	100.0	100.0	100.0	100.0	100.0	100.0	100.0	100.0	100.0	-	-
300.	60.0	84.0	93.7	97.5	99.0	100.0	100.0	100.0	100.0	100.0	100.0	100.0	100.0	100.0	100.0	100.0	100.0	-	-
325.	65.0	87.8	95.8	98.5	99.5	100.0	100.0	100.0	100.0	100.0	100.0	100.0	100.0	100.0	100.0	100.0	100.0	-	-
350.	70.0	91.0	97.3	99.2	99.8	100.0	100.0	100.0	100.0	100.0	100.0	100.0	100.0	100.0	100.0	100.0	100.0	-	-
375.	75.0	93.8	98.5	99.6	99.9	100.0	100.0	100.0	100.0	100.0	100.0	100.0	100.0	100.0	100.0	100.0	100.0	-	-
400.	80.0	96.0	99.2	99.8	100.0	100.0	100.0	100.0	100.0	100.0	100.0	100.0	100.0	100.0	100.0	100.0	100.0	-	-
425.	85.0	97.8	99.7	100.0	100.0	100.0	100.0	100.0	100.0	100.0	100.0	100.0	100.0	100.0	100.0	100.0	100.0	-	-
450.	90.0	99.0	99.9	100.0	100.0	100.0	100.0	100.0	100.0	100.0	100.0	100.0	100.0	100.0	100.0	100.0	100.0	-	-
475.	95.0	99.8	100.0	100.0	100.0	100.0	100.0	100.0	100.0	100.0	100.0	100.0	100.0	100.0	100.0	100.0	100.0	-	-
500.	100.0	100.0	100.0	100.0	100.0	100.0	100.0	100.0	100.0	100.0	100.0	100.0	100.0	100.0	100.0	100.0	100.0	-	-
550.	-	-	-	-	-	-	-	-	-	-	-	-	-	-	-	-	-	-	-
600.	-	-	-	-	-	-	-	-	-	-	-	-	-	-	-	-	-	-	-
650.	-	-	-	-	-	-	-	-	-	-	-	-	-	-	-	-	-	-	-
700.	-	-	-	-	-	-	-	-	-	-	-	-	-	-	-	-	-	-	-

Page 3 PROBABILITY, IN PER CENT, OF FINDING AT LEAST ONE ERROR IF TOTAL NO. OF ERRORS IN UNIVERSE IS AS INDICATED

SAMPLE	TOTAL ERRORS IN UNIVERSE SIZE OF 1000.																		
SIZE	1	2	3	4	5	10	15	20	25	30	40	50	75	100	200	300	500	1000	2000
5.	0.5	1.0	1.5	2.0	2.5	4.9	7.3	9.6	11.9	14.2	18.5	22.7	32.3	41.0	67.3	83.3	96.9	100.0	-
10.	1.0	2.0	3.0	3.9	4.9	9.6	14.1	18.4	22.5	26.4	33.6	40.3	54.3	65.3	89.4	97.2	99.9	100.0	-
15.	1.5	3.0	4.4	5.9	7.3	14.1	20.4	26.3	31.8	36.9	46.0	53.9	69.2	79.7	96.6	99.5	100.0	100.0	-
20.	2.0	4.0	5.9	7.8	9.6	18.4	26.3	33.5	40.0	45.9	56.2	64.5	79.3	88.1	98.9	99.9	100.0	100.0	-
25.	2.5	4.9	7.3	9.6	11.9	22.5	31.8	40.0	47.3	53.7	64.4	72.7	86.1	93.1	99.7	100.0	100.0	100.0	-
30.	3.0	5.9	8.7	11.5	14.2	26.4	36.9	45.9	53.7	60.4	71.2	79.0	90.7	96.0	99.9	100.0	100.0	100.0	-
35.	3.5	6.9	10.1	13.3	16.3	30.1	41.6	51.3	59.4	66.2	76.6	83.9	93.8	97.7	100.0	100.0	100.0	100.0	-
40.	4.0	7.8	11.5	15.1	18.5	33.6	46.0	56.2	64.4	71.2	81.1	87.7	95.9	98.6	100.0	100.0	100.0	100.0	-
45.	4.5	8.8	12.9	16.8	20.6	37.0	50.1	60.5	68.8	75.4	84.7	90.6	97.2	99.2	100.0	100.0	100.0	100.0	-
50.	5.0	9.8	14.3	18.6	22.7	40.3	53.9	64.5	72.7	79.0	87.7	92.8	98.2	99.6	100.0	100.0	100.0	100.0	-
55.	5.5	10.7	15.6	20.3	24.7	43.4	57.5	68.1	76.1	82.1	90.1	94.5	98.8	99.7	100.0	100.0	100.0	100.0	-
60.	6.0	11.6	17.0	22.0	26.7	46.3	60.7	71.3	79.1	84.8	92.0	95.8	99.2	99.9	100.0	100.0	100.0	100.0	-
65.	6.5	12.6	18.3	23.6	28.6	49.1	63.8	74.3	81.8	87.1	93.6	96.8	99.5	99.9	100.0	100.0	100.0	100.0	-
70.	7.0	13.5	19.6	25.2	30.5	51.8	66.6	76.9	84.1	89.0	94.8	97.6	99.7	100.0	100.0	100.0	100.0	100.0	-
75.	7.5	14.4	20.9	26.8	32.3	54.3	69.2	79.3	86.1	90.7	95.9	98.2	99.8	100.0	100.0	100.0	100.0	100.0	-
80.	8.0	15.4	22.2	28.4	34.1	56.7	71.6	81.4	87.9	92.1	96.7	98.6	99.9	100.0	100.0	100.0	100.0	100.0	-
85.	8.5	16.3	23.4	29.9	35.9	59.0	73.9	83.4	89.5	93.3	97.3	99.0	99.9	100.0	100.0	100.0	100.0	100.0	-
90.	9.0	17.2	24.7	31.5	37.7	61.2	76.0	85.1	90.8	94.3	97.9	99.2	99.9	100.0	100.0	100.0	100.0	100.0	-
95.	9.5	18.1	25.9	33.0	39.4	63.3	77.9	86.7	92.0	95.2	98.3	99.4	100.0	100.0	100.0	100.0	100.0	100.0	-
100.	10.0	19.0	27.1	34.4	41.0	65.3	79.7	88.1	93.1	96.0	98.6	99.6	100.0	100.0	100.0	100.0	100.0	100.0	-
125.	12.5	23.4	33.0	41.4	48.8	73.9	86.7	93.3	96.6	98.3	99.6	99.9	100.0	100.0	100.0	100.0	100.0	100.0	-
150.	15.0	27.8	38.6	47.9	55.7	80.5	91.4	96.3	98.4	99.3	99.9	100.0	100.0	100.0	100.0	100.0	100.0	100.0	-
175.	17.5	32.0	43.9	53.7	61.9	85.5	94.5	98.0	99.2	99.7	100.0	100.0	100.0	100.0	100.0	100.0	100.0	100.0	-
200.	20.0	36.0	48.8	59.1	67.3	89.4	96.6	98.9	99.7	99.9	100.0	100.0	100.0	100.0	100.0	100.0	100.0	100.0	-
225.	22.5	40.0	53.5	64.0	72.1	92.3	97.9	99.4	99.8	100.0	100.0	100.0	100.0	100.0	100.0	100.0	100.0	100.0	-
250.	25.0	43.8	57.9	68.4	76.3	94.5	98.7	99.7	99.9	100.0	100.0	100.0	100.0	100.0	100.0	100.0	100.0	100.0	-
275.	27.5	47.5	61.9	72.4	80.0	96.1	99.2	99.9	100.0	100.0	100.0	100.0	100.0	100.0	100.0	100.0	100.0	100.0	-
300.	30.0	51.0	65.7	76.1	83.3	97.2	99.5	99.9	100.0	100.0	100.0	100.0	100.0	100.0	100.0	100.0	100.0	100.0	-
325.	32.5	54.5	69.3	79.3	86.1	98.1	99.7	100.0	100.0	100.0	100.0	100.0	100.0	100.0	100.0	100.0	100.0	100.0	-
350.	35.0	57.8	72.6	82.2	88.5	98.7	99.9	100.0	100.0	100.0	100.0	100.0	100.0	100.0	100.0	100.0	100.0	100.0	-
375.	37.5	61.0	75.6	84.8	90.5	99.1	99.9	100.0	100.0	100.0	100.0	100.0	100.0	100.0	100.0	100.0	100.0	100.0	-
400.	40.0	64.0	78.4	87.1	92.3	99.4	100.0	100.0	100.0	100.0	100.0	100.0	100.0	100.0	100.0	100.0	100.0	100.0	-
425.	42.5	67.0	81.0	89.1	93.8	99.6	100.0	100.0	100.0	100.0	100.0	100.0	100.0	100.0	100.0	100.0	100.0	100.0	-
450.	45.0	69.8	83.4	90.9	95.0	99.8	100.0	100.0	100.0	100.0	100.0	100.0	100.0	100.0	100.0	100.0	100.0	100.0	-
475.	47.5	72.5	85.6	92.4	96.0	99.8	100.0	100.0	100.0	100.0	100.0	100.0	100.0	100.0	100.0	100.0	100.0	100.0	-
500.	50.0	75.0	87.5	93.8	96.9	99.9	100.0	100.0	100.0	100.0	100.0	100.0	100.0	100.0	100.0	100.0	100.0	100.0	-
550.	55.0	79.8	90.9	95.9	98.2	100.0	100.0	100.0	100.0	100.0	100.0	100.0	100.0	100.0	100.0	100.0	100.0	100.0	-
600.	60.0	84.0	93.6	97.5	99.0	100.0	100.0	100.0	100.0	100.0	100.0	100.0	100.0	100.0	100.0	100.0	100.0	100.0	-
650.	65.0	87.8	95.7	98.5	99.5	100.0	100.0	100.0	100.0	100.0	100.0	100.0	100.0	100.0	100.0	100.0	100.0	100.0	-
700.	70.0	91.0	97.3	99.2	99.8	100.0	100.0	100.0	100.0	100.0	100.0	100.0	100.0	100.0	100.0	100.0	100.0	100.0	-

Page 4 PROBABILITY, IN PER CENT, OF FINDING AT LEAST ONE ERROR IF TOTAL NO. OF ERRORS IN UNIVERSE IS AS INDICATED

TOTAL ERRORS IN UNIVERSE SIZE OF 1600.

SAMPLE SIZE	1	2	3	4	5	10	15	20	25	30	40	50	75	100	200	300	500	1000	2000
5.	0.3	0.6	0.9	1.2	1.6	3.1	4.6	6.1	7.6	9.0	11.9	14.7	21.4	27.6	48.8	64.6	84.7	99.3	-
10.	0.6	1.2	1.9	2.5	3.1	6.1	9.0	11.9	14.6	17.3	22.4	27.3	38.2	47.7	73.8	87.5	97.7	100.0	-
15.	0.9	1.9	2.8	3.7	4.6	9.0	13.2	17.3	21.1	24.8	31.7	38.0	51.5	62.2	86.6	95.6	99.6	100.0	-
20.	1.3	2.5	3.7	4.9	6.1	11.9	17.3	22.4	27.2	31.7	39.9	47.2	61.9	72.7	93.2	98.5	99.9	100.0	-
25.	1.6	3.1	4.6	6.1	7.6	14.6	21.1	27.2	32.7	37.9	47.2	55.1	70.2	80.3	96.5	99.5	100.0	100.0	-
30.	1.9	3.7	5.5	7.3	9.0	17.3	24.8	31.7	37.9	43.6	53.5	61.8	76.6	85.8	98.2	99.8	100.0	100.0	-
35.	2.2	4.3	6.4	8.5	10.5	19.9	28.3	35.9	42.7	48.8	59.2	67.5	81.7	89.8	99.1	99.9	100.0	100.0	-
40.	2.5	4.9	7.3	9.6	11.9	22.4	31.7	39.9	47.2	53.5	64.1	72.4	85.7	92.7	99.6	100.0	100.0	100.0	-
45.	2.8	5.5	8.2	10.8	13.3	24.9	34.9	43.7	51.3	57.8	68.5	76.5	88.8	94.7	99.8	100.0	100.0	100.0	-
50.	3.1	6.2	9.1	11.9	14.7	27.3	38.0	47.2	55.1	61.8	72.4	80.1	91.3	96.2	99.9	100.0	100.0	100.0	-
55.	3.4	6.8	10.0	13.1	16.1	29.6	41.0	50.5	58.6	65.3	75.8	83.1	93.2	97.3	99.9	100.0	100.0	100.0	-
60.	3.8	7.4	10.8	14.2	17.4	31.8	43.8	53.7	61.8	68.6	78.7	85.7	94.7	98.1	100.0	100.0	100.0	100.0	-
65.	4.1	8.0	11.7	15.3	18.7	34.0	46.5	56.6	64.8	71.5	81.4	87.8	95.9	98.6	100.0	100.0	100.0	100.0	-
70.	4.4	8.6	12.6	16.4	20.1	36.2	49.0	59.4	67.6	74.2	83.7	89.7	96.8	99.0	100.0	100.0	100.0	100.0	-
75.	4.7	9.2	13.4	17.5	21.4	38.2	51.5	61.9	70.2	76.6	85.7	91.3	97.5	99.3	100.0	100.0	100.0	100.0	-
80.	5.0	9.8	14.3	18.6	22.6	40.2	53.8	64.4	72.5	78.8	87.5	92.6	98.1	99.5	100.0	100.0	100.0	100.0	-
85.	5.3	10.3	15.1	19.6	23.9	42.2	56.1	66.7	74.7	80.9	89.0	93.8	98.5	99.6	100.0	100.0	100.0	100.0	-
90.	5.6	10.9	16.0	20.7	25.2	44.0	58.2	68.8	76.7	82.7	90.4	94.7	98.8	99.7	100.0	100.0	100.0	100.0	-
95.	5.9	11.5	16.8	21.7	26.4	45.9	60.2	70.8	78.6	84.3	91.6	95.5	99.1	99.8	100.0	100.0	100.0	100.0	-
100.	6.2	12.1	17.6	22.8	27.6	47.7	62.2	72.7	80.3	85.8	92.7	96.2	99.3	99.9	100.0	100.0	100.0	100.0	-
125.	7.8	15.0	21.7	27.8	33.5	55.8	70.6	80.5	87.1	91.5	96.3	98.4	99.8	100.0	100.0	100.0	100.0	100.0	-
150.	9.4	17.9	25.6	32.6	38.9	62.7	77.3	86.2	91.6	94.9	98.1	99.3	99.9	100.0	100.0	100.0	100.0	100.0	-
175.	10.9	20.7	29.4	37.1	44.0	68.7	82.5	90.3	94.6	97.0	99.1	99.7	100.0	100.0	100.0	100.0	100.0	100.0	-
200.	12.5	23.4	33.0	41.4	48.8	73.8	86.6	93.2	96.5	98.2	99.6	99.9	100.0	100.0	100.0	100.0	100.0	100.0	-
225.	14.1	26.2	36.6	45.5	53.2	78.1	89.8	95.3	97.8	99.0	99.8	100.0	100.0	100.0	100.0	100.0	100.0	100.0	-
250.	15.6	28.8	40.0	49.4	57.3	81.8	92.3	96.7	98.6	99.4	99.9	100.0	100.0	100.0	100.0	100.0	100.0	100.0	-
275.	17.2	31.4	43.2	53.0	61.1	84.9	94.2	97.8	99.1	99.7	100.0	100.0	100.0	100.0	100.0	100.0	100.0	100.0	-
300.	18.7	34.0	46.4	56.5	64.6	87.5	95.6	98.5	99.5	99.8	100.0	100.0	100.0	100.0	100.0	100.0	100.0	100.0	-
325.	20.3	36.5	49.4	59.7	67.9	89.7	96.7	99.0	99.7	99.9	100.0	100.0	100.0	100.0	100.0	100.0	100.0	100.0	-
350.	21.9	39.0	52.3	62.8	70.9	91.6	97.6	99.3	99.8	99.9	100.0	100.0	100.0	100.0	100.0	100.0	100.0	100.0	-
375.	23.4	41.4	55.1	65.7	73.7	93.1	98.2	99.5	99.9	100.0	100.0	100.0	100.0	100.0	100.0	100.0	100.0	100.0	-
400.	25.0	43.8	57.8	68.4	76.3	94.4	98.7	99.7	99.9	100.0	100.0	100.0	100.0	100.0	100.0	100.0	100.0	100.0	-
425.	26.6	46.1	60.4	71.0	78.7	95.5	99.0	99.8	100.0	100.0	100.0	100.0	100.0	100.0	100.0	100.0	100.0	100.0	-
450.	28.1	48.4	62.9	73.4	80.9	96.4	99.3	99.9	100.0	100.0	100.0	100.0	100.0	100.0	100.0	100.0	100.0	100.0	-
475.	29.7	50.6	65.3	75.6	82.9	97.1	99.5	99.9	100.0	100.0	100.0	100.0	100.0	100.0	100.0	100.0	100.0	100.0	-
500.	31.2	52.7	67.5	77.7	84.7	97.7	99.6	99.9	100.0	100.0	100.0	100.0	100.0	100.0	100.0	100.0	100.0	100.0	-
550.	34.4	56.9	71.8	81.5	87.9	98.5	99.8	100.0	100.0	100.0	100.0	100.0	100.0	100.0	100.0	100.0	100.0	100.0	-
600.	37.5	61.0	75.6	84.8	90.5	99.1	99.9	100.0	100.0	100.0	100.0	100.0	100.0	100.0	100.0	100.0	100.0	100.0	-
650.	40.6	64.8	79.1	87.6	92.7	99.5	100.0	100.0	100.0	100.0	100.0	100.0	100.0	100.0	100.0	100.0	100.0	100.0	-
700.	43.7	68.4	82.2	90.0	94.4	99.7	100.0	100.0	100.0	100.0	100.0	100.0	100.0	100.0	100.0	100.0	100.0	100.0	-

Page 5

PROBABILITY, IN PER CENT, OF FINDING AT LEAST ONE ERROR IF TOTAL NO. OF ERRORS IN UNIVERSE IS AS INDICATED

SAMPLE SIZE	TOTAL ERRORS IN UNIVERSE SIZE OF 2000.																		
	1	2	3	4	5	10	15	20	25	30	40	50	75	100	200	300	500	1000	2000
5.	0.3	0.5	0.7	1.0	1.2	2.5	3.7	4.9	6.1	7.3	9.6	11.9	17.4	22.6	41.0	55.7	76.3	96.9	100.0
10.	0.5	1.0	1.5	2.0	2.5	4.9	7.3	9.6	11.8	14.1	18.3	22.4	31.8	40.2	65.2	80.4	94.4	99.9	100.0
15.	0.8	1.5	2.2	3.0	3.7	7.3	10.7	14.0	17.3	20.3	26.2	31.7	43.8	53.8	79.5	91.3	98.7	100.0	100.0
20.	1.0	2.0	3.0	3.9	4.9	9.6	14.0	18.3	22.3	26.2	33.4	39.9	53.6	64.3	88.0	96.2	99.7	100.0	100.0
25.	1.3	2.5	3.7	4.9	6.1	11.8	17.3	22.3	27.1	31.6	39.8	47.1	61.8	72.5	92.9	98.3	99.9	100.0	100.0
30.	1.5	3.0	4.4	5.9	7.3	14.1	20.3	26.2	31.6	36.7	45.7	53.5	68.5	78.8	95.9	99.3	100.0	100.0	100.0
35.	1.8	3.5	5.2	6.8	8.5	16.2	23.3	29.9	35.9	41.3	51.0	59.1	74.1	83.7	97.6	99.7	100.0	100.0	100.0
40.	2.0	4.0	5.9	7.8	9.6	18.3	26.2	33.4	39.8	45.7	55.8	64.0	78.7	87.4	98.6	99.9	100.0	100.0	100.0
45.	2.3	4.5	6.6	8.7	10.8	20.4	29.0	36.7	43.6	49.7	60.1	68.4	82.4	90.3	99.2	99.9	100.0	100.0	100.0
50.	2.5	4.9	7.3	9.6	11.9	22.4	31.7	39.9	47.1	53.5	64.0	72.2	85.6	92.6	99.5	100.0	100.0	100.0	100.0
55.	2.8	5.4	8.0	10.6	13.0	24.4	34.3	42.9	50.4	56.9	67.6	75.6	88.1	94.3	99.7	100.0	100.0	100.0	100.0
60.	3.0	5.9	8.7	11.5	14.1	26.3	36.8	45.8	53.5	60.2	70.8	78.6	90.3	95.6	99.8	100.0	100.0	100.0	100.0
65.	3.3	6.4	9.4	12.4	15.2	28.2	39.2	48.5	56.4	63.2	73.7	81.2	92.0	96.6	99.9	100.0	100.0	100.0	100.0
70.	3.5	6.9	10.1	13.3	16.3	30.0	41.5	51.1	59.2	65.9	76.3	83.5	93.4	97.4	99.9	100.0	100.0	100.0	100.0
75.	3.8	7.4	10.8	14.2	17.4	31.8	43.8	53.6	61.8	68.5	78.7	85.6	94.6	98.0	100.0	100.0	100.0	100.0	100.0
80.	4.0	7.8	11.5	15.1	18.5	33.6	45.9	56.0	64.2	70.9	80.8	87.3	95.6	98.5	100.0	100.0	100.0	100.0	100.0
85.	4.3	8.3	12.2	16.0	19.5	35.3	48.0	58.2	66.5	73.1	82.7	88.9	96.4	98.8	100.0	100.0	100.0	100.0	100.0
90.	4.5	8.8	12.9	16.8	20.6	37.0	50.0	60.4	68.6	75.1	84.4	90.3	97.0	99.1	100.0	100.0	100.0	100.0	100.0
95.	4.8	9.3	13.6	17.7	21.6	38.6	51.9	62.4	70.6	77.0	86.0	91.5	97.6	99.3	100.0	100.0	100.0	100.0	100.0
100.	5.0	9.8	14.3	18.6	22.6	40.2	53.8	64.3	72.5	78.8	87.4	92.6	98.0	99.5	100.0	100.0	100.0	100.0	100.0
125.	6.2	12.1	17.6	22.8	27.6	47.6	62.2	72.7	80.3	85.8	92.6	96.2	99.3	99.9	100.0	100.0	100.0	100.0	100.0
150.	7.5	14.4	20.9	26.8	32.3	54.2	69.1	79.1	85.9	90.5	95.7	98.1	99.7	100.0	100.0	100.0	100.0	100.0	100.0
175.	8.8	16.7	24.0	30.7	36.8	60.1	74.8	84.1	90.0	93.7	97.5	99.0	99.9	100.0	100.0	100.0	100.0	100.0	100.0
200.	10.0	19.0	27.1	34.4	41.0	65.2	79.5	88.0	92.9	95.9	98.6	99.5	100.0	100.0	100.0	100.0	100.0	100.0	100.0
225.	11.3	21.2	30.1	38.0	45.0	69.8	83.4	90.9	95.0	97.3	99.2	99.8	100.0	100.0	100.0	100.0	100.0	100.0	100.0
250.	12.5	23.4	33.0	41.4	48.7	73.8	86.6	93.2	96.5	98.2	99.5	99.9	100.0	100.0	100.0	100.0	100.0	100.0	100.0
275.	13.8	25.6	35.9	44.7	52.3	77.3	89.2	94.9	97.6	98.9	99.7	99.9	100.0	100.0	100.0	100.0	100.0	100.0	100.0
300.	15.0	27.8	38.6	47.8	55.7	80.4	91.3	96.2	98.3	99.3	99.9	100.0	100.0	100.0	100.0	100.0	100.0	100.0	100.0
325.	16.2	29.9	41.3	50.8	58.8	83.1	93.1	97.2	98.8	99.5	99.9	100.0	100.0	100.0	100.0	100.0	100.0	100.0	100.0
350.	17.5	31.9	43.9	53.7	61.8	85.5	94.5	97.9	99.2	99.7	100.0	100.0	100.0	100.0	100.0	100.0	100.0	100.0	100.0
375.	18.7	34.0	46.4	56.4	64.6	87.5	95.6	98.5	99.5	99.8	100.0	100.0	100.0	100.0	100.0	100.0	100.0	100.0	100.0
400.	20.0	36.0	48.8	59.1	67.3	89.3	96.5	98.9	99.6	99.9	100.0	100.0	100.0	100.0	100.0	100.0	100.0	100.0	100.0
425.	21.3	38.0	51.2	61.6	69.8	90.9	97.3	99.2	99.8	99.9	100.0	100.0	100.0	100.0	100.0	100.0	100.0	100.0	100.0
450.	22.5	39.9	53.5	64.0	72.1	92.2	97.8	99.4	99.8	100.0	100.0	100.0	100.0	100.0	100.0	100.0	100.0	100.0	100.0
475.	23.8	41.9	55.7	66.2	74.3	93.4	98.3	99.6	99.9	100.0	100.0	100.0	100.0	100.0	100.0	100.0	100.0	100.0	100.0
500.	25.0	43.8	57.8	68.4	76.3	94.4	98.7	99.7	99.9	100.0	100.0	100.0	100.0	100.0	100.0	100.0	100.0	100.0	100.0
550.	27.5	47.4	61.9	72.4	80.0	96.0	99.2	99.8	100.0	100.0	100.0	100.0	100.0	100.0	100.0	100.0	100.0	100.0	100.0
600.	30.0	51.0	65.7	76.0	83.2	97.2	99.5	99.9	100.0	100.0	100.0	100.0	100.0	100.0	100.0	100.0	100.0	100.0	100.0
650.	32.5	54.4	69.3	79.3	86.0	98.1	99.7	100.0	100.0	100.0	100.0	100.0	100.0	100.0	100.0	100.0	100.0	100.0	100.0
700.	35.0	57.8	72.6	82.2	88.4	98.7	99.8	100.0	100.0	100.0	100.0	100.0	100.0	100.0	100.0	100.0	100.0	100.0	100.0

Page 6 PROBABILITY, IN PER CENT, OF FINDING AT LEAST ONE ERROR IF TOTAL NO. OF ERRORS IN UNIVERSE IS AS INDICATED

SAMPLE SIZE	TOTAL ERRORS IN UNIVERSE SIZE OF 3000. 1	2	3	4	5	10	15	20	25	30	40	50	75	100	200	300	500	1000	2000
5.	0.2	0.3	0.5	0.7	0.8	1.7	2.5	3.3	4.1	4.9	6.5	8.1	11.9	15.6	29.2	41.0	59.8	86.9	99.6
10.	0.3	0.7	1.0	1.3	1.7	3.3	4.9	6.5	8.0	9.6	12.6	15.5	22.4	28.8	49.9	65.2	83.9	98.3	100.0
15.	0.5	1.0	1.5	2.0	2.5	4.9	7.3	9.6	11.8	14.0	18.3	22.3	31.7	39.9	64.6	79.5	93.6	99.8	100.0
20.	0.7	1.3	2.0	2.6	3.3	6.5	9.6	12.6	15.5	18.3	23.6	28.6	39.8	49.3	75.0	87.9	97.4	100.0	100.0
25.	0.8	1.7	2.5	3.3	4.1	8.0	11.8	15.5	18.9	22.3	28.6	34.4	47.0	57.3	82.3	92.9	99.0	100.0	100.0
30.	1.0	2.0	3.0	3.9	4.9	9.6	14.0	18.3	22.3	26.1	33.3	39.8	53.4	64.0	87.5	95.8	99.6	100.0	100.0
35.	1.2	2.3	3.5	4.6	5.7	11.1	16.2	21.0	25.5	29.8	37.7	44.7	59.0	69.7	91.2	97.6	99.8	100.0	100.0
40.	1.3	2.6	3.9	5.2	6.5	12.6	18.3	23.6	28.6	33.3	41.8	49.2	63.9	74.5	93.8	98.6	99.9	100.0	100.0
45.	1.5	3.0	4.4	5.9	7.3	14.0	20.3	26.2	31.6	36.6	45.6	53.3	68.3	78.5	95.6	99.2	100.0	100.0	100.0
50.	1.7	3.3	4.9	6.5	8.1	15.5	22.3	28.6	34.4	39.8	49.2	57.1	72.1	81.9	96.9	99.5	100.0	100.0	100.0
55.	1.8	3.6	5.4	7.1	8.8	16.9	24.3	31.0	37.2	42.8	52.5	60.7	75.5	84.8	97.8	99.7	100.0	100.0	100.0
60.	2.0	4.0	5.9	7.8	9.6	18.3	26.2	33.3	39.8	45.6	55.7	63.9	78.4	87.2	98.5	99.8	100.0	100.0	100.0
65.	2.2	4.3	6.4	8.4	10.4	19.7	28.1	35.6	42.3	48.3	58.6	66.9	81.1	89.2	98.9	99.9	100.0	100.0	100.0
70.	2.3	4.6	6.8	9.0	11.1	21.1	29.9	37.7	44.7	50.9	61.4	69.6	83.4	90.9	99.2	99.9	100.0	100.0	100.0
75.	2.5	4.9	7.3	9.6	11.9	22.4	31.7	39.8	47.0	53.4	63.9	72.1	85.4	92.4	99.5	100.0	100 0	100.0	100.0
80.	2.7	5.3	7.8	10.3	12.6	23.7	33.4	41.9	49.3	55.7	66.3	74.4	87.2	93.6	99.6	100.0	100.0	100.0	100.0
85.	2.8	5.6	8.3	10.9	13.4	25.0	35.1	43.8	51.4	58.0	68.6	76.5	88.7	94.6	99.7	100.0	100.0	100.0	100.0
90.	3.0	5.9	8.7	11.5	14.1	26.3	36.7	45.7	53.4	60.1	70.7	78.5	90.1	95.5	99.8	100.0	100.0	100.0	100.0
95.	3.2	6.2	9.2	12.1	14.9	27.6	38.4	47.6	55.4	62.1	72.6	80.3	91.3	96.2	99.9	100.0	100.0	100.0	100.0
100.	3.3	6.6	9.7	12.7	15.6	28.8	39.9	49.3	57.3	64.0	74.5	81.9	92.4	96.8	99.9	100.0	100.0	100.0	100.0
125.	4.2	8.2	12.0	15.7	19.2	34.7	47.3	57.4	65.6	72.3	82.0	88.3	96.1	98.7	100.0	100.0	100.0	100.0	100.0
150.	5.0	9.8	14.3	18.6	22.6	40.2	53.8	64.3	72.4	78.7	87.3	92.5	98.0	99.5	100.0	100.0	100.0	100.0	100.0
175.	5.8	11.3	16.5	21.4	26.0	45.2	59.5	70.1	77.9	83.7	91.1	95.2	99.0	99.8	100.0	100.0	100.0	100.0	100.0
200.	6.7	12.9	18.7	24.1	29.2	49.9	64.6	75.0	82.3	87.5	93.8	96.9	99.5	99.9	100.0	100.0	100.0	100.0	100.0
225.	7.5	14.4	20.9	26.8	32.3	54.2	69.0	79.1	85.9	90.5	95.7	98.0	99.7	100.0	100.0	100.0	100.0	100.0	100.0
250.	8.3	16.0	23.0	29.4	35.3	58.2	73.0	82.6	88.7	92.7	97.0	98.8	99.9	100.0	100.0	100.0	100.0	100.0	100.0
275.	9.2	17.5	25.1	31.9	38.2	61.8	76.4	85.5	91.1	94.5	97.9	99.2	99.9	100.0	100.0	100.0	100.0	100.0	100.0
300.	10.0	19.0	27.1	34.4	41.0	65.2	79.5	87.9	92.9	95.8	98.6	99.5	100.0	100.0	100.0	100.0	100.0	100.0	100.0
325.	10.8	20.5	29.1	36.8	43.7	68.3	82.2	90.0	94.4	96.8	99.0	99.7	100.0	100.0	100.0	100.0	100.0	100.0	100.0
350.	11.7	22.0	31.1	39.1	46.2	71.1	84.5	91.7	95.6	97.6	99.3	99.8	100.0	100.0	100.0	100.0	100.0	100.0	100.0
375.	12.5	23.4	33.0	41.4	48.7	73.7	86.6	93.1	96.5	98.2	99.5	99.9	100.0	100.0	100.0	100.0	100.0	100.0	100.0
400.	13.3	24.9	34.9	43.6	51.1	76.1	88.4	94.3	97.2	98.7	99.7	99.9	100.0	100.0	100.0	100.0	100.0	100.0	100.0
425.	14.2	26.3	36.8	45.7	53.4	78.3	89.9	95.3	97.8	99.0	99.8	100.0	100.0	100.0	100.0	100.0	100.0	100.0	100.0
450.	15.0	27.8	38.6	47.8	55.7	80.4	91.3	96.2	98.3	99.3	99.9	100.0	100.0	100.0	100.0	100.0	100.0	100.0	100.0
475.	15.8	29.2	40.4	49.8	57.8	82.2	92.5	96.9	98.7	99.4	99.9	100.0	100.0	100.0	100.0	100.0	100.0	100.0	100.0
500.	16.7	30.6	42.1	51.8	59.8	83.9	93.6	97.4	99.0	99.6	99.9	100.0	100.0	100.0	100.0	100.0	100.0	100.0	100.0
550.	18.3	33.3	45.5	55.5	63.7	86.8	95.2	98.3	99.4	99.8	100.0	100.0	100.0	100.0	100.0	100.0	100.0	100.0	100.0
600.	20.0	36.0	48.8	59.1	67.3	89.3	96.5	98.9	99.6	99.9	100.0	100.0	100.0	100.0	100.0	100.0	100.0	100.0	100.0
650.	21.7	38.6	51.9	62.4	70.5	91.3	97.5	99.3	99.8	99.9	100.0	100.0	100.0	100.0	100.0	100.0	100.0	100.0	100.0
700.	23.3	41.2	55.0	65.5	73.5	93.0	98.2	99.5	99.9	100.0	100.0	100.0	100.0	100.0	100.0	100.0	100.0	100.0	100.0

Page 7

PROBABILITY, IN PER CENT, OF FINDING AT LEAST ONE ERROR IF TOTAL NO. OF ERRORS IN UNIVERSE IS AS INDICATED

SAMPLE SIZE	TOTAL ERRORS IN UNIVERSE SIZE OF 4000.																		
	1	2	3	4	5	10	15	20	25	30	40	50	75	100	200	300	500	1000	2000
5.	0.1	0.2	0.4	0.5	0.6	1.2	1.9	2.5	3.1	3.7	4.9	6.1	9.0	11.9	22.6	32.3	48.7	76.3	96.9
10.	0.3	0.5	0.7	1.0	1.2	2.5	3.7	4.9	6.1	7.3	9.6	11.8	17.3	22.4	40.2	54.2	73.7	94.4	99.9
15.	0.4	0.7	1.1	1.5	1.9	3.7	5.5	7.3	9.0	10.7	14.0	17.2	24.8	31.6	53.7	69.0	86.6	98.7	100.0
20.	0.5	1.0	1.5	2.0	2.5	4.9	7.3	9.6	11.8	14.0	18.2	22.3	31.6	39.8	64.2	79.1	93.1	99.7	100.0
25.	0.6	1.2	1.9	2.5	3.1	6.1	9.0	11.8	14.5	17.2	22.3	27.1	37.8	47.0	72.4	85.8	96.5	99.9	100.0
30.	0.8	1.5	2.2	3.0	3.7	7.3	10.7	14.0	17.2	20.3	26.1	31.5	43.4	53.3	78.7	90.4	98.2	100.0	100.0
35.	0.9	1.7	2.6	3.5	4.3	8.4	12.4	16.2	19.8	23.3	29.8	35.7	48.6	58.9	83.5	93.5	99.1	100.0	100.0
40.	1.0	2.0	3.0	3.9	4.9	9.6	14.0	18.2	22.3	26.1	33.2	39.7	53.3	63.9	87.3	95.6	99.5	100.0	100.0
45.	1.1	2.2	3.3	4.4	5.5	10.7	15.6	20.3	24.7	28.9	36.5	43.4	57.5	68.2	90.2	97.1	99.8	100.0	100.0
50.	1.3	2.5	3.7	4.9	6.1	11.8	17.2	22.3	27.1	31.5	39.7	46.9	61.4	72.0	92.4	98.0	99.9	100.0	100.0
55.	1.4	2.7	4.1	5.4	6.7	12.9	18.8	24.2	29.3	34.1	42.7	50.2	64.9	75.4	94.2	98.7	99.9	100.0	100.0
60.	1.5	3.0	4.4	5.9	7.3	14.0	20.3	26.1	31.5	36.6	45.5	53.3	68.2	78.4	95.5	99.1	100.0	100.0	100.0
65.	1.6	3.2	4.8	6.3	7.9	15.1	21.8	28.0	33.7	38.9	48.2	56.1	71.1	81.0	96.5	99.4	100.0	100.0	100.0
70.	1.8	3.5	5.2	6.8	8.5	16.2	23.3	29.8	35.8	41.2	50.8	58.9	73.7	83.3	97.3	99.6	100.0	100.0	100.0
75.	1.9	3.7	5.5	7.3	9.0	17.3	24.8	31.6	37.8	43.4	53.3	61.4	76.1	85.3	97.9	99.7	100.0	100.0	100.0
80.	2.0	4.0	5.9	7.8	9.6	18.3	26.2	33.3	39.7	45.6	55.6	63.8	78.3	87.1	98.4	99.8	100.0	100.0	100.0
85.	2.1	4.2	6.2	8.2	10.2	19.3	27.6	35.0	41.6	47.6	57.8	66.1	80.3	88.6	98.8	99.9	100.0	100.0	100.0
90.	2.3	4.4	6.6	8.7	10.8	20.4	29.0	36.6	43.5	49.6	59.9	68.2	82.1	90.0	99.1	99.9	100.0	100.0	100.0
95.	2.4	4.7	7.0	9.2	11.3	21.4	30.3	38.2	45.3	51.5	61.9	70.2	83.8	91.2	99.3	99.9	100.0	100.0	100.0
100.	2.5	4.9	7.3	9.6	11.9	22.4	31.6	39.8	47.0	53.3	63.9	72.0	85.3	92.3	99.4	100.0	100.0	100.0	100.0
125.	3.1	6.2	9.1	11.9	14.7	27.2	37.9	47.1	54.9	61.6	72.1	79.8	91.0	96.0	99.9	100.0	100.0	100.0	100.0
150.	3.8	7.4	10.8	14.2	17.4	31.8	43.7	53.5	61.7	68.4	78.5	85.4	94.5	97.9	100.0	100.0	100.0	100.0	100.0
175.	4.4	8.6	12.6	16.4	20.1	36.1	48.9	59.2	67.4	74.0	83.4	89.5	96.6	98.9	100.0	100.0	100.0	100.0	100.0
200.	5.0	9.8	14.3	18.6	22.6	40.2	53.7	64.2	72.4	78.7	87.3	92.4	97.9	99.4	100.0	100.0	100.0	100.0	100.0
225.	5.6	10.9	15.9	20.7	25.1	44.0	58.1	68.7	76.6	82.5	90.2	94.6	98.8	99.7	100.0	100.0	100.0	100.0	100.0
250.	6.2	12.1	17.6	22.8	27.6	47.6	62.1	72.6	80.2	85.7	92.5	96.1	99.2	99.9	100.0	100.0	100.0	100.0	100.0
275.	6.9	13.3	19.2	24.8	30.0	51.0	65.7	76.0	83.2	88.3	94.3	97.2	99.5	99.9	100.0	100.0	100.0	100.0	100.0
300.	7.5	14.4	20.9	26.8	32.3	54.2	69.0	79.1	85.8	90.4	95.6	98.0	99.7	100.0	100.0	100.0	100.0	100.0	100.0
325.	8.1	15.6	22.5	28.8	34.6	57.2	72.0	81.7	88.1	92.2	96.7	98.6	99.8	100.0	100.0	100.0	100.0	100.0	100.0
350.	8.8	16.7	24.0	30.7	36.8	60.0	74.7	84.1	89.9	93.7	97.5	99.0	99.9	100.0	100.0	100.0	100.0	100.0	100.0
375.	9.4	17.9	25.6	32.6	38.9	62.7	77.2	86.1	91.5	94.8	98.1	99.3	99.9	100.0	100.0	100.0	100.0	100.0	100.0
400.	10.0	19.0	27.1	34.4	41.0	65.2	79.5	87.9	92.9	95.8	98.6	99.5	100.0	100.0	100.0	100.0	100.0	100.0	100.0
425.	10.6	20.1	28.6	36.2	43.0	67.5	81.5	89.5	94.0	96.6	98.9	99.6	100.0	100.0	100.0	100.0	100.0	100.0	100.0
450.	11.3	21.2	30.1	38.0	45.0	69.7	83.4	90.9	95.0	97.3	99.2	99.8	100.0	100.0	100.0	100.0	100.0	100.0	100.0
475.	11.9	22.3	31.6	39.7	46.9	71.8	85.0	92.1	95.8	97.8	99.4	99.8	100.0	100.0	100.0	100.0	100.0	100.0	100.0
500.	12.5	23.4	33.0	41.4	48.7	73.7	86.6	93.1	96.5	98.2	99.5	99.9	100.0	100.0	100.0	100.0	100.0	100.0	100.0
550.	13.8	25.6	35.8	44.7	52.3	77.3	89.2	94.8	97.6	98.8	99.7	99.9	100.0	100.0	100.0	100.0	100.0	100.0	100.0
600.	15.0	27.8	38.6	47.8	55.6	80.4	91.3	96.2	98.3	99.3	99.9	100.0	100.0	100.0	100.0	100.0	100.0	100.0	100.0
650.	16.2	29.9	41.3	50.8	58.8	83.1	93.0	97.1	98.8	99.5	99.9	100.0	100.0	100.0	100.0	100.0	100.0	100.0	100.0
700.	17.5	31.9	43.9	53.7	61.8	85.4	94.4	97.9	99.2	99.7	100.0	100.0	100.0	100.0	100.0	100.0	100.0	100.0	100.0

Page 8 PROBABILITY, IN PER CENT, OF FINDING AT LEAST ONE ERROR IF TOTAL NO. OF ERRORS IN UNIVERSE IS AS INDICATED

SAMPLE SIZE	TOTAL ERRORS IN UNIVERSE SIZE OF 5000.																		
	1	2	3	4	5	10	15	20	25	30	40	50	75	100	200	300	500	1000	2000
5.	0.1	0.2	0.3	0.4	0.5	1.0	1.5	2.0	2.5	3.0	3.9	4.9	7.3	9.6	18.5	26.6	41.0	67.2	92.2
10.	0.2	0.4	0.6	0.8	1.0	2.0	3.0	3.9	4.9	5.8	7.7	9.6	14.0	18.3	33.5	46.2	65.2	89.3	99.4
15.	0.3	0.6	0.9	1.2	1.5	3.0	4.4	5.8	7.3	8.6	11.4	14.0	20.3	26.2	45.8	60.5	79.5	96.5	100.0
20.	0.4	0.8	1.2	1.6	2.0	3.9	5.8	7.7	9.6	11.4	14.9	18.2	26.1	33.3	55.9	71.1	87.9	98.9	100.0
25.	0.5	1.0	1.5	2.0	2.5	4.9	7.3	9.6	11.8	14.0	18.2	22.3	31.5	39.7	64.1	78.8	92.9	99.6	100.0
30.	0.6	1.2	1.8	2.4	3.0	5.8	8.6	11.4	14.0	16.6	21.5	26.1	36.5	45.5	70.7	84.5	95.8	99.9	100.0
35.	0.7	1.4	2.1	2.8	3.5	6.8	10.0	13.1	16.1	19.1	24.6	29.7	41.2	50.8	76.2	88.6	97.5	100.0	100.0
40.	0.8	1.6	2.4	3.2	3.9	7.7	11.4	14.9	18.2	21.5	27.6	33.2	45.5	55.6	80.6	91.7	98.5	100.0	100.0
45.	0.9	1.8	2.7	3.6	4.4	8.7	12.7	16.6	20.3	23.8	30.4	36.5	49.5	59.9	84.2	93.9	99.1	100.0	100.0
50.	1.0	2.0	3.0	3.9	4.9	9.6	14.0	18.2	22.3	26.1	33.2	39.6	53.2	63.8	87.1	95.5	99.5	100.0	100.0
55.	1.1	2.2	3.3	4.3	5.4	10.5	15.3	19.9	24.2	28.3	35.9	42.6	56.6	67.3	89.5	96.7	99.7	100.0	100.0
60.	1.2	2.4	3.6	4.7	5.9	11.4	16.6	21.5	26.1	30.5	38.4	45.5	59.8	70.5	91.5	97.6	99.8	100.0	100.0
65.	1.3	2.6	3.9	5.1	6.3	12.3	17.8	23.1	28.0	32.5	40.9	48.2	62.8	73.3	93.1	98.3	99.9	100.0	100.0
70.	1.4	2.8	4.1	5.5	6.8	13.2	19.1	24.6	29.8	34.6	43.2	50.8	65.5	75.9	94.4	98.7	99.9	100.0	100.0
75.	1.5	3.0	4.4	5.9	7.3	14.0	20.3	26.1	31.5	36.5	45.5	53.2	68.1	78.3	95.4	99.1	100.0	100.0	100.0
80.	1.6	3.2	4.7	6.2	7.8	14.9	21.5	27.6	33.2	38.4	47.7	55.5	70.4	80.4	96.3	99.3	100.0	100.0	100.0
85.	1.7	3.4	5.0	6.6	8.2	15.8	22.7	29.1	34.9	40.3	49.8	57.8	72.6	82.3	97.0	99.5	100.0	100.0	100.0
90.	1.8	3.6	5.3	7.0	8.7	16.6	23.9	30.5	36.6	42.1	51.8	59.9	74.7	84.0	97.5	99.6	100.0	100.0	100.0
95.	1.9	3.8	5.6	7.4	9.1	17.5	25.0	31.9	38.2	43.9	53.7	61.9	76.5	85.6	98.0	99.7	100.0	100.0	100.0
100.	2.0	4.0	5.9	7.8	9.6	18.3	26.2	33.3	39.7	45.5	55.6	63.8	78.3	87.0	98.4	99.8	100.0	100.0	100.0
125.	2.5	4.9	7.3	9.6	11.9	22.4	31.6	39.8	47.0	53.3	63.8	72.0	85.2	92.3	99.4	100.0	100.0	100.0	100.0
150.	3.0	5.9	8.7	11.5	14.1	26.3	36.7	45.7	53.4	60.0	70.6	78.4	90.0	95.4	99.8	100.0	100.0	100.0	100.0
175.	3.5	6.9	10.1	13.3	16.3	30.0	41.4	51.0	59.1	65.8	76.1	83.3	93.2	97.3	99.9	100.0	100.0	100.0	100.0
200.	4.0	7.8	11.5	15.1	18.5	33.5	45.8	55.9	64.1	70.7	80.6	87.1	95.4	98.4	100.0	100.0	100.0	100.0	100.0
225.	4.5	8.8	12.9	16.8	20.6	36.9	49.9	60.3	68.5	75.0	84.3	90.1	96.9	99.0	100.0	100.0	100.0	100.0	100.0
250.	5.0	9.8	14.3	18.6	22.6	40.2	53.7	64.2	72.3	78.6	87.3	92.4	97.9	99.4	100.0	100.0	100.0	100.0	100.0
275.	5.5	10.7	15.6	20.3	24.6	43.2	57.2	67.8	75.8	81.8	89.7	94.2	98.6	99.7	100.0	100.0	100.0	100.0	100.0
300.	6.0	11.6	16.9	21.9	26.6	46.2	60.5	71.1	78.8	84.5	91.7	95.5	99.1	99.8	100.0	100.0	100.0	100.0	100.0
325.	6.5	12.6	18.3	23.6	28.6	49.0	63.6	74.0	81.4	86.8	93.3	96.6	99.4	99.9	100.0	100.0	100.0	100.0	100.0
350.	7.0	13.5	19.6	25.2	30.4	51.6	66.4	76.6	83.8	88.7	94.6	97.4	99.6	99.9	100.0	100.0	100.0	100.0	100.0
375.	7.5	14.4	20.9	26.8	32.3	54.2	69.0	79.0	85.8	90.4	95.6	98.0	99.7	100.0	100.0	100.0	100.0	100.0	100.0
400.	8.0	15.4	22.1	28.4	34.1	56.6	71.4	81.2	87.6	91.9	96.5	98.5	99.8	100.0	100.0	100.0	100.0	100.0	100.0
425.	8.5	16.3	23.4	29.9	35.9	58.9	73.7	83.1	89.2	93.1	97.2	98.8	99.9	100.0	100.0	100.0	100.0	100.0	100.0
450.	9.0	17.2	24.6	31.4	37.6	61.1	75.7	84.9	90.6	94.1	97.7	99.1	99.9	100.0	100.0	100.0	100.0	100.0	100.0
475.	9.5	18.1	25.9	32.9	39.3	63.2	77.7	86.5	91.8	95.0	98.2	99.3	99.9	100.0	100.0	100.0	100.0	100.0	100.0
500.	10.0	19.0	27.1	34.4	41.0	65.2	79.5	87.9	92.9	95.8	98.5	99.5	100.0	100.0	100.0	100.0	100.0	100.0	100.0
550.	11.0	20.8	29.5	37.3	44.2	68.9	82.6	90.3	94.6	97.0	99.1	99.7	100.0	100.0	100.0	100.0	100.0	100.0	100.0
600.	12.0	22.6	31.9	40.0	47.2	72.2	85.3	92.3	95.9	97.9	99.4	99.8	100.0	100.0	100.0	100.0	100.0	100.0	100.0
650.	13.0	24.3	34.2	42.7	50.2	75.2	87.7	93.9	97.0	98.5	99.6	99.9	100.0	100.0	100.0	100.0	100.0	100.0	100.0
700.	14.0	26.0	36.4	45.3	53.0	77.9	89.6	95.1	97.7	98.9	99.8	99.9	100.0	100.0	100.0	100.0	100.0	100.0	100.0

Page 9 PROBABILITY, IN PER CENT, OF FINDING AT LEAST ONE ERROR IF TOTAL NO. OF ERRORS IN UNIVERSE IS AS INDICATED

SAMPLE	TOTAL ERRORS IN UNIVERSE SIZE OF 10000.																		
SIZE	1	2	3	4	5	10	15	20	25	30	40	50	75	100	200	300	500	1000	2000
5.	0.1	0.1	0.1	0.2	0.2	0.5	0.7	1.0	1.2	1.5	2.0	2.5	3.7	4.9	9.6	14.1	22.6	41.0	67.2
10.	0.1	0.2	0.3	0.4	0.5	1.0	1.5	2.0	2.5	3.0	3.9	4.9	7.3	9.6	18.3	26.3	40.1	65.1	89.3
15.	0.2	0.3	0.4	0.6	0.7	1.5	2.2	3.0	3.7	4.4	5.8	7.2	10.7	14.0	26.2	36.7	53.7	79.4	96.5
20.	0.2	0.4	0.6	0.8	1.0	2.0	3.0	3.9	4.9	5.8	7.7	9.5	14.0	18.2	33.3	45.7	64.2	87.9	98.9
25.	0.3	0.5	0.7	1.0	1.2	2.5	3.7	4.9	6.1	7.2	9.5	11.8	17.2	22.2	39.7	53.3	72.3	92.8	99.6
30.	0.3	0.6	0.9	1.2	1.5	3.0	4.4	5.8	7.2	8.6	11.3	14.0	20.2	26.1	45.5	60.0	78.6	95.8	99.9
35.	0.4	0.7	1.0	1.4	1.7	3.4	5.1	6.8	8.4	10.0	13.1	16.1	23.2	29.7	50.8	65.6	83.4	97.5	100.0
40.	0.4	0.8	1.2	1.6	2.0	3.9	5.8	7.7	9.5	11.3	14.8	18.2	26.0	33.2	55.5	70.5	87.2	98.5	100.0
45.	0.5	0.9	1.3	1.8	2.2	4.4	6.5	8.6	10.7	12.7	16.5	20.2	28.8	36.4	59.8	74.7	90.1	99.1	100.0
50.	0.5	1.0	1.5	2.0	2.5	4.9	7.2	9.5	11.8	14.0	18.2	22.2	31.4	39.6	63.7	78.3	92.4	99.5	100.0
55.	0.6	1.1	1.6	2.2	2.7	5.4	7.9	10.5	12.9	15.3	19.8	24.2	34.0	42.6	67.2	81.4	94.1	99.7	100.0
60.	0.6	1.2	1.8	2.4	3.0	5.8	8.6	11.4	14.0	16.5	21.4	26.0	36.4	45.4	70.4	84.0	95.4	99.8	100.0
65.	0.7	1.3	1.9	2.6	3.2	6.3	9.3	12.2	15.1	17.8	23.0	27.9	38.8	48.1	73.2	86.3	96.5	99.9	100.0
70.	0.7	1.4	2.1	2.8	3.5	6.8	10.0	13.1	16.1	19.0	24.5	29.7	41.1	50.6	75.8	88.2	97.3	99.9	100.0
75.	0.8	1.5	2.2	3.0	3.7	7.3	10.7	14.0	17.2	20.2	26.0	31.4	43.3	53.1	78.1	89.9	97.9	100.0	100.0
80.	0.8	1.6	2.4	3.2	3.9	7.7	11.4	14.9	18.2	21.4	27.5	33.1	45.4	55.4	80.3	91.3	98.4	100.0	100.0
85.	0.9	1.7	2.5	3.4	4.2	8.2	12.0	15.7	19.2	22.6	29.0	34.8	47.4	57.6	82.2	92.6	98.7	100.0	100.0
90.	0.9	1.8	2.7	3.6	4.4	8.6	12.7	16.6	20.3	23.8	30.4	36.4	49.4	59.7	83.9	93.6	99.0	100.0	100.0
95.	1.0	1.9	2.8	3.7	4.7	9.1	13.3	17.4	21.3	24.9	31.8	38.0	51.3	61.7	85.5	94.5	99.3	100.0	100.0
100.	1.0	2.0	3.0	3.9	4.9	9.6	14.0	18.2	22.2	26.1	33.2	39.6	53.1	63.6	86.9	95.3	99.4	100.0	100.0
125.	1.3	2.5	3.7	4.9	6.1	11.8	17.2	22.3	27.0	31.5	39.6	46.8	61.2	71.8	92.1	97.8	99.8	100.0	100.0
150.	1.5	3.0	4.4	5.9	7.3	14.0	20.3	26.1	31.5	36.5	45.4	53.1	67.9	78.1	95.3	99.0	100.0	100.0	100.0
175.	1.8	3.5	5.2	6.8	8.5	16.2	23.3	29.8	35.7	41.2	50.7	58.7	73.5	83.0	97.2	99.5	100.0	100.0	100.0
200.	2.0	4.0	5.9	7.8	9.6	18.3	26.2	33.3	39.7	45.5	55.5	63.7	78.1	86.9	98.3	99.8	100.0	100.0	100.0
225.	2.3	4.4	6.6	8.7	10.8	20.4	28.9	36.6	43.4	49.5	59.8	68.0	82.0	89.8	99.0	99.9	100.0	100.0	100.0
250.	2.5	4.9	7.3	9.6	11.9	22.4	31.6	39.8	46.9	53.3	63.7	71.9	85.1	92.1	99.4	100.0	100.0	100.0	100.0
275.	2.8	5.4	8.0	10.6	13.0	24.3	34.2	42.8	50.2	56.7	67.3	75.3	87.7	93.9	99.6	100.0	100.0	100.0	100.0
300.	3.0	5.9	8.7	11.5	14.1	26.3	36.7	45.7	53.3	60.0	70.5	78.3	89.9	95.3	99.8	100.0	100.0	100.0	100.0
325.	3.3	6.4	9.4	12.4	15.2	28.1	39.1	48.4	56.3	62.9	73.4	80.9	91.7	96.4	99.9	100.0	100.0	100.0	100.0
350.	3.5	6.9	10.1	13.3	16.3	30.0	41.4	51.0	59.0	65.7	76.0	83.2	93.2	97.2	99.9	100.0	100.0	100.0	100.0
375.	3.8	7.4	10.8	14.2	17.4	31.8	43.7	53.5	61.6	68.3	78.4	85.3	94.4	97.9	100.0	100.0	100.0	100.0	100.0
400.	4.0	7.8	11.5	15.1	18.5	33.5	45.8	55.8	64.0	70.7	80.5	87.1	95.4	98.3	100.0	100.0	100.0	100.0	100.0
425.	4.3	8.3	12.2	15.9	19.5	35.2	47.9	58.1	66.3	72.9	82.5	88.7	96.2	98.7	100.0	100.0	100.0	100.0	100.0
450.	4.5	8.8	12.9	16.8	20.6	36.9	49.9	60.2	68.4	74.9	84.2	90.1	96.9	99.0	100.0	100.0	100.0	100.0	100.0
475.	4.8	9.3	13.6	17.7	21.6	38.5	51.8	62.3	70.4	76.8	85.8	91.3	97.4	99.2	100.0	100.0	100.0	100.0	100.0
500.	5.0	9.8	14.3	18.6	22.6	40.1	53.7	64.2	72.3	78.6	87.2	92.4	97.9	99.4	100.0	100.0	100.0	100.0	100.0
550.	5.5	10.7	15.6	20.3	24.6	43.2	57.2	67.8	75.7	81.7	89.6	94.1	98.6	99.7	100.0	100.0	100.0	100.0	100.0
600.	6.0	11.6	16.9	21.9	26.6	46.2	60.5	71.0	78.7	84.4	91.6	95.5	99.1	99.8	100.0	100.0	100.0	100.0	100.0
650.	6.5	12.6	18.3	23.6	28.5	49.0	63.5	74.0	81.4	86.7	93.2	96.6	99.4	99.9	100.0	100.0	100.0	100.0	100.0
700.	7.0	13.5	19.6	25.2	30.4	51.6	66.4	76.6	83.7	88.7	94.5	97.4	99.6	99.9	100.0	100.0	100.0	100.0	100.0

Page 10

PROBABILITY, IN PER CENT, OF FINDING AT LEAST ONE ERROR IF TOTAL NO. OF ERRORS IN UNIVERSE IS AS INDICATED

SAMPLE SIZE	TOTAL ERRORS IN UNIVERSE SIZE OF 25000.																	
	1	2	3	4	5	10	15	20	25	30	40	50	75	100	200	300	500	1000	2000
5.	0.0	0.0	0.1	0.1	0.1	0.2	0.3	0.4	0.5	0.6	0.8	1.0	1.5	2.0	3.9	5.9	9.6	18.5	34.1
10.	0.0	0.1	0.1	0.2	0.2	0.4	0.6	0.8	1.0	1.2	1.6	2.0	3.0	3.9	7.7	11.4	18.3	33.5	56.6
15.	0.1	0.1	0.2	0.2	0.3	0.6	0.9	1.2	1.5	1.8	2.4	3.0	4.4	5.8	11.4	16.6	26.1	45.8	71.4
20.	0.1	0.2	0.2	0.3	0.4	0.8	1.2	1.6	2.0	2.4	3.2	3.9	5.8	7.7	14.8	21.5	33.2	55.8	81.1
25.	0.1	0.2	0.3	0.4	0.5	1.0	1.5	2.0	2.5	3.0	3.9	4.9	7.2	9.5	18.2	26.1	39.7	64.0	87.6
30.	0.1	0.2	0.4	0.5	0.6	1.2	1.8	2.4	3.0	3.5	4.7	5.8	8.6	11.3	21.4	30.4	45.5	70.6	91.8
35.	0.1	0.3	0.4	0.6	0.7	1.4	2.1	2.8	3.4	4.1	5.5	6.8	10.0	13.1	24.5	34.5	50.7	76.1	94.6
40.	0.2	0.3	0.5	0.6	0.8	1.6	2.4	3.2	3.9	4.7	6.2	7.7	11.3	14.8	27.5	38.3	55.5	80.5	96.4
45.	0.2	0.4	0.5	0.7	0.9	1.8	2.7	3.5	4.4	5.3	7.0	8.6	12.7	16.5	30.4	41.9	59.7	84.1	97.7
50.	0.2	0.4	0.6	0.8	1.0	2.0	3.0	3.9	4.9	5.8	7.7	9.5	14.0	18.2	33.1	45.4	63.6	87.0	98.5
55.	0.2	0.4	0.7	0.9	1.1	2.2	3.3	4.3	5.4	6.4	8.4	10.4	15.2	19.8	35.7	48.6	67.1	89.4	99.0
60.	0.2	0.5	0.7	1.0	1.2	2.4	3.5	4.7	5.8	7.0	9.2	11.3	16.5	21.4	38.3	51.6	70.3	91.4	99.3
65.	0.3	0.5	0.8	1.0	1.3	2.6	3.8	5.1	6.3	7.5	9.9	12.2	17.8	23.0	40.7	54.4	73.1	93.0	99.6
70.	0.3	0.6	0.8	1.1	1.4	2.8	4.1	5.5	6.8	8.1	10.6	13.1	19.0	24.5	43.1	57.1	75.7	94.3	99.7
75.	0.3	0.6	0.9	1.2	1.5	3.0	4.4	5.8	7.2	8.6	11.3	14.0	20.2	26.0	45.3	59.6	78.1	95.3	99.8
80.	0.3	0.6	1.0	1.3	1.6	3.2	4.7	6.2	7.7	9.2	12.0	14.8	21.4	27.5	47.5	62.0	80.2	96.2	99.9
85.	0.3	0.7	1.0	1.4	1.7	3.3	5.0	6.6	8.2	9.7	12.7	15.7	22.6	28.9	49.5	64.2	82.1	96.9	99.9
90.	0.4	0.7	1.1	1.4	1.8	3.5	5.3	7.0	8.6	10.3	13.4	16.5	23.7	30.3	51.5	66.3	83.8	97.5	99.9
95.	0.4	0.8	1.1	1.5	1.9	3.7	5.6	7.3	9.1	10.8	14.1	17.3	24.9	31.7	53.4	68.3	85.4	97.9	100.0
100.	0.4	0.8	1.2	1.6	2.0	3.9	5.8	7.7	9.5	11.3	14.8	18.2	26.0	33.1	55.3	70.2	86.8	98.3	100.0
125.	0.5	1.0	1.5	2.0	2.5	4.9	7.2	9.5	11.8	14.0	18.2	22.2	31.4	39.5	63.5	78.0	92.0	99.4	100.0
150.	0.6	1.2	1.8	2.4	3.0	5.8	8.6	11.3	14.0	16.5	21.4	26.0	36.4	45.3	70.1	83.7	95.2	99.8	100.0
175.	0.7	1.4	2.1	2.8	3.5	6.8	10.0	13.1	16.1	19.0	24.5	29.6	41.0	50.5	75.6	88.0	97.1	99.9	100.0
200.	0.8	1.6	2.4	3.2	3.9	7.7	11.4	14.8	18.2	21.4	27.5	33.1	45.3	55.3	80.1	91.1	98.3	100.0	100.0
225.	0.9	1.8	2.7	3.6	4.4	8.6	12.7	16.5	20.2	23.8	30.4	36.4	49.3	59.6	83.7	93.5	99.0	100.0	100.0
250.	1.0	2.0	3.0	3.9	4.9	9.6	14.0	18.2	22.2	26.0	33.1	39.5	53.0	63.5	86.7	95.2	99.4	100.0	100.0
275.	1.1	2.2	3.3	4.3	5.4	10.5	15.3	19.9	24.2	28.3	35.8	42.5	56.4	67.0	89.2	96.5	99.6	100.0	100.0
300.	1.2	2.4	3.6	4.7	5.9	11.4	16.6	21.5	26.1	30.4	38.3	45.4	59.6	70.2	91.1	97.4	99.8	100.0	100.0
325.	1.3	2.6	3.8	5.1	6.3	12.3	17.8	23.0	27.9	32.5	40.8	48.1	62.6	73.0	92.8	98.1	99.9	100.0	100.0
350.	1.4	2.8	4.1	5.5	6.8	13.2	19.1	24.6	29.7	34.5	43.1	50.6	65.3	75.7	94.1	98.6	99.9	100.0	100.0
375.	1.5	3.0	4.4	5.9	7.3	14.0	20.3	26.1	31.5	36.5	45.4	53.1	67.9	78.0	95.2	99.0	100.0	100.0	100.0
400.	1.6	3.2	4.7	6.2	7.7	14.9	21.5	27.6	33.2	38.4	47.6	55.4	70.2	80.1	96.1	99.2	100.0	100.0	100.0
425.	1.7	3.4	5.0	6.6	8.2	15.8	22.7	29.0	34.9	40.2	49.7	57.6	72.4	82.1	96.8	99.4	100.0	100.0	100.0
450.	1.8	3.6	5.3	7.0	8.7	16.6	23.9	30.5	36.5	42.0	51.7	59.7	74.4	83.8	97.4	99.6	100.0	100.0	100.0
475.	1.9	3.8	5.6	7.4	9.1	17.5	25.0	31.9	38.1	43.8	53.6	61.7	76.3	85.4	97.9	99.7	100.0	100.0	100.0
500.	2.0	4.0	5.9	7.8	9.6	18.3	26.1	33.2	39.7	45.5	55.5	63.6	78.1	86.8	98.3	99.8	100.0	100.0	100.0
550.	2.2	4.4	6.5	8.5	10.5	19.9	28.4	35.9	42.7	48.7	59.0	67.2	81.2	89.2	98.9	99.9	100.0	100.0	100.0
600.	2.4	4.7	7.0	9.3	11.4	21.6	30.5	38.5	45.5	51.8	62.2	70.4	83.9	91.2	99.2	99.9	100.0	100.0	100.0
650.	2.6	5.1	7.6	10.0	12.3	23.2	32.7	41.0	48.3	54.7	65.2	73.2	86.2	92.9	99.5	100.0	100.0	100.0	100.0
700.	2.8	5.5	8.2	10.7	13.2	24.7	34.7	43.3	50.9	57.4	67.9	75.9	88.2	94.2	99.7	100.0	100.0	100.0	100.0

Page 11 PROBABILITY, IN PER CENT, OF FINDING AT LEAST ONE ERROR IF TOTAL NO. OF ERRORS IN UNIVERSE IS AS INDICATED

SAMPLE SIZE	TOTAL ERRORS IN UNIVERSE SIZE OF 50000.																		
	1	2	3	4	5	10	15	20	25	30	40	50	75	100	200	300	500	1000	2000
5.	0.0	0.0	0.0	0.0	0.0	0.1	0.1	0.2	0.2	0.3	0.4	0.5	0.7	1.0	2.0	3.0	4.9	9.6	18.5
10.	0.0	0.0	0.1	0.1	0.1	0.2	0.3	0.4	0.5	0.6	0.8	1.0	1.5	2.0	3.9	5.8	9.6	18.3	33.5
15.	0.0	0.1	0.1	0.1	0.1	0.3	0.4	0.6	0.7	0.9	1.2	1.5	2.2	3.0	5.8	8.6	14.0	26.1	45.8
20.	0.0	0.1	0.1	0.2	0.2	0.4	0.6	0.8	1.0	1.2	1.6	2.0	3.0	3.9	7.7	11.3	18.2	33.2	55.8
25.	0.1	0.1	0.1	0.2	0.2	0.5	0.7	1.0	1.2	1.5	2.0	2.5	3.7	4.9	9.5	14.0	22.2	39.7	64.0
30.	0.1	0.1	0.2	0.2	0.3	0.6	0.9	1.2	1.5	1.8	2.4	3.0	4.4	5.8	11.3	16.5	26.0	45.5	70.6
35.	0.1	0.1	0.2	0.3	0.3	0.7	1.0	1.4	1.7	2.1	2.8	3.4	5.1	6.8	13.1	19.0	29.7	50.7	76.1
40.	0.1	0.2	0.2	0.3	0.4	0.8	1.2	1.6	2.0	2.4	3.2	3.9	5.8	7.7	14.8	21.4	33.1	55.4	80.5
45.	0.1	0.2	0.3	0.4	0.4	0.9	1.3	1.8	2.2	2.7	3.5	4.4	6.5	8.6	16.5	23.7	36.4	59.7	84.1
50.	0.1	0.2	0.3	0.4	0.5	1.0	1.5	2.0	2.5	3.0	3.9	4.9	7.2	9.5	18.2	26.0	39.5	63.6	87.0
55.	0.1	0.2	0.3	0.4	0.5	1.1	1.6	2.2	2.7	3.2	4.3	5.4	7.9	10.4	19.8	28.2	42.5	67.1	89.4
60.	0.1	0.2	0.4	0.5	0.6	1.2	1.8	2.4	3.0	3.5	4.7	5.8	8.6	11.3	21.4	30.3	45.3	70.3	91.4
65.	0.1	0.3	0.4	0.5	0.6	1.3	1.9	2.6	3.2	3.8	5.1	6.3	9.3	12.2	22.9	32.4	48.0	73.1	93.0
70.	0.1	0.3	0.4	0.6	0.7	1.4	2.1	2.8	3.4	4.1	5.5	6.8	10.0	13.1	24.5	34.4	50.5	75.7	94.3
75.	0.2	0.3	0.4	0.6	0.7	1.5	2.2	3.0	3.7	4.4	5.8	7.2	10.7	14.0	26.0	36.3	53.0	78.0	95.3
80.	0.2	0.3	0.5	0.6	0.8	1.6	2.4	3.2	3.9	4.7	6.2	7.7	11.3	14.8	27.5	38.2	55.3	80.2	96.2
85.	0.2	0.3	0.5	0.7	0.8	1.7	2.5	3.3	4.2	5.0	6.6	8.2	12.0	15.7	28.9	40.1	57.5	82.1	96.9
90.	0.2	0.4	0.5	0.7	0.9	1.8	2.7	3.5	4.4	5.3	7.0	8.6	12.6	16.5	30.3	41.8	59.6	83.8	97.5
95.	0.2	0.4	0.6	0.8	0.9	1.9	2.8	3.7	4.6	5.5	7.3	9.1	13.3	17.3	31.7	43.6	61.5	85.4	97.9
100.	0.2	0.4	0.6	0.8	1.0	2.0	3.0	3.9	4.9	5.8	7.7	9.5	14.0	18.2	33.0	45.3	63.4	86.8	98.3
125.	0.3	0.5	0.7	1.0	1.2	2.5	3.7	4.9	6.1	7.2	9.5	11.8	17.1	22.2	39.4	52.9	71.6	92.0	99.4
150.	0.3	0.6	0.9	1.2	1.5	3.0	4.4	5.8	7.2	8.6	11.3	14.0	20.2	26.0	45.2	59.5	77.9	95.2	99.8
175.	0.4	0.7	1.0	1.4	1.7	3.4	5.1	6.8	8.4	10.0	13.1	16.1	23.1	29.6	50.5	65.2	82.8	97.1	99.9
200.	0.4	0.8	1.2	1.6	2.0	3.9	5.8	7.7	9.5	11.3	14.8	18.2	26.0	33.0	55.2	70.1	86.7	98.3	100.0
225.	0.5	0.9	1.3	1.8	2.2	4.4	6.5	8.6	10.7	12.7	16.5	20.2	28.7	36.3	59.5	74.3	89.6	98.9	100.0
250.	0.5	1.0	1.5	2.0	2.5	4.9	7.2	9.5	11.8	14.0	18.2	22.2	31.4	39.5	63.4	77.9	91.9	99.4	100.0
275.	0.6	1.1	1.6	2.2	2.7	5.4	7.9	10.4	12.9	15.3	19.8	24.1	33.9	42.4	66.9	81.0	93.7	99.6	100.0
300.	0.6	1.2	1.8	2.4	3.0	5.8	8.6	11.3	14.0	16.5	21.4	26.0	36.3	45.3	70.1	83.6	95.1	99.8	100.0
325.	0.7	1.3	1.9	2.6	3.2	6.3	9.3	12.2	15.0	17.8	23.0	27.8	38.7	47.9	72.9	85.9	96.2	99.9	100.0
350.	0.7	1.4	2.1	2.8	3.5	6.8	10.0	13.1	16.1	19.0	24.5	29.6	41.0	50.5	75.5	87.9	97.1	99.9	100.0
375.	0.8	1.5	2.2	3.0	3.7	7.3	10.7	14.0	17.2	20.2	26.0	31.4	43.2	52.9	77.9	89.6	97.7	100.0	100.0
400.	0.8	1.6	2.4	3.2	3.9	7.7	11.4	14.8	18.2	21.4	27.5	33.1	45.3	55.2	80.0	91.1	98.2	100.0	100.0
425.	0.9	1.7	2.5	3.4	4.2	8.2	12.0	15.7	19.2	22.6	28.9	34.8	47.3	57.4	81.9	92.3	98.6	100.0	100.0
450.	0.9	1.8	2.7	3.6	4.4	8.6	12.7	16.5	20.2	23.8	30.4	36.4	49.3	59.5	83.7	93.4	98.9	100.0	100.0
475.	1.0	1.9	2.8	3.7	4.7	9.1	13.3	17.4	21.2	24.9	31.7	38.0	51.2	61.5	85.2	94.3	99.2	100.0	100.0
500.	1.0	2.0	3.0	3.9	4.9	9.6	14.0	18.2	22.2	26.0	33.1	39.5	53.0	63.4	86.7	95.1	99.4	100.0	100.0
550.	1.1	2.2	3.3	4.3	5.4	10.5	15.3	19.8	24.2	28.2	35.8	42.5	56.4	67.0	89.1	96.4	99.6	100.0	100.0
600.	1.2	2.4	3.6	4.7	5.9	11.4	16.6	21.5	26.1	30.4	38.3	45.3	59.6	70.1	91.1	97.4	99.8	100.0	100.0
650.	1.3	2.6	3.8	5.1	6.3	12.3	17.8	23.0	27.9	32.5	40.8	48.0	62.5	73.0	92.7	98.1	99.9	100.0	100.0
700.	1.4	2.8	4.1	5.5	6.8	13.2	19.1	24.6	29.7	34.5	43.1	50.6	65.3	75.6	94.1	98.6	99.9	100.0	100.0

Page 12 PROBABILITY, IN PER CENT, OF FINDING AT LEAST ONE ERROR IF TOTAL NO. OF ERRORS IN UNIVERSE IS AS INDICATED

SAMPLE SIZE	TOTAL ERRORS IN UNIVERSE SIZE OF 100000. 1	2	3	4	5	10	15	20	25	30	40	50	75	100	200	300	500	1000	2000
5.	0.0	0.0	0.0	0.0	0.0	0.1	0.1	0.1	0.1	0.1	0.2	0.2	0.4	0.5	1.0	1.5	2.5	4.9	9.6
10.	0.0	0.0	0.0	0.0	0.0	0.1	0.1	0.2	0.2	0.3	0.4	0.5	0.7	1.0	2.0	3.0	4.9	9.6	18.3
15.	0.0	0.0	0.0	0.1	0.1	0.1	0.2	0.3	0.4	0.4	0.6	0.7	1.1	1.5	3.0	4.4	7.2	14.0	26.1
20.	0.0	0.0	0.1	0.1	0.1	0.2	0.3	0.4	0.5	0.6	0.8	1.0	1.5	2.0	3.9	5.8	9.5	18.2	33.2
25.	0.0	0.0	0.1	0.1	0.1	0.2	0.4	0.5	0.6	0.7	1.0	1.2	1.9	2.5	4.9	7.2	11.8	22.2	39.7
30.	0.0	0.1	0.1	0.1	0.1	0.3	0.4	0.6	0.7	0.9	1.2	1.5	2.2	3.0	5.8	8.6	14.0	26.0	45.5
35.	0.0	0.1	0.1	0.1	0.2	0.3	0.5	0.7	0.9	1.0	1.4	1.7	2.6	3.4	6.8	10.0	16.1	29.7	50.7
40.	0.0	0.1	0.1	0.2	0.2	0.4	0.6	0.8	1.0	1.2	1.6	2.0	3.0	3.9	7.7	11.3	18.2	33.1	55.4
45.	0.0	0.1	0.1	0.2	0.2	0.4	0.7	0.9	1.1	1.3	1.8	2.2	3.3	4.4	8.6	12.6	20.2	36.4	59.7
50.	0.1	0.1	0.1	0.2	0.2	0.5	0.7	1.0	1.2	1.5	2.0	2.5	3.7	4.9	9.5	14.0	22.2	39.5	63.6
55.	0.1	0.1	0.2	0.2	0.3	0.5	0.8	1.1	1.4	1.6	2.2	2.7	4.0	5.4	10.4	15.2	24.1	42.5	67.1
60.	0.1	0.1	0.2	0.2	0.3	0.6	0.9	1.2	1.5	1.8	2.4	3.0	4.4	5.8	11.3	16.5	26.0	45.3	70.3
65.	0.1	0.1	0.2	0.3	0.3	0.6	1.0	1.3	1.6	1.9	2.6	3.2	4.8	6.3	12.2	17.7	27.8	48.0	73.1
70.	0.1	0.1	0.2	0.3	0.3	0.7	1.0	1.4	1.7	2.1	2.8	3.4	5.1	6.8	13.1	19.0	29.6	50.5	75.7
75.	0.1	0.1	0.2	0.3	0.4	0.7	1.1	1.5	1.9	2.2	3.0	3.7	5.5	7.2	13.9	20.2	31.3	53.0	78.0
80.	0.1	0.2	0.2	0.3	0.4	0.8	1.2	1.6	2.0	2.4	3.2	3.9	5.8	7.7	14.8	21.4	33.0	55.3	80.1
85.	0.1	0.2	0.3	0.3	0.4	0.8	1.3	1.7	2.1	2.5	3.3	4.2	6.2	8.2	15.7	22.5	34.7	57.5	82.1
90.	0.1	0.2	0.3	0.4	0.4	0.9	1.3	1.8	2.2	2.7	3.5	4.4	6.5	8.6	16.5	23.7	36.3	59.5	83.8
95.	0.1	0.2	0.3	0.4	0.5	0.9	1.4	1.9	2.3	2.8	3.7	4.6	6.9	9.1	17.3	24.8	37.9	61.5	85.3
100.	0.1	0.2	0.3	0.4	0.5	1.0	1.5	2.0	2.5	3.0	3.9	4.9	7.2	9.5	18.2	26.0	39.4	63.4	86.8
125.	0.1	0.2	0.4	0.5	0.6	1.2	1.9	2.5	3.1	3.7	4.9	6.1	9.0	11.8	22.2	31.3	46.6	71.6	92.0
150.	0.2	0.3	0.4	0.6	0.7	1.5	2.2	3.0	3.7	4.4	5.8	7.2	10.7	13.9	26.0	36.3	52.9	77.9	95.2
175.	0.2	0.3	0.5	0.7	0.9	1.7	2.6	3.4	4.3	5.1	6.8	8.4	12.3	16.1	29.6	40.9	58.4	82.8	97.1
200.	0.2	0.4	0.6	0.8	1.0	2.0	3.0	3.9	4.9	5.8	7.7	9.5	13.9	18.2	33.0	45.2	63.3	86.6	98.2
225.	0.2	0.4	0.7	0.9	1.1	2.2	3.3	4.4	5.5	6.5	8.6	10.7	15.5	20.2	36.3	49.2	67.7	89.6	98.9
250.	0.3	0.5	0.7	1.0	1.2	2.5	3.7	4.9	6.1	7.2	9.5	11.8	17.1	22.2	39.4	52.9	71.5	91.9	99.4
275.	0.3	0.5	0.8	1.1	1.4	2.7	4.0	5.4	6.7	7.9	10.4	12.9	18.7	24.1	42.4	56.3	74.9	93.7	99.6
300.	0.3	0.6	0.9	1.2	1.5	3.0	4.4	5.8	7.2	8.6	11.3	14.0	20.2	26.0	45.2	59.5	77.8	95.1	99.8
325.	0.3	0.6	1.0	1.3	1.6	3.2	4.8	6.3	7.8	9.3	12.2	15.0	21.7	27.8	47.9	62.4	80.4	96.2	99.9
350.	0.4	0.7	1.0	1.4	1.7	3.4	5.1	6.8	8.4	10.0	13.1	16.1	23.1	29.6	50.4	65.1	82.8	97.1	99.9
375.	0.4	0.7	1.1	1.5	1.9	3.7	5.5	7.2	9.0	10.7	14.0	17.1	24.6	31.3	52.9	67.7	84.8	97.7	99.9
400.	0.4	0.8	1.2	1.6	2.0	3.9	5.8	7.7	9.5	11.3	14.8	18.2	26.0	33.0	55.2	70.0	86.6	98.2	100.0
425.	0.4	0.8	1.3	1.7	2.1	4.2	6.2	8.2	10.1	12.0	15.7	19.2	27.4	34.7	57.4	72.2	88.2	98.6	100.0
450.	0.5	0.9	1.3	1.8	2.2	4.4	6.5	8.6	10.7	12.7	16.5	20.2	28.7	36.3	59.5	74.2	89.6	98.9	100.0
475.	0.5	0.9	1.4	1.9	2.4	4.6	6.9	9.1	11.2	13.3	17.3	21.2	30.0	37.9	61.5	76.1	90.8	99.2	100.0
500.	0.5	1.0	1.5	2.0	2.5	4.9	7.2	9.5	11.8	14.0	18.2	22.2	31.3	39.4	63.3	77.8	91.9	99.4	100.0
550.	0.6	1.1	1.6	2.2	2.7	5.4	7.9	10.4	12.9	15.3	19.8	24.1	33.9	42.4	66.9	80.9	93.7	99.6	100.0
600.	0.6	1.2	1.8	2.4	3.0	5.8	8.6	11.3	14.0	16.5	21.4	26.0	36.3	45.2	70.0	83.6	95.1	99.8	100.0
650.	0.7	1.3	1.9	2.6	3.2	6.3	9.3	12.2	15.0	17.8	23.0	27.8	38.7	47.9	72.9	85.9	96.2	99.9	100.0
700.	0.7	1.4	2.1	2.8	3.5	6.8	10.0	13.1	16.1	19.0	24.5	29.6	41.0	50.5	75.5	87.9	97.0	99.9	100.0

Page 13 PROBABILITY, IN PER CENT, OF FINDING AT LEAST ONE ERROR IF TOTAL NO. OF ERRORS IN UNIVERSE IS AS INDICATED

SAMPLE SIZE	TOTAL ERRORS IN UNIVERSE SIZE OF 150000.																		
	1	2	3	4	5	10	15	20	25	30	40	50	75	100	200	300	500	1000	2000
5.	0.0	0.0	0.0	0.0	0.0	0.0	0.1	0.1	0.1	0.1	0.1	0.2	0.2	0.3	0.7	1.0	1.7	3.3	6.5
10.	0.0	0.0	0.0	0.0	0.0	0.1	0.1	0.1	0.2	0.2	0.3	0.3	0.5	0.7	1.3	2.0	3.3	6.5	12.6
15.	0.0	0.0	0.0	0.0	0.0	0.1	0.1	0.2	0.2	0.3	0.4	0.5	0.7	1.0	2.0	3.0	4.9	9.5	18.2
20.	0.0	0.0	0.0	0.1	0.1	0.1	0.2	0.3	0.3	0.4	0.5	0.7	1.0	1.3	2.6	3.9	6.5	12.5	23.5
25.	0.0	0.0	0.0	0.1	0.1	0.2	0.2	0.3	0.4	0.5	0.7	0.8	1.2	1.7	3.3	4.9	8.0	15.4	28.5
30.	0.0	0.0	0.1	0.1	0.1	0.2	0.3	0.4	0.5	0.6	0.8	1.0	1.5	2.0	3.9	5.8	9.5	18.2	33.2
35.	0.0	0.0	0.1	0.1	0.1	0.2	0.3	0.5	0.6	0.7	0.9	1.2	1.7	2.3	4.6	6.8	11.0	20.9	37.5
40.	0.0	0.1	0.1	0.1	0.1	0.3	0.4	0.5	0.7	0.8	1.1	1.3	2.0	2.6	5.2	7.7	12.5	23.5	41.6
45.	0.0	0.1	0.1	0.1	0.1	0.3	0.4	0.6	0.7	0.9	1.2	1.5	2.2	3.0	5.8	8.6	14.0	26.0	45.3
50.	0.0	0.1	0.1	0.1	0.2	0.3	0.5	0.7	0.8	1.0	1.3	1.7	2.5	3.3	6.5	9.5	15.4	28.4	48.9
55.	0.0	0.1	0.1	0.1	0.2	0.4	0.5	0.7	0.9	1.1	1.5	1.8	2.7	3.6	7.1	10.4	16.8	30.8	52.2
60.	0.0	0.1	0.1	0.2	0.2	0.4	0.6	0.8	1.0	1.2	1.6	2.0	3.0	3.9	7.7	11.3	18.2	33.1	55.3
65.	0.0	0.1	0.1	0.2	0.2	0.4	0.6	0.9	1.1	1.3	1.7	2.1	3.2	4.2	8.3	12.2	19.5	35.3	58.2
70.	0.0	0.1	0.1	0.2	0.2	0.5	0.7	0.9	1.2	1.4	1.9	2.3	3.4	4.6	8.9	13.1	20.8	37.4	60.9
75.	0.1	0.1	0.1	0.2	0.2	0.5	0.7	1.0	1.2	1.5	2.0	2.5	3.7	4.9	9.5	13.9	22.2	39.5	63.5
80.	0.1	0.1	0.2	0.2	0.3	0.5	0.8	1.1	1.3	1.6	2.1	2.6	3.9	5.2	10.1	14.8	23.4	41.4	65.8
85.	0.1	0.1	0.2	0.2	0.3	0.6	0.8	1.1	1.4	1.7	2.2	2.8	4.2	5.5	10.7	15.7	24.7	43.4	68.1
90.	0.1	0.1	0.2	0.2	0.3	0.6	0.9	1.2	1.5	1.8	2.4	3.0	4.4	5.8	11.3	16.5	26.0	45.2	70.1
95.	0.1	0.1	0.2	0.3	0.3	0.6	0.9	1.3	1.6	1.9	2.5	3.1	4.6	6.1	11.9	17.3	27.2	47.0	72.1
100.	0.1	0.1	0.2	0.3	0.3	0.7	1.0	1.3	1.7	2.0	2.6	3.3	4.9	6.5	12.5	18.1	28.4	48.8	73.9
125.	0.1	0.2	0.2	0.3	0.4	0.8	1.2	1.7	2.1	2.5	3.3	4.1	6.1	8.0	15.4	22.1	34.1	56.7	81.3
150.	0.1	0.2	0.3	0.4	0.5	1.0	1.5	2.0	2.5	3.0	3.9	4.9	7.2	9.5	18.1	26.0	39.4	63.4	86.7
175.	0.1	0.2	0.3	0.5	0.6	1.2	1.7	2.3	2.9	3.4	4.6	5.7	8.4	11.0	20.8	29.6	44.3	69.0	90.5
200.	0.1	0.3	0.4	0.5	0.7	1.3	2.0	2.6	3.3	3.9	5.2	6.5	9.5	12.5	23.4	33.0	48.7	73.8	93.2
225.	0.2	0.3	0.4	0.6	0.7	1.5	2.2	3.0	3.7	4.4	5.8	7.2	10.7	13.9	25.9	36.3	52.8	77.8	95.1
250.	0.2	0.3	0.5	0.7	0.8	1.7	2.5	3.3	4.1	4.9	6.5	8.0	11.8	15.4	28.4	39.4	56.6	81.2	96.5
275.	0.2	0.4	0.5	0.7	0.9	1.8	2.7	3.6	4.5	5.4	7.1	8.8	12.9	16.8	30.7	42.4	60.1	84.1	97.5
300.	0.2	0.4	0.6	0.8	1.0	2.0	3.0	3.9	4.9	5.8	7.7	9.5	13.9	18.1	33.0	45.2	63.3	86.6	98.2
325.	0.2	0.4	0.6	0.9	1.1	2.1	3.2	4.2	5.3	6.3	8.3	10.3	15.0	19.5	35.2	47.9	66.3	88.7	98.7
350.	0.2	0.5	0.7	0.9	1.2	2.3	3.4	4.6	5.7	6.8	8.9	11.0	16.1	20.8	37.3	50.4	69.0	90.4	99.1
375.	0.3	0.5	0.7	1.0	1.2	2.5	3.7	4.9	6.1	7.2	9.5	11.8	17.1	22.2	39.4	52.8	71.5	91.9	99.4
400.	0.3	0.5	0.8	1.1	1.3	2.6	3.9	5.2	6.5	7.7	10.1	12.5	18.2	23.4	41.4	55.2	73.7	93.1	99.5
425.	0.3	0.6	0.8	1.1	1.4	2.8	4.2	5.5	6.8	8.2	10.7	13.2	19.2	24.7	43.3	57.3	75.9	94.2	99.7
450.	0.3	0.6	0.9	1.2	1.5	3.0	4.4	5.8	7.2	8.6	11.3	14.0	20.2	26.0	45.2	59.4	77.8	95.1	99.8
475.	0.3	0.6	0.9	1.3	1.6	3.1	4.6	6.1	7.6	9.1	11.9	14.7	21.2	27.2	47.0	61.4	79.6	95.9	99.8
500.	0.3	0.7	1.0	1.3	1.7	3.3	4.9	6.5	8.0	9.5	12.5	15.4	22.2	28.4	48.7	63.3	81.2	96.5	99.9
550.	0.4	0.7	1.1	1.5	1.8	3.6	5.4	7.1	8.8	10.4	13.7	16.8	24.1	30.8	52.1	66.8	84.1	97.5	99.9
600.	0.4	0.8	1.2	1.6	2.0	3.9	5.8	7.7	9.5	11.3	14.8	18.2	26.0	33.0	55.2	70.0	86.6	98.2	100.0
650.	0.4	0.9	1.3	1.7	2.1	4.2	6.3	8.3	10.3	12.2	15.9	19.5	27.8	35.2	58.1	72.9	88.6	98.7	100.0
700.	0.5	0.9	1.4	1.9	2.3	4.6	6.8	8.9	11.0	13.1	17.1	20.9	29.6	37.4	60.8	75.5	90.4	99.1	100.0

Page 14

PROBABILITY, IN PER CENT, OF FINDING AT LEAST ONE ERROR IF TOTAL NO. OF ERRORS IN UNIVERSE IS AS INDICATED

SAMPLE	TOTAL ERRORS IN UNIVERSE SIZE OF 200000.																		
SIZE	1	2	3	4	5	10	15	20	25	30	40	50	75	100	200	300	500	1000	2000
5.	0.0	0.0	0.0	0.0	0.0	0.0	0.0	0.1	0.1	0.1	0.1	0.1	0.2	0.2	0.5	0.7	1.2	2.5	4.9
10.	0.0	0.0	0.0	0.0	0.0	0.0	0.1	0.1	0.1	0.1	0.2	0.2	0.4	0.5	1.0	1.5	2.5	4.9	9.6
15.	0.0	0.0	0.0	0.0	0.0	0.1	0.1	0.1	0.2	0.2	0.3	0.4	0.6	0.7	1.5	2.2	3.7	7.2	14.0
20.	0.0	0.0	0.0	0.0	0.0	0.1	0.1	0.2	0.2	0.3	0.4	0.5	0.7	1.0	2.0	3.0	4.9	9.5	18.2
25.	0.0	0.0	0.0	0.0	0.1	0.1	0.2	0.2	0.3	0.4	0.5	0.6	0.9	1.2	2.5	3.7	6.1	11.8	22.2
30.	0.0	0.0	0.0	0.1	0.1	0.1	0.2	0.3	0.4	0.4	0.6	0.7	1.1	1.5	3.0	4.4	7.2	14.0	26.0
35.	0.0	0.0	0.1	0.1	0.1	0.2	0.3	0.3	0.4	0.5	0.7	0.9	1.3	1.7	3.4	5.1	8.4	16.1	29.7
40.	0.0	0.0	0.1	0.1	0.1	0.2	0.3	0.4	0.5	0.6	0.8	1.0	1.5	2.0	3.9	5.8	9.5	18.2	33.1
45.	0.0	0.0	0.1	0.1	0.1	0.2	0.3	0.4	0.6	0.7	0.9	1.1	1.7	2.2	4.4	6.5	10.7	20.2	36.4
50.	0.0	0.0	0.1	0.1	0.1	0.2	0.4	0.5	0.6	0.7	1.0	1.2	1.9	2.5	4.9	7.2	11.8	22.2	39.5
55.	0.0	0.1	0.1	0.1	0.1	0.3	0.4	0.5	0.7	0.8	1.1	1.4	2.0	2.7	5.4	7.9	12.9	24.1	42.5
60.	0.0	0.1	0.1	0.1	0.1	0.3	0.4	0.6	0.7	0.9	1.2	1.5	2.2	3.0	5.8	8.6	13.9	26.0	45.3
65.	0.0	0.1	0.1	0.1	0.2	0.3	0.5	0.6	0.8	1.0	1.3	1.6	2.4	3.2	6.3	9.3	15.0	27.8	48.0
70.	0.0	0.1	0.1	0.1	0.2	0.3	0.5	0.7	0.9	1.0	1.4	1.7	2.6	3.4	6.8	10.0	16.1	29.6	50.5
75.	0.0	0.1	0.1	0.1	0.2	0.4	0.6	0.7	0.9	1.1	1.5	1.9	2.8	3.7	7.2	10.6	17.1	31.3	52.9
80.	0.0	0.1	0.1	0.2	0.2	0.4	0.6	0.8	1.0	1.2	1.6	2.0	3.0	3.9	7.7	11.3	18.2	33.0	55.3
85.	0.0	0.1	0.1	0.2	0.2	0.4	0.6	0.8	1.1	1.3	1.7	2.1	3.1	4.2	8.2	12.0	19.2	34.7	57.4
90.	0.0	0.1	0.1	0.2	0.2	0.4	0.7	0.9	1.1	1.3	1.8	2.2	3.3	4.4	8.6	12.6	20.2	36.3	59.5
95.	0.0	0.1	0.1	0.2	0.2	0.5	0.7	0.9	1.2	1.4	1.9	2.3	3.5	4.6	9.1	13.3	21.2	37.9	61.5
100.	0.1	0.1	0.1	0.2	0.2	0.5	0.7	1.0	1.2	1.5	2.0	2.5	3.7	4.9	9.5	13.9	22.1	39.4	63.4
125.	0.1	0.1	0.2	0.2	0.3	0.6	0.9	1.2	1.6	1.9	2.5	3.1	4.6	6.1	11.8	17.1	26.9	46.6	71.5
150.	0.1	0.1	0.2	0.3	0.4	0.7	1.1	1.5	1.9	2.2	3.0	3.7	5.5	7.2	13.9	20.2	31.3	52.9	77.9
175.	0.1	0.2	0.3	0.3	0.4	0.9	1.3	1.7	2.2	2.6	3.4	4.3	6.4	8.4	16.1	23.1	35.5	58.4	82.8
200.	0.1	0.2	0.3	0.4	0.5	1.0	1.5	2.0	2.5	3.0	3.9	4.9	7.2	9.5	18.1	25.9	39.4	63.3	86.6
225.	0.1	0.2	0.3	0.4	0.6	1.1	1.7	2.2	2.8	3.3	4.4	5.5	8.1	10.6	20.2	28.7	43.1	67.6	89.6
250.	0.1	0.2	0.4	0.5	0.6	1.2	1.9	2.5	3.1	3.7	4.9	6.1	9.0	11.8	22.1	31.3	46.5	71.5	91.9
275.	0.1	0.3	0.4	0.5	0.7	1.4	2.0	2.7	3.4	4.0	5.4	6.6	9.8	12.9	24.1	33.8	49.8	74.8	93.7
300.	0.2	0.3	0.4	0.6	0.7	1.5	2.2	3.0	3.7	4.4	5.8	7.2	10.6	13.9	25.9	36.3	52.8	77.8	95.1
325.	0.2	0.3	0.5	0.6	0.8	1.6	2.4	3.2	4.0	4.8	6.3	7.8	11.5	15.0	27.8	38.6	55.7	80.4	96.2
350.	0.2	0.3	0.5	0.7	0.9	1.7	2.6	3.4	4.3	5.1	6.8	8.4	12.3	16.1	29.6	40.9	58.4	82.7	97.0
375.	0.2	0.4	0.6	0.7	0.9	1.9	2.8	3.7	4.6	5.5	7.2	9.0	13.1	17.1	31.3	43.1	60.9	84.8	97.7
400.	0.2	0.4	0.6	0.8	1.0	2.0	3.0	3.9	4.9	5.8	7.7	9.5	13.9	18.1	33.0	45.2	63.3	86.6	98.2
425.	0.2	0.4	0.6	0.8	1.1	2.1	3.1	4.2	5.2	6.2	8.2	10.1	14.7	19.2	34.7	47.2	65.5	88.1	98.6
450.	0.2	0.4	0.7	0.9	1.1	2.2	3.3	4.4	5.5	6.5	8.6	10.7	15.5	20.2	36.3	49.1	67.6	89.5	98.9
475.	0.2	0.5	0.7	0.9	1.2	2.3	3.5	4.6	5.8	6.9	9.1	11.2	16.3	21.2	37.9	51.0	69.6	90.8	99.2
500.	0.3	0.5	0.7	1.0	1.2	2.5	3.7	4.9	6.1	7.2	9.5	11.8	17.1	22.1	39.4	52.8	71.4	91.9	99.3
550.	0.3	0.5	0.8	1.1	1.4	2.7	4.0	5.4	6.7	7.9	10.4	12.9	18.7	24.1	42.4	56.3	74.8	93.7	99.6
600.	0.3	0.6	0.9	1.2	1.5	3.0	4.4	5.8	7.2	8.6	11.3	14.0	20.2	26.0	45.2	59.4	77.8	95.1	99.8
650.	0.3	0.6	1.0	1.3	1.6	3.2	4.8	6.3	7.8	9.3	12.2	15.0	21.7	27.8	47.9	62.4	80.4	96.2	99.9
700.	0.4	0.7	1.0	1.4	1.7	3.4	5.1	6.8	8.4	10.0	13.1	16.1	23.1	29.6	50.4	65.1	82.7	97.0	99.9

Appendix D: Table for Determining Minimum Sample Sizes and for Evaluating Attributes Sample Results

Page 1

PROBABILITY THAT ERROR RATE IN UNIVERSE
100–199 **IS LESS THAN:**

SIZE OF SAMPLE EXAMINED	NO. OF ERRORS FOUND	1%	2%	3%	4%	5%	6%	7%	8%	9%	10%	12%	14%	16%	18%	20%
40	0	36.08	59.35	74.28	83.82	89.87	93.70	96.10	97.60	98.53	99.11	99.68	99.89	99.96	99.99	100.00
	1	3.92	17.92	34.46	49.97	63.04	78.39	81.24	87.01	91.14	94.05	97.41	98.92	99.57	99.84	99.94
	2		2.59	9.57	19.97	32.06	44.35	55.78	65.78	74.13	80.84	90.05	95.15	97.77	99.02	99.59
	3		0.14	1.54	5.26	11.54	19.98	29.81	40.18	50.35	59.78	75.29	85.94	92.52	96.26	98.23
	4			0.13	0.89	2.91	6.64	12.20	19.38	27.75	36.79	54.80	70.28	81.87	89.68	94.48
	5				0.09	0.51	1.62	3.80	7.31	12.25	18.52	33.88	50.52	65.73	77.92	86.70
50	0	43.84	68.68	82.65	90.46	94.79	97.18	98.48	99.19	99.57	99.77	99.94	99.98	100.00	100.00	100.00
	1	6.16	26.06	46.76	63.77	76.32	84.99	90.72	94.39	96.67	98.06	99.37	99.81	99.94	99.98	100.00
	2		4.90	16.64	31.98	47.63	61.53	72.82	81.42	87.65	92.00	96.85	98.86	99.61	99.88	99.96
	3		0.36	3.53	10.94	21.96	34.92	48.09	60.21	70.55	78.88	89.99	95.69	98.29	99.37	99.78
	4			0.40	2.47	7.29	15.09	25.26	36.74	48.39	59.32	76.97	88.31	94.61	97.72	99.11
	5			0.02	0.35	1.71	4.89	10.37	18.09	27.56	38.00	58.68	75.49	86.90	93.62	97.15
	6				0.03	0.27	1.17	3.29	7.10	12.84	20.36	38.94	58.20	74.30	85.69	92.73
	7					0.03	0.20	0.80	2.21	4.85	9.03	22.15	39.67	57.83	73.33	84.66
60	0	51.11	76.30	88.62	94.58	97.44	98.81	99.45	99.75	99.89	99.95	99.99	100.00	100.00	100.00	100.00
	1	8.89	34.78	58.26	75.02	85.74	92.15	95.80	97.81	98.88	99.44	99.87	99.97	99.99	100.00	100.00
	2		8.16	25.33	44.89	62.20	75.52	84.86	91.00	94.82	97.10	99.16	99.78	99.95	99.99	100.00
	3		0.75	6.75	18.99	34.83	50.99	65.20	76.48	84.77	90.50	96.64	98.94	99.70	99.92	99.98
	4			0.99	5.42	14.45	27.15	41.49	55.47	67.71	77.59	90.36	96.87	98.78	99.63	99.90
	5			0.06	0.99	4.31	11.09	21.22	33.58	46.66	59.08	78.74	90.46	96.23	98.67	99.58
	6				0.10	0.89	3.40	8.55	16.60	27.05	38.88	62.14	79.89	90.69	96.19	98.61
	7					0.12	0.76	2.67	6.60	12.99	21.71	43.41	64.83	81.03	91.02	96.22
	8					0.01	0.12	0.63	2.09	5.10	10.15	26.39	47.37	67.24	82.15	91.40
	9						0.01	0.11	0.52	1.62	3.94	13.80	30.69	50.90	69.44	83.23
70	0	57.86	82.44	92.76	97.05	98.81	99.53	99.81	99.93	99.97	99.99	100.00	100.00	100.00	100.00	100.00
	1	12.14	43.72	68.45	83.63	91.95	96.20	98.26	99.23	99.67	99.86	99.98	100.00	100.00	100.00	100.00
	2		12.42	35.16	57.49	74.48	85.66	92.36	96.11	98.09	99.09	99.81	99.97	99.99	100.00	100.00
	3		1.42	11.40	29.06	48.72	65.93	78.88	87.64	93.12	96.33	99.07	99.79	99.96	99.99	100.00
	4			2.07	10.15	24.37	41.55	58.18	72.02	82.35	89.42	96.68	99.10	99.79	99.96	99.99
	5			0.16	2.30	8.94	20.62	35.59	51.23	65.34	76.74	90.99	97.07	99.18	99.80	99.96
	6				0.30	2.30	7.81	17.54	30.58	45.04	58.98	80.55	92.37	97.46	99.27	99.81
	7				0.02	0.39	2.20	6.81	14.97	26.32	39.54	65.34	83.69	93.56	97.82	99.36
	8					0.04	0.44	2.03	5.91	12.81	22.68	47.48	70.66	86.33	94.59	98.16
	9						0.06	0.46	1.85	5.12	10.97	30.36	54.47	75.15	88.55	95.49
	10							0.07	0.45	1.66	4.42	16.86	37.73	60.63	78.97	90.43

Source: USAF Auditor General, 1961

PROBABILITY THAT ERROR RATE IN UNIVERSE

Page 2 100–199 **IS LESS THAN:**

SIZE OF SAMPLE EXAMINED	NO. OF ERRORS FOUND	1%	2%	3%	4%	5%	6%	7%	8%	9%	10%	12%	14%	16%	18%	20%
80	0	64.12	87.30	95.57	98.47	99.48	99.83	99.94	99.98	99.99	100.00	100.00	100.00	100.00	100.00	100.00
	1	15.88	52.57	77.07	89.84	95.76	98.31	99.35	99.76	99.91	99.97	100.00	100.00	100.00	100.00	100.00
	2		17.69	45.57	68.89	88.93	92.30	96.53	98.51	99.39	99.76	99.97	100.00	100.00	100.00	100.00
	3		2.45	17.57	40.47	62.16	78.24	88.44	94.25	97.30	98.79	99.79	99.97	100.00	100.00	100.00
	4			3.87	16.89	36.43	56.48	72.88	84.40	91.61	95.75	99.06	99.83	99.97	100.00	100.00
	5			0.36	4.64	16.02	33.06	51.58	67.90	[illegible]	88.70	96.88	99.29	99.86	99.98	100.00
	6				0.74	5.03	15.09	30.17	47.28	63.28	76.23	91.79	97.73	99.48	99.90	99.98
	7				0.05	1.05	5.18	14.16	27.64	43.47	58.99	82.40	94.05	98.36	99.62	99.93
	8					0.13	1.28	5.18	13.24	25.39	40.05	68.39	87.00	95.70	98.83	99.73
	9					0.01	0.21	1.44	5.09	12.36	23.37	51.31	75.84	90.45	96.92	99.18
	10						0.02	0.29	1.53	4.93	11.52	34.15	61.07	81.69	93.03	97.81
	11							0.04	0.35	1.58	4.73	19.86	44.65	69.30	86.24	94.95
	12								0.06	0.40	1.60	9.97	29.22	54.34	76.09	89.77
90	0	69.87	91.07	97.40	99.26	99.79	99.94	99.98	100.00	100.00	100.00	100.00	100.00	100.00	100.00	100.00
	1	20.13	61.05	84.03	94.06	97.98	99.31	99.78	99.93	99.98	99.99	100.00	100.00	100.00	100.00	100.00
	2		23.93	55.98	78.49	90.62	96.24	98.59	99.50	99.83	99.94	99.99	100.00	100.00	100.00	100.00
	3		3.95	25.21	52.39	73.97	87.31	94.34	97.66	99.09	99.66	99.96	100.00	100.00	100.00	100.00
	4			6.62	25.62	49.59	70.17	84.19	92.34	96.57	98.56	99.79	99.97	100.00	100.00	100.00
	5			0.76	8.41	25.63	47.27	66.89	81.27	90.31	95.36	99.12	99.87	99.98	100.00	100.00
	6				1.63	9.62	25.44	45.27	64.00	78.54	88.27	97.19	99.48	99.92	99.99	100.00
	7				0.14	2.44	10.46	25.11	43.50	61.43	75.99	92.70	98.32	99.69	99.95	99.99
	8					0.37	3.14	11.04	24.72	41.90	59.09	84.28	95.51	99.00	99.82	99.97
	9					0.03	0.64	3.72	11.43	24.28	40.43	71.35	89.88	97.27	99.42	99.90
	10						0.08	0.92	4.19	11.69	23.81	54.98	80.46	93.59	98.36	99.67
	11							0.16	1.18	4.57	11.83	37.81	67.21	86.99	96.02	99.04
	12							0.02	0.24	1.42	4.88	22.81	51.43	76.85	91.53	97.57
	13								0.04	0.34	1.64	11.90	35.49	63.47	84.10	94.61
	14									0.06	0.44	5.30	21.78	48.28	73.43	89.37

Page 3

PROBABILITY THAT ERROR RATE IN UNIVERSE 200 - 399 IS LESS THAN:

SIZE OF SAMPLE EXAMINED	NO. OF ERRORS FOUND	1%	2%	3%	4%	5%	6%	7%	8%	9%	10%	12%	14%	16%	18%	20%
40	0	34.50	57.29	72.28	82.09	88.49	92.64	95.31	97.03	98.13	98.83	99.55	99.83	99.94	99.98	99.99
	1	5.14	18.58	34.16	48.88	61.48	71.66	79.55	85.49	89.85	92.99	96.78	98.58	99.40	99.75	99.90
	2	0.35	3.65	10.79	20.80	32.23	43.83	54.70	64.34	72.51	79.21	88.69	94.19	97.17	98.68	99.41
	3	0.01	0.45	2.36	6.45	12.78	20.95	30.28	40.06	49.68	58.67	73.72	84.42	91.32	95.42	97.71
	4		0.04	0.37	1.49	3.89	7.88	13.46	20.44	28.43	36.97	54.01	68.87	80.31	88.30	98.44
	5			0.04	0.26	0.92	2.36	4.84	8.57	13.55	19.67	34.31	50.02	64.52	76.40	85.23
50	0	41.51	65.99	80.34	88.70	93.55	96.34	97.94	98.84	99.36	99.65	99.89	99.97	99.99	100.00	100.00
	1	7.80	26.32	45.54	61.72	74.06	82.91	88.99	93.05	95.69	97.36	99.06	99.68	99.90	99.97	99.99
	2	0.67	6.54	17.95	32.21	46.73	59.83	70.74	79.30	85.72	90.37	95.88	98.35	99.38	99.78	99.93
	3	0.02	1.04	5.01	12.60	23.10	35.14	47.37	58.73	68.62	76.80	88.21	94.50	97.61	99.03	99.63
	4		0.11	1.01	3.74	8.97	16.68	26.31	36.97	47.78	58.00	74.96	86.44	93.25	96.89	98.66
	5		0.01	0.15	0.85	2.76	6.42	12.09	19.61	28.54	38.24	57.48	73.54	84.97	92.14	96.19
	6			0.02	0.15	0.68	2.02	4.61	8.76	14.56	21.82	39.19	57.10	72.42	83.74	91.15
60	0	47.94	73.09	86.19	92.97	96.45	98.22	99.11	99.56	99.79	99.90	99.98	99.99	100.00	100.00	100.00
	1	10.86	34.30	55.98	72.20	83.16	90.13	94.36	96.85	98.27	99.07	99.74	99.93	99.98	100.00	100.00
	2	1.15	10.32	26.26	44.01	60.04	72.86	82.29	88.84	93.16	95.92	98.64	99.58	99.88	99.97	99.99
	3	0.05	2.01	8.91	20.67	35.13	49.83	63.04	73.89	82.22	88.29	95.33	98.31	99.44	99.83	99.95
	4		0.25	2.20	7.49	16.50	28.30	41.34	54.10	65.52	75.04	88.14	94.97	98.07	99.32	99.78
	5		0.02	0.40	2.11	6.23	13.28	22.93	34.27	46.15	57.54	76.22	88.24	94.79	97.90	99.23
	6			0.05	0.46	1.90	5.15	10.73	18.65	28.40	39.18	60.45	77.39	88.49	94.73	97.80
	7			0.01	0.08	0.47	1.66	4.24	8.69	15.18	23.50	43.36	63.01	78.54	88.84	94.76
	8				0.01	0.09	0.45	1.42	3.48	7.05	12.37	27.89	47.02	65.81	79.67	89.26
70	0	53.82	78.86	90.41	95.69	98.08	99.15	99.63	99.84	99.93	99.97	99.99	100.00	100.00	100.00	100.00
	1	14.28	42.21	65.15	80.35	89.43	94.51	97.23	98.64	99.34	99.69	99.93	99.99	100.00	100.00	100.00
	2	1.81	14.89	35.19	55.24	71.25	82.55	89.88	94.35	96.95	98.40	99.59	99.90	99.98	100.00	100.00
	3	0.09	3.46	14.04	30.09	47.56	63.26	75.69	84.68	90.74	94.60	98.33	99.54	99.88	99.97	99.99
	4		0.52	4.14	12.88	26.08	41.37	56.26	69.11	79.21	86.60	95.02	98.37	99.52	99.87	99.97
	5		0.05	0.90	4.33	11.67	22.78	36.22	50.12	62.93	73.71	88.32	95.51	98.47	99.54	99.87
	6			0.14	1.15	4.27	10.52	20.00	31.85	44.71	57.20	77.48	89.85	96.00	98.61	99.56
	7			0.02	0.24	1.28	4.07	9.44	17.62	28.10	39.92	63.08	80.65	91.21	96.49	98.75
	8				0.04	0.31	1.32	3.81	8.45	15.55	24.85	47.04	68.05	83.35	92.41	96.94
	9				0.01	0.06	0.36	1.32	3.52	7.55	13.74	31.86	53.34	72.32	85.68	93.47
	10					0.01	0.08	0.39	1.27	3.22	6.73	19.48	38.48	58.92	76.02	87.70

PROBABILITY THAT ERROR RATE IN UNIVERSE

Page 4 200 - 399 **IS LESS THAN:**

SIZE OF SAMPLE EXAMINED	NO. OF ERRORS FOUND	1%	2%	3%	4%	5%	6%	7%	8%	9%	10%	12%	14%	16%	18%	20%
80	0	59.19	83.52	93.41	97.89	98.98	99.61	99.85	99.94	99.98	99.99	100.00	100.00	100.00	100.00	100.00
	1	18.00	49.82	72.94	86.46	98.56	97.06	98.70	99.44	99.76	99.90	99.98	100.00	100.00	100.00	100.00
	2	2.65	20.14	44.25	65.29	80.10	89.29	94.52	97.31	98.72	99.41	99.89	99.98	100.00	100.00	100.00
	3	0.15	5.44	20.27	40.18	59.30	74.42	84.93	91.59	95.52	97.71	99.46	99.89	99.98	100.00	100.00
	4		0.97	6.95	19.79	36.92	54.89	69.36	80.67	88.46	93.44	98.12	99.53	99.90	99.98	100.00
	5		0.11	1.78	7.75	19.07	34.16	50.22	64.81	76.58	85.23	94.90	98.50	99.62	99.91	99.98
	6		0.01	0.34	2.41	8.18	18.25	31.76	46.60	60.70	72.67	88.66	96.05	98.82	99.69	99.93
	7			0.05	0.59	2.86	8.26	17.41	29.63	43.39	56.95	78.73	91.28	96.97	99.09	99.76
	8				0.12	0.88	3.17	8.23	16.55	27.70	40.51	65.46	83.44	93.31	97.68	99.30
	9				0.02	0.20	1.03	3.36	8.10	15.71	25.95	50.35	72.41	87.14	94.89	98.25
	10					0.04	0.28	1.18	3.47	7.90	14.89	35.51	58.98	78.07	90.05	96.12
	11					0.01	0.07	0.36	1.30	3.51	7.64	22.83	44.68	66.41	82.67	92.34
90	0	64.08	87.25	95.53	98.45	99.47	99.82	99.94	99.98	99.99	100.00	100.00	100.00	100.00	100.00	100.00
	1	21.96	56.96	79.38	90.89	96.20	98.48	99.41	99.78	99.92	99.97	100.00	100.00	100.00	100.00	100.00
	2	3.71	25.92	53.04	73.85	86.71	93.70	97.17	98.78	99.50	99.80	99.97	100.00	100.00	100.00	100.00
	3	0.24	8.01	27.36	50.29	69.63	83.02	91.17	95.67	97.98	99.10	99.84	99.98	100.00	100.00	100.00
	4		1.64	10.72	27.93	48.13	66.21	79.76	88.70	94.07	97.04	99.36	99.88	99.98	100.00	100.00
	5		0.21	3.16	12.49	28.09	46.34	63.35	76.85	86.34	92.41	98.00	99.56	99.92	99.99	100.00
	6		0.02	0.69	4.47	13.69	28.02	44.79	60.89	74.23	84.10	94.90	98.65	99.70	99.94	99.99
	7			0.11	1.28	5.55	14.52	27.83	43.41	58.71	71.85	89.09	96.58	99.11	99.80	99.96
	8			0.01	0.29	1.87	6.42	15.09	27.56	42.15	56.76	79.95	92.55	97.72	99.41	99.87
	9				0.05	0.52	2.42	7.12	15.48	27.28	41.00	67.65	85.86	94.95	98.50	99.62
	10				0.01	0.12	0.77	2.91	7.67	15.78	26.86	53.37	76.21	90.13	96.61	99.02
	11					0.02	0.21	1.03	3.34	8.11	15.88	38.91	64.03	82.77	93.19	97.75
	12						0.05	0.32	1.28	3.72	8.45	26.06	50.45	72.81	87.71	95.86
	13						0.01	0.08	0.43	1.52	4.04	15.96	37.04	60.82	79.84	91.36

Page 5

PROBABILITY THAT ERROR RATE IN UNIVERSE
200 - 399 **IS LESS THAN:**

SIZE OF SAMPLE EXAMINED	NO. OF ERRORS FOUND	1%	2%	3%	4%	5%	6%	7%	8%	9%	10%	12%	14%	16%	18%	20%
100	0	68.52	90.22	97.01	99.10	99.73	99.92	99.98	99.99	100.00	100.00	100.00	100.00	100.00	100.00	100.00
	1	26.12	63.53	84.57	94.02	97.82	99.24	99.74	99.92	99.97	99.99	100.00	100.00	100.00	100.00	100.00
	2	4.99	32.07	61.28	80.84	91.43	96.44	98.61	99.48	99.81	99.94	99.99	100.00	100.00	100.00	100.00
	3	0.87	11.16	35.02	59.85	78.16	89.23	95.09	97.90	99.15	99.67	99.96	100.00	100.00	100.00	100.00
	4		2.60	15.44	36.86	58.91	76.13	87.37	93.81	97.16	98.77	99.80	99.97	100.00	100.00	100.00
	5		0.39	5.17	18.54	38.15	58.22	74.52	85.76	92.62	96.42	99.29	99.88	99.98	100.00	100.00
	6		0.03	1.30	7.54	20.91	39.11	57.69	73.20	84.38	91.53	97.94	99.59	99.93	99.99	100.00
	7			0.24	2.46	9.62	22.79	39.86	57.26	72.11	83.16	94.99	98.82	99.77	99.96	99.99
	8			0.03	0.64	3.69	11.44	24.32	40.47	56.91	71.19	89.60	97.07	99.33	99.87	99.98
	9				0.18	1.18	4.92	13.02	25.60	40.98	56.62	81.15	93.68	98.31	99.63	99.93
	10				0.02	0.31	1.81	6.09	14.41	26.69	41.41	69.70	87.97	96.23	99.04	99.80
	11					0.07	0.57	2.48	7.19	15.64	27.68	56.17	79.54	92.50	97.79	99.46
	12					0.01	0.15	0.88	3.17	8.21	16.73	42.10	68.55	86.59	95.42	98.73
	13						0.08	0.27	1.23	3.86	9.16	29.16	55.82	78.24	91.44	97.27
	14						0.01	0.07	0.42	1.62	4.53	18.58	42.63	67.65	85.41	94.68
	15							0.02	0.13	0.61	2.02	10.86	30.87	55.55	77.15	90.50
110	0	72.53	92.57	98.02	99.48	99.87	99.97	99.99	100.00	100.00	100.00	100.00	100.00	100.00	100.00	100.00
	1	30.41	69.47	88.66	96.17	98.78	99.63	99.89	99.97	99.99	100.00	100.00	100.00	100.00	100.00	100.00
	2	6.51	38.43	68.63	86.34	94.66	98.07	99.34	99.79	99.93	99.98	100.00	100.00	100.00	100.00	100.00
	3	0.55	14.91	42.96	68.47	84.87	93.47	97.41	99.04	99.66	99.89	99.99	100.00	100.00	100.00	100.00
	4		3.91	21.04	46.12	68.62	83.89	92.53	96.81	98.73	99.53	99.94	99.99	100.00	100.00	100.00
	5		0.66	7.91	25.74	48.58	68.91	83.24	91.80	96.30	98.44	99.77	99.97	100.00	100.00	100.00
	6		0.06	2.24	11.73	29.52	50.60	69.29	82.81	91.22	95.86	99.25	99.89	99.99	100.00	100.00
	7			0.47	4.32	15.20	32.67	52.34	69.72	82.54	90.77	97.93	99.64	99.95	99.99	100.00
	8			0.07	1.27	6.58	18.34	35.37	53.88	70.18	82.39	95.16	98.99	99.83	99.98	100.00
	9			0.01	0.30	2.38	8.89	21.17	37.74	55.26	70.65	90.15	97.52	99.51	99.92	99.99
	10				0.05	0.72	3.70	11.16	23.75	39.85	56.52	82.32	94.67	98.76	99.77	99.97
	11				0.01	0.18	1.32	5.15	13.35	26.10	41.76	71.63	89.81	97.21	99.40	99.90
	12					0.04	0.40	2.08	6.67	15.45	28.27	58.78	82.48	94.35	98.58	99.71
	13					0.01	0.10	0.73	2.96	8.22	17.45	45.11	72.62	89.67	96.98	99.30
	14						0.02	0.22	1.16	3.93	9.78	32.16	60.79	82.77	94.16	98.44
	15							0.06	0.40	1.68	4.97	21.19	48.01	73.62	89.67	96.82
	16							0.01	0.12	0.64	2.28	12.86	35.57	62.61	83.17	94.07
	17								0.03	0.22	0.95	7.18	24.61	50.59	74.61	89.77

Page 6

PROBABILITY THAT ERROR RATE IN UNIVERSE 400 - 1000 IS LESS THAN:

SIZE OF SAMPLE EXAMINED	NO. OF ERRORS FOUND	1%	2%	3%	4%	5%	6%	7%	8%	9%	10%	12%	14%	16%	18%	20%
40	0	33.64	56.15	71.15	81.11	87.68	92.00	94.83	96.68	97.88	98.65	99.46	99.79	99.92	99.97	99.99
	1	5.73	18.88	33.97	48.28	60.63	70.70	78.60	84.62	89.10	92.37	96.38	98.35	99.28	99.69	99.87
	2	0.59	4.21	11.41	21.21	32.29	43.54	54.12	63.55	71.62	78.30	87.90	93.62	96.80	98.46	99.29
	3	0.04	0.67	2.83	7.09	13.42	21.43	30.50	39.98	49.31	58.07	72.86	83.57	90.62	94.92	97.38
	4		0.08	0.54	1.86	4.45	8.54	14.13	20.99	28.77	37.05	53.58	68.10	79.45	87.51	92.82
50	0	40.27	64.51	79.03	87.68	92.80	95.82	97.59	98.61	99.21	99.55	99.86	99.96	99.99	100.00	100.00
	1	8.53	26.40	44.88	60.63	72.83	81.74	88.00	92.25	95.08	96.92	98.84	99.58	99.86	99.95	99.99
	2	1.10	7.36	18.57	32.29	46.25	58.93	69.62	78.14	84.64	89.44	95.28	98.02	99.22	99.71	99.90
	3	0.09	1.48	5.79	13.42	23.64	35.23	46.97	57.95	67.60	75.67	87.20	93.78	97.18	98.80	99.52
	4	0.01	0.22	1.41	4.45	9.83	17.47	26.81	37.06	47.44	57.31	73.88	85.39	92.45	96.36	98.86
	5		0.03	0.27	1.20	3.38	7.25	12.97	20.36	29.01	38.34	56.85	72.50	83.90	91.28	95.60
	6			0.04	0.27	0.97	2.54	5.36	9.64	15.43	22.55	39.29	56.52	71.42	82.67	90.25
60	0	46.29	71.34	84.81	92.00	95.82	97.83	98.88	99.43	99.71	99.85	99.96	99.99	100.00	100.00	100.00
	1	11.68	34.01	54.80	70.70	81.74	88.96	93.49	96.24	97.87	98.81	99.65	99.90	99.97	99.99	100.00
	2	1.82	11.31	26.65	43.54	58.93	71.46	80.90	87.61	92.19	95.19	98.29	99.43	99.82	99.95	99.99
	3	0.19	2.75	9.96	21.43	35.23	49.24	61.94	72.54	80.86	87.05	94.52	97.89	99.25	99.75	99.92
	4	0.01	0.51	2.93	8.54	17.47	28.82	41.23	53.41	64.41	73.72	86.91	94.13	97.59	99.09	99.68
	5		0.07	0.69	2.80	7.25	14.35	23.73	34.58	45.88	56.76	74.91	87.01	93.92	97.40	98.97
	6		0.01	0.13	0.76	2.54	6.11	11.83	19.62	29.02	39.30	59.60	76.09	87.27	93.85	97.28
	7			0.02	0.18	0.76	2.25	5.13	9.78	16.26	24.35	43.32	62.10	77.24	87.64	93.88
70	0	51.77	76.91	89.04	94.83	97.59	98.88	99.49	99.77	99.89	99.95	99.99	100.00	100.00	100.00	100.00
	1	15.11	41.43	63.48	78.60	88.00	93.49	96.56	98.23	99.10	99.55	99.90	99.98	100.00	100.00	100.00
	2	2.76	15.93	35.14	54.12	69.62	80.90	88.49	93.30	96.22	97.92	99.41	99.85	99.96	99.99	100.00
	3	0.34	4.54	15.21	30.50	46.97	61.94	74.06	83.09	89.39	93.56	97.83	99.34	99.82	99.95	99.99
	4	0.03	0.99	5.25	14.13	26.81	41.23	55.33	67.67	77.60	85.08	94.02	97.87	99.31	99.80	99.95
	5		0.17	1.47	5.44	12.97	23.73	36.46	49.57	61.76	72.19	86.86	94.55	97.98	99.33	99.79
	6		0.02	0.34	1.76	5.36	11.83	21.11	32.40	44.52	56.33	75.93	88.46	95.10	98.14	99.36
	7			0.07	0.48	1.90	5.13	10.76	18.84	28.89	40.05	61.98	79.08	89.88	95.64	98.31
	8			0.01	0.11	0.59	1.95	4.84	9.75	16.84	25.82	46.81	66.78	81.80	91.16	96.16
	9				0.02	0.16	0.65	1.93	4.51	8.82	15.07	32.51	52.79	70.94	84.16	92.30
80	0	56.73	81.44	92.11	96.68	98.61	99.43	99.77	99.91	99.96	99.98	100.00	100.00	100.00	100.00	100.00
	1	18.74	48.49	70.87	84.62	92.25	96.24	98.23	99.18	99.63	99.84	99.97	99.99	100.00	100.00	100.00
	2	3.92	21.08	43.59	63.55	78.14	87.61	93.30	96.51	98.24	99.14	99.81	99.96	99.99	100.00	100.00
	3	0.55	6.87	21.35	39.98	57.95	72.54	83.09	90.08	94.42	96.98	99.19	99.81	99.96	99.99	100.00

Page 7

PROBABILITY THAT ERROR RATE IN UNIVERSE
400 - 1000 IS LESS THAN:

SIZE OF SAMPLE EXAMINED	NO. OF ERRORS FOUND	1%	2%	3%	4%	5%	6%	7%	8%	9%	10%	12%	14%	16%	18%	20%
80 (Cont)	4	0.05	1.72	8.43	20.99	37.06	53.41	67.67	78.79	86.76	92.08	97.47	99.29	99.82	99.96	99.99
	5		0.34	2.72	9.25	20.36	34.58	49.57	63.35	74.75	83.42	93.70	97.93	99.40	99.84	99.96
	6		0.05	0.73	3.45	9.64	19.62	32.40	46.25	59.49	70.96	86.97	95.00	98.33	99.51	99.87
	7		0.01	0.16	1.10	3.96	9.78	18.84	30.45	43.30	56.00	76.92	89.74	96.06	98.67	99.60
	8			0.08	0.30	1.42	4.30	9.75	18.04	28.67	40.66	64.09	81.63	91.93	96.91	98.95
	9			0.01	0.07	0.44	1.67	4.51	9.62	17.24	27.05	49.88	70.78	85.41	93.67	97.59
	10				0.01	0.12	0.58	1.87	4.62	9.40	16.45	36.06	58.01	76.32	88.44	95.06
	11					0.03	0.18	0.69	2.01	4.66	9.14	24.13	44.65	65.07	80.89	90.87
90	0	61.23	85.12	94.35	97.88	99.21	99.71	99.89	99.96	99.99	100.00	100.00	100.00	100.00	100.00	100.00
	1	22.51	55.06	77.03	89.10	95.08	97.87	99.10	99.63	99.85	99.94	99.99	100.00	100.00	100.00	100.00
	2	5.31	26.57	51.69	71.62	84.64	92.19	96.22	98.24	99.21	99.66	99.94	99.99	100.00	100.00	100.00
	3	0.85	9.73	28.11	49.31	67.60	80.86	89.39	94.42	97.20	98.65	99.72	99.95	99.99	100.00	100.00
	4	0.09	2.76	12.46	28.77	47.44	64.41	77.60	86.76	92.58	96.04	99.00	99.78	99.96	99.99	100.00
	5	0.01	0.62	4.54	14.23	29.01	45.88	61.76	74.75	84.30	90.74	97.20	99.27	99.84	99.97	99.99
	6		0.11	1.38	6.00	15.43	29.02	44.52	59.49	72.22	82.01	98.51	98.08	99.48	99.88	99.98
	7		0.02	0.35	2.17	7.17	16.26	28.89	43.30	57.50	69.95	87.18	95.45	98.63	99.64	99.92
	8			0.08	0.68	2.92	8.09	16.84	28.67	42.20	55.78	77.87	90.89	96.83	99.05	99.75
	9			0.01	0.19	1.05	3.58	8.82	17.24	28.40	41.17	66.00	83.83	93.55	97.81	99.35
	10				0.04	0.33	1.42	4.16	9.40	17.50	28.09	52.64	74.24	88.28	95.47	98.49
	11				0.01	0.09	0.50	1.77	4.66	9.86	17.66	39.31	62.62	80.72	91.58	96.85
	12					0.02	0.16	0.68	2.11	5.09	10.22	27.38	50.00	70.99	85.74	94.01
	13					0.01	0.05	0.24	0.87	2.41	5.45	17.75	37.63	59.66	77.82	89.58
100	0	65.31	88.10	95.97	98.65	99.55	99.85	99.95	99.98	100.00	100.00	100.00	100.00	100.00	100.00	100.00
	1	26.87	61.08	82.07	92.37	96.92	98.81	99.55	99.84	99.94	99.98	100.00	100.00	100.00	100.00	100.00
	2	6.92	32.28	59.19	78.30	89.44	95.19	97.92	99.14	99.66	99.87	99.98	100.00	100.00	100.00	100.00
	3	1.23	13.09	35.22	58.07	75.67	87.05	93.56	96.98	98.65	99.42	99.90	99.99	100.00	100.00	100.00
	4	0.15	4.15	17.26	37.05	57.31	73.72	85.08	92.08	96.04	98.11	99.62	99.94	99.99	100.00	100.00
	5	0.01	1.04	7.02	20.26	38.34	56.76	72.19	83.42	90.74	95.12	98.83	99.76	99.96	99.99	100.00
	6		0.21	2.39	9.52	22.54	39.30	56.33	70.96	82.01	89.53	96.99	99.28	99.85	99.97	100.00
	7		0.03	0.69	3.86	11.66	24.35	40.05	56.00	69.95	80.79	93.42	98.16	99.57	99.91	99.98
	8			0.17	1.36	5.31	13.49	25.82	40.66	55.73	69.10	87.45	95.89	98.88	99.74	99.95
	9			0.04	0.42	2.14	6.69	15.07	27.05	41.17	55.50	78.80	91.91	97.46	99.33	99.85
	10			0.01	0.11	0.77	2.98	7.97	16.45	28.09	41.60	67.74	85.74	94.85	98.44	99.60
	11				0.03	0.25	1.19	3.82	9.14	17.66	28.98	55.15	77.24	90.58	96.76	99.06

Page 8

PROBABILITY THAT ERROR RATE IN UNIVERSE
400 - 1000 IS LESS THAN:

SIZE OF SAMPLE EXAMINED	NO. OF ERRORS FOUND	1%	2%	3%	4%	5%	6%	7%	8%	9%	10%	12%	14%	16%	18%	20%
100 (Cont)	12				0.01	0.07	0.43	1.66	4.65	10.22	18.73	42.30	66.70	84.32	93.89	98.00
	13					0.02	0.14	0.66	2.17	5.45	11.21	30.46	54.89	75.98	89.43	96.10
	14						0.04	0.24	0.93	2.68	6.22	20.54	42.84	65.88	83.12	93.02
	15						0.01	0.08	0.36	1.22	3.20	12.95	31.61	54.69	74.95	88.42
120	0	72.32	92.44	97.97	99.46	99.86	99.96	99.99	100.00	100.00	100.00	100.00	100.00	100.00	100.00	100.00
	1	34.19	71.39	89.37	96.38	98.84	99.65	99.90	99.97	99.99	100.00	100.00	100.00	100.00	100.00	100.00
	2	10.77	43.77	71.95	87.90	95.28	98.29	99.41	99.81	99.94	99.98	100.00	100.00	100.00	100.00	100.00
	3	2.33	21.11	49.46	72.86	87.20	94.52	97.83	99.19	99.72	99.90	99.99	100.00	100.00	100.00	100.00
	4	0.35	8.07	28.66	53.58	73.88	86.91	94.02	97.47	99.00	99.62	99.95	100.00	100.00	100.00	100.00
	5	0.04	2.47	13.99	34.52	56.85	74.91	86.86	93.70	97.20	98.83	99.83	99.98	100.00	100.00	100.00
	6		0.61	5.77	19.41	39.29	59.60	75.93	86.97	93.51	96.99	99.46	99.92	99.99	100.00	100.00
	7		0.12	2.03	9.53	24.26	43.32	61.98	76.92	87.18	93.42	98.58	99.76	99.97	100.00	100.00
	8		0.02	0.61	4.09	13.36	28.60	46.81	64.09	77.87	87.45	96.75	99.34	99.89	99.99	100.00
	9			0.16	1.55	6.57	17.11	32.51	49.88	66.00	78.80	93.43	98.44	99.70	99.95	99.99
	10			0.04	0.52	2.89	9.27	20.71	36.06	52.64	67.74	88.13	96.68	99.27	99.87	99.98
	11			0.01	0.15	1.14	4.56	12.08	24.13	39.31	55.15	80.55	93.63	98.37	99.66	99.94
	12				0.04	0.41	2.03	6.46	14.92	27.38	42.30	70.84	88.88	96.71	99.22	99.85
	13				0.01	0.13	0.83	3.17	8.53	17.75	30.45	59.56	82.18	93,93	98.36	99.64
	14					0.04	0.31	1.42	4.50	10.70	20.54	47.65	73.55	89.67	96.82	99.21
	15					0.01	0.10	0.59	2.20	6.00	12.95	36.13	63.37	83.69	94.29	98.39
	16						0.03	0.22	0.99	3.13	7.64	25.91	52.33	75.96	90.46	96.97
	17						0.01	0.08	0.42	1.52	4.21	17.54	41.26	66.73	85.10	94.68
	18							0.03	0.16	0.69	2.17	11.19	30.99	56.49	78.15	91.23
150	0	80.47	96.25	99.29	99.87	99.98	100.00	100.00	100.00	100.00	100.00	100.00	100.00	100.00	100.00	100.00
	1	45.63	82.73	95.43	98.91	99.76	99.95	99.99	100.00	100.00	100.00	100.00	100.00	100.00	100.00	100.00
	2	17.90	59.72	85.26	95.45	98.75	99.68	99.92	99.98	100.00	100.00	100.00	100.00	100.00	100.00	100.00
	3	4.91	35.19	68.21	87.50	95.78	98.73	99.65	99.91	99.98	99.99	100.00	100.00	100.00	100.00	100.00
	4	0.95	16.82	47.66	74.20	89.40	96.20	98.78	99.64	99.90	99.97	100.00	100.00	100.00	100.00	100.00
	5	0.13	6.53	28.76	57.05	78.75	91.01	96.64	98.87	99.65	99.90	99.99	100.00	100.00	100.00	100.00
	6	0.01	2.08	14.93	39.28	64.38	82.33	92.37	97.06	98.97	99.67	99.97	100.00	100.00	100.00	100.00
	7		0.54	6.68	24.06	48.27	70.25	85.23	93.53	97.45	99.09	99.91	99.99	100.00	100.00	100.00
	8		0.12	2.58	13.09	32.94	55.90	75.02	87.61	94.52	97.81	99.73	99.97	100.00	100.00	100.00
	9		0.02	0.86	6.33	20.38	41.16	62.37	78.97	89.58	95.36	99.30	99.92	99.99	100.00	100.00
	10			0.25	2.72	11.42	27.91	48.59	67.88	82.25	91.22	98.39	99.79	99.98	100.00	100.00
	11			0.06	1.04	5.80	17.39	35.29	55.22	72.57	85.00	96.68	99.48	99.94	99.99	100.00

PROBABILITY THAT ERROR RATE IN UNIVERSE

Page 9 400 - 1000 IS LESS THAN:

SIZE OF SAMPLE EXAMINED	NO. OF ERRORS FOUND	1%	2%	3%	4%	5%	6%	7%	8%	9%	10%	12%	14%	16%	18%	20%
150 (Cont)	12			0.01	0.36	2.67	9.94	23.81	42.29	61.10	76.58	93.77	98.84	99.84	99.98	100.00
	13				0.11	1.11	5.22	14.90	30.37	48.81	66.27	89.27	97.64	99.62	99.95	100.00
	14				0.03	0.42	2.51	8.64	20.41	36.84	54.78	82.95	95.59	99.18	99.88	99.99
	15				0.01	0.15	1.11	4.65	12.81	26.21	43.07	74.79	92.34	98.34	99.73	99.97
	16					0.05	0.45	2.32	7.50	17.53	32.11	65.10	87.63	96.89	99.42	99.92
	17					0.01	0.17	1.07	4.11	11.02	22.64	54.46	81.30	94.56	98.84	99.81
	18						0.06	0.46	2.10	6.51	15.09	43.64	73.40	91.08	97.82	99.60
	19						0.02	0.18	1.00	3.61	9.49	33.41	64.21	86.25	96.16	99.20
	20						0.01	0.07	0.45	1.88	5.63	24.38	54.23	79.95	93.62	98.49
	21							0.02	0.19	0.92	3.15	16.94	44.09	72.28	89.97	97.31
	22							0.01	0.07	0.42	1.67	11.19	34.42	63.51	85.06	95.47
	23								0.03	0.18	0.83	7.03	25.76	54.05	78.83	92.76
	24								0.01	0.07	0.39	4.20	18.46	44.45	71.38	89.00
180	0	86.39	98.19	99.76	99.97	100.00	100.00	100.00	100.00	100.00	100.00	100.00	100.00	100.00	100.00	100.00
	1	56.19	90.05	98.16	99.69	99.95	99.99	100.00	100.00	100.00	100.00	100.00	100.00	100.00	100.00	100.00
	2	26.22	72.79	92.89	98.46	99.71	99.95	99.99	100.00	100.00	100.00	100.00	100.00	100.00	100.00	100.00
	3	8.73	49.84	81.85	94.92	98.80	99.75	99.95	99.99	100.00	100.00	100.00	100.00	100.00	100.00	100.00
	4	2.07	28.37	65.24	87.51	96.36	99.09	99.80	99.96	99.99	100.00	100.00	100.00	100.00	100.00	100.00
	5	0.35	13.34	46.12	75.56	91.28	97.40	99.33	88.84	99.97	99.99	100.00	100.00	100.00	100.00	100.00
	6	0.04	5.19	28.60	60.04	82.67	93.85	98.14	99.51	99.88	99.97	100.00	100.00	100.00	100.00	100.00
	7		1.67	15.49	43.37	70.55	87.64	95.64	98.67	99.64	99.91	100.00	100.00	100.00	100.00	100.00
	8		0.45	7.32	28.28	56.07	78.37	91.16	96.91	99.05	99.74	99.99	100.00	100.00	100.00	100.00
	9		0.10	3.02	16.58	41.15	66.40	84.16	93.67	97.81	99.33	99.95	100.00	100.00	100.00	100.00
	10		0.02	1.09	8.73	27.73	52.83	74.56	88.44	95.47	98.44	99.87	99.99	100.00	100.00	100.00
	11			0.34	4.13	17.12	39.24	62.85	80.89	91.58	96.76	99.66	99.98	100.00	100.00	100.00
	12			0.10	1.75	9.66	27.09	50.09	71.14	85.74	93.89	99.22	99.93	100.00	100.00	100.00
	13			0.02	0.67	4.98	17.34	37.55	59.76	77.82	89.43	98.36	99.83	99.99	100.00	100.00
	14				0.23	2.35	10.27	26.40	47.70	68.05	83.12	96.82	99.62	99.97	100.00	100.00
	15				0.07	1.01	5.63	17.36	36.03	57.01	74.95	94.29	99.18	99.92	99.99	100.00
	16				0.02	0.40	2.86	10.66	25.68	45.57	65.23	90.46	98.37	99.81	99.99	100.00
	17				0.01	0.15	1.34	6.12	17.25	34.63	54.53	85.10	96.98	99.60	99.96	100.00
	18					0.05	0.58	3.28	10.90	24.97	43.64	78.15	94.76	99.18	99.91	99.99
	19					0.01	0.24	1.64	6.47	17.05	33.33	69.73	91.46	98.44	99.81	99.98
	20						0.09	0.77	3.61	11.01	24.25	60.23	86.87	97.20	99.60	99.96
	21						0.03	0.33	1.90	6.72	16.78	50.18	80.88	95.27	99.21	99.91
	22						0.01	0.14	0.94	3.88	11.04	40.23	73.54	92.41	98.55	99.81

Page 10

PROBABILITY THAT ERROR RATE IN UNIVERSE 400 - 1000 IS LESS THAN:

SIZE OF SAMPLE EXAMINED	NO. OF ERRORS FOUND	1%	2%	3%	4%	5%	6%	7%	8%	9%	10%	12%	14%	16%	18%	20%
180 (Cont)	23							0.05	0.44	2.12	6.98	30.97	65.08	88.45	97.46	99.61
	24							0.02	0.19	1.09	4.08	22.85	55.89	83.26	95.77	99.27
	25							0.01	0.08	0.53	2.30	16.13	46.44	76.83	93.30	98.67
210	0	90.65	99.15	99.92	99.99	100.00	100.00	100.00	100.00	100.00	100.00	100.00	100.00	100.00	100.00	100.00
	1	65.49	94.51	99.30	99.92	99.99	100.00	100.00	100.00	100.00	100.00	100.00	100.00	100.00	100.00	100.00
	2	35.24	82.58	96.82	99.53	99.94	99.99	100.00	100.00	100.00	100.00	100.00	100.00	100.00	100.00	100.00
	3	13.81	63.31	90.50	98.15	99.70	99.96	99.99	100.00	100.00	100.00	100.00	100.00	100.00	100.00	100.00
	4	3.91	41.42	78.95	94.66	98.92	99.82	99.97	100.00	100.00	100.00	100.00	100.00	100.00	100.00	100.00
	5	0.79	22.79	62.77	87.79	96.95	99.37	99.89	99.98	100.00	100.00	100.00	100.00	100.00	100.00	100.00
	6	0.11	10.49	44.73	76.93	92.88	98.25	99.64	99.93	99.99	100.00	100.00	100.00	100.00	100.00	100.00
	7	0.01	4.03	28.28	62.71	85.88	95.83	98.99	99.79	99.96	99.99	100.00	100.00	100.00	100.00	100.00
	8		1.29	15.78	47.00	75.69	91.44	97.57	99.42	99.88	99.98	100.00	100.00	100.00	100.00	100.00
	9		0.34	7.75	32.14	62.85	84.50	94.85	98.58	99.67	99.93	100.00	100.00	100.00	100.00	100.00
	10		0.08	3.35	19.95	48.75	74.89	90.28	96.92	99.18	99.81	99.99	100.00	100.00	100.00	100.00
	11		0.01	1.27	11.22	35.09	63.10	83.47	93.99	98.17	99.52	99.98	100.00	100.00	100.00	100.00
	12			0.43	5.70	23.34	50.18	74.37	89.35	96.33	98.92	99.94	100.00	100.00	100.00	100.00
	13			0.13	2.62	14.31	37.47	63.40	82.70	93.25	97.78	99.84	99.99	100.00	100.00	100.00
	14			0.03	1.09	8.08	26.18	51.41	74.03	88.59	95.79	99.63	99.98	100.00	100.00	100.00
	15			0.01	0.41	4.20	17.06	39.47	63.73	82.11	92.62	99.21	99.95	100.00	100.00	100.00
	16				0.14	2.00	10.36	28.60	52.49	73.83	87.96	98.42	99.88	99.99	100.00	100.00
	17				0.04	0.88	5.86	19.51	41.20	64.09	81.66	97.06	99.73	99.98	100.00	100.00
	18				0.01	0.36	3.08	12.51	30.72	53.46	73.72	94.88	99.44	99.96	100.00	100.00
	19					0.13	1.51	7.53	21.71	42.71	64.45	91.61	98.90	99.91	100.00	100.00
	20					0.05	0.69	4.26	14.52	32.59	54.36	87.04	97.97	99.81	99.99	100.00
	21					0.01	0.29	2.26	9.18	23.70	44.08	81.06	96.46	99.61	99.97	100.00
	22						0.12	1.12	5.49	16.40	34.28	73.71	94.17	99.24	99.94	100.00
	23						0.04	0.53	3.10	10.78	25.51	65.21	90.88	98.61	99.87	99.99
	24						0.01	0.23	1.65	6.74	18.15	55.96	86.44	97.57	99.73	99.98
	25							0.09	0.83	4.00	12.33	46.45	80.75	95.87	99.48	99.96
	26							0.04	0.40	2.25	7.99	37.21	73.88	93.62	99.05	99.91
	27							0.01	0.18	1.20	4.93	28.72	65.98	90.36	98.34	99.82
	28								0.08	0.61	2.91	21.33	57.38	86.06	97.23	99.65
	29								0.03	0.29	1.63	15.22	48.48	80.64	95.58	99.37
	30								0.01	0.13	0.87	10.43	39.72	74.16	93.21	98.89
	31									0.06	0.44	6.86	31.50	66.76	90.01	98.12

PROBABILITY THAT ERROR RATE IN UNIVERSE

Page 11 SIZE OF OVER 1000 IS LESS THAN:

SIZE OF SAMPLE EXAMINED	NO. OF ERRORS FOUND	1%	2%	3%	4%	5%	6%	7%	8%	9%	10%	12%	14%	16%	18%	20%
40	0	33.10	55.43	70.43	80.46	87.15	91.58	94.51	96.44	97.70	98.52	99.40	99.76	99.91	99.96	99.99
	1	6.07	19.05	33.85	47.90	60.10	70.10	77.99	84.06	88.60	91.95	96.12	98.20	99.19	99.65	99.85
	2	0.75	4.57	11.78	21.45	32.33	43.35	53.75	63.06	71.06	77.72	87.39	93.24	96.55	98.31	99.21
	3	0.07	0.82	3.14	7.48	13.82	21.73	30.63	39.93	45.08	57.69	72.32	83.02	90.16	94.58	97.15
	4	0.01	0.12	0.67	2.10	4.80	8.96	14.54	21.32	28.97	37.10	53.31	67.62	78.90	87.01	92.44
	5		0.01	0.12	0.49	1.39	3.09	5.82	9.67	14.65	20.63	34.64	49.58	63.46	75.04	83.87
	6			0.02	0.10	0.34	0.91	1.99	3.76	6.39	9.95	19.80	32.45	46.31	59.71	71.41
50	0	39.50	63.58	78.19	87.01	92.31	95.47	97.34	98.45	99.10	99.49	99.83	99.95	99.98	100.00	100.00
	1	8.94	26.42	44.47	59.95	72.06	81.00	87.35	91.73	94.68	96.62	98.69	99.52	99.83	99.94	99.98
	2	1.38	7.84	18.92	32.33	45.95	58.38	68.92	77.40	83.95	88.83	94.87	97.79	99.10	99.65	99.87
	3	0.16	1.78	6.28	13.91	23.96	35.27	46.73	57.47	66.97	74.97	86.55	93.30	96.88	98.64	99.43
	4	0.02	0.32	1.68	4.90	10.36	17.94	27.10	37.11	47.23	56.88	73.21	84.72	91.92	96.01	98.15
	5		0.05	0.37	1.44	3.78	7.76	13.51	20.81	29.28	38.39	56.47	71.86	83.23	90.71	95.20
	6		0.01	0.07	0.36	1.18	2.89	5.83	10.19	15.96	22.98	39.35	56.16	70.81	81.99	89.66
60	0	45.28	70.25	83.92	91.37	95.39	97.56	98.72	99.33	99.65	99.82	99.95	99.99	100.00	100.00	100.00
	1	12.12	33.81	54.08	69.78	80.85	88.21	92.91	95.92	97.58	98.62	99.57	99.87	99.96	99.99	100.00
	2	2.24	11.87	26.85	43.24	58.26	70.60	80.02	86.83	91.54	94.70	98.04	99.32	99.79	99.93	99.98
	3	0.31	3.22	10.57	21.87	35.27	48.87	61.27	71.71	80.00	86.26	93.99	97.59	99.10	99.69	99.90
	4		0.73	3.40	9.17	18.03	29.11	41.15	52.98	63.73	72.90	86.12	93.57	97.27	98.93	99.61
	5		0.13	0.91	3.25	7.87	14.98	24.20	34.74	45.71	56.28	74.10	86.23	93.35	97.05	98.79
	6			0.21	0.99	2.97	6.71	12.50	20.20	29.37	39.36	59.08	75.29	86.50	93.27	96.92
70	0	50.52	75.69	88.14	94.26	97.24	98.69	99.38	99.71	99.86	99.94	99.99	100.00	100.00	100.00	100.00
	1	15.53	40.96	62.47	77.51	87.03	92.81	96.10	97.93	98.92	99.45	99.86	99.97	99.99	100.00	100.00
	2	3.34	16.50	35.08	53.44	68.63	79.87	87.59	92.60	95.72	97.58	99.28	99.80	99.95	99.99	100.00
	3	0.54	5.19	15.87	30.71	46.61	61.15	73.07	82.10	88.53	92.88	97.48	99.19	99.76	99.93	99.98
	4	0.07	1.32	5.93	14.85	27.21	41.13	54.77	66.80	76.61	84.12	93.36	97.51	99.16	99.74	99.92
	5		0.28	1.86	6.12	13.72	24.27	36.58	49.24	61.06	71.28	85.94	93.92	97.64	99.17	99.73
	6		0.50	0.50	2.18	6.04	12.61	21.75	32.70	44.40	55.82	74.98	87.57	94.50	97.81	99.20

PROBABILITY THAT ERROR RATE IN UNIVERSE

Page 12 **SIZE OF** OVER 1000 IS LESS THAN:

SIZE OF SAMPLE EXAMINED	NO. OF ERRORS FOUND	1%	2%	3%	4%	5%	6%	7%	8%	9%	10%	12%	14%	16%	18%	20%
70 (Cont)	7			0.12	0.68	2.34	5.80	11.54	19.54	29.33	40.12	61.33	78.13	89.04	95.08	98.00
	8			0.02	0.19	0.80	2.38	5.49	10.54	17.59	26.37	46.66	66.03	80.85	90.36	95.63
	9				0.05	0.25	0.88	2.36	5.14	9.60	15.86	32.88	52.46	70.10	83.23	91.55
80	0	55.25	80.14	91.26	96.18	98.35	99.29	99.70	99.87	99.95	99.98	100.00	100.00	100.00	100.00	100.00
	1	19.08	47.70	69.62	83.46	91.40	95.68	97.89	98.99	99.53	99.78	99.96	99.99	100.00	100.00	100.00
	2	4.66	21.56	43.19	62.52	76.94	86.56	92.50	95.96	97.89	98.93	99.74	99.94	100.00	100.00	100.00
	3	0.87	7.69	21.93	39.84	57.16	71.42	81.50	89.11	94.69	96.47	98.99	99.74	99.99	99.99	100.00
	4	0.13	2.24	9.28	21.64	37.11	52.83	66.67	77.65	85.69	91.20	97.01	99.10	99.94	99.94	99.99
	5		0.55	3.33	10.12	21.08	34.78	49.18	62.50	73.66	82.31	92.91	97.52	99.76	99.78	99.95
	6		0.11	1.03	4.12	10.53	20.39	32.73	46.03	58.79	69.96	85.92	94.30	99.23	99.36	99.82
	7			0.28	1.47	4.66	10.68	19.64	30.89	43.24	55.44	75.84	88.77	97.97	98.36	99.47
	8				0.47	1.84	5.02	10.65	18.88	29.21	40.73	63.29	80.54	95.44	96.37	98.69
	9				0.13	0.65	2.13	5.24	10.51	18.10	27.66	49.61	69.83	84.34	92.88	97.13
	10					0.21	0.82	2.35	5.36	10.31	17.34	36.36	57.45	75.28	87.43	94.35
90	0	59.53	83.77	93.55	97.46	99.01	99.62	99.85	99.95	99.98	99.99	100.00	100.00	100.00	100.00	100.00
	1	22.73	53.96	75.60	87.95	94.33	97.43	98.87	99.51	99.80	99.92	99.97	99.98	100.00	100.00	100.00
	2	6.19	26.88	50.90	70.30	83.36	91.20	95.56	97.85	98.99	99.54	99.91	99.92	99.99	100.00	100.00
	3	1.29	10.68	28.49	48.74	66.42	79.55	88.26	93.59	96.65	98.31	99.61	99.69	99.93	99.99	100.00
	4	0.22	3.48	13.41	29.20	47.03	63.36	76.31	85.55	91.61	95.35	98.72	99.05	99.93	99.99	100.00
	5		0.96	5.39	15.19	29.48	45.60	60.84	73.52	83.05	89.68	96.63	97.58	99.77	99.97	99.99
	6		0.23	1.88	6.92	16.39	29.53	44.35	58.69	71.05	80.75	92.60	94.70	99.31	99.90	99.96
	7			0.57	2.79	8.13	17.23	29.45	43.22	50.81	68.85	85.99	89.83	98.27	99.70	99.87
	8			0.16	1.00	3.62	9.08	17.81	24.27	42.20	55.13	76.65	82.61	96.21	99.22	99.65
	9				0.32	1.45	4.32	9.83	18.21	29.03	41.25	65.05	73.09	92.64	98.22	99.14
	10					0.53	1.89	4.97	10.43	18.48	28.75	52.23	61.82	87.13	96.34	98.10
100	0	63.40	86.74	95.25	98.31	99.41	99.80	99.93	99.98	99.99	100.00	100.00	100.00	100.00	100.00	100.00
	1	26.42	59.67	80.54	91.28	96.29	98.48	99.40	99.77	99.91	99.97	100.00	100.00	100.00	100.00	100.00
	2	7.94	32.33	58.02	76.79	88.17	94.34	97.42	98.87	99.52	99.81	99.97	100.00	100.00	100.00	100.00
	3	1.84	14.10	35.28	57.05	74.22	85.70	92.56	96.33	98.27	99.22	99.86	99.98	100.00	100.00	100.00

PROBABILITY THAT ERROR RATE IN UNIVERSE

Page 13 SIZE OF OVER 1000 IS LESS THAN:

SIZE OF SAMPLE EXAMINED	NO. OF ERRORS FOUND	1%	2%	3%	4%	5%	6%	7%	8%	9%	10%	12%	14%	16%	18%	20%
100 (Cont)	4	0.34	5.08	18.22	37.11	56.40	72.32	83.68	90.97	95.26	97.63	99.47	99.90	99.98	100.00	100.00
	5	0.05	1.55	8.08	21.16	38.40	55.93	70.86	82.01	89.55	94.24	98.48	99.66	99.93	99.99	100.00
	6	0.01	0.41	3.12	10.64	23.40	39.37	55.57	69.68	80.60	88.28	96.33	99.03	99.78	99.96	99.99
	7		0.09	1.06	4.75	12.80	25.17	40.12	55.29	68.72	79.40	92.39	97.67	99.39	99.86	99.97
	8		0.02	0.32	1.90	6.31	14.63	26.60	40.74	55.06	67.91	86.14	95.08	98.53	99.62	99.91
	9			0.09	0.68	2.82	7.75	16.20	27.80	41.25	54.87	77.44	90.78	96.84	99.08	99.77
	10			0.02	0.22	1.15	3.76	9.08	17.57	28.82	41.68	66.63	84.40	93.93	98.00	99.43
	11				0.07	0.43	1.68	4.69	10.29	18.76	29.70	54.58	75.91	89.39	96.05	98.74
	12				0.02	0.15	0.69	2.24	5.59	11.38	19.82	42.39	65.66	82.97	92.89	97.47
	13					0.05	0.26	0.99	2.82	6.45	12.39	31.14	54.36	74.69	88.19	95.31
	14					0.01	0.09	0.41	1.33	3.41	7.26	21.60	42.94	64.90	81.77	91.96
	15						0.03	0.16	0.59	1.69	3.99	14.15	32.27	54.20	72.70	87.15

Page 14

PROBABILITY THAT ERROR RATE IN UNIVERSE SIZE OF OVER 1000 IS LESS THAN:

SIZE OF SAMPLE EXAMINED	NO. OF ERRORS FOUND	1%	2%	3%	4%	5%	6%	7%	8%	9%	10%	12%	14%	16%	18%	20%
120	0	70.06	91.15	97.41	99.25	99.79	99.94	99.98	100.00	100.00	100.00	100.00	100.00	100.00	100.00	100.00
	1	33.77	69.46	87.82	95.53	98.45	99.48	99.83	99.95	99.98	100.00	100.00	100.00	100.00	100.00	100.00
	2	11.96	43.13	70.16	86.28	94.25	97.75	99.17	99.71	99.90	99.97	100.00	100.00	100.00	100.00	100.00
	3	3.30	22.00	48.67	71.13	85.56	93.40	97.19	98.87	99.60	99.84	99.98	100.00	100.00	100.00	100.00
	4	0.74	9.38	29.24	52.67	72.18	85.27	92.83	96.75	98.61	99.44	99.92	99.99	100.00	100.00	100.00
	5	0.14	3.41	15.29	34.83	55.85	73.23	85.23	92.47	96.42	98.40	99.72	99.96	99.99	100.00	100.00
	6	0.02	1.07	7.03	20.57	39.37	58.50	74.26	85.35	92.26	96.18	99.21	99.87	99.98	100.00	100.00
	7		0.30	2.86	10.90	25.24	43.20	60.81	75.25	85.57	92.16	98.08	99.62	99.94	99.99	100.00
	8		0.07	1.04	5.21	14.74	29.39	46.51	62.85	76.21	85.86	95.89	99.05	99.82	99.97	100.00
	9		0.02	0.34	2.26	7.86	18.43	33.12	49.44	64.70	77.14	92.18	97.89	99.53	99.91	99.99
	10			0.10	0.89	3.85	10.66	21.93	36.49	52.06	66.39	86.56	95.79	98.94	99.78	99.96
	11			0.03	0.32	1.73	5.70	13.50	25.23	39.56	54.45	78.90	92.39	97.80	99.48	99.90
	12			0.01	0.11	0.72	2.83	7.75	16.33	28.33	42.39	69.41	87.35	95.83	95.88	99.75
	13				0.03	0.28	1.31	4.15	9.91	19.11	31.27	58.66	80.53	92.71	97.78	99.44
	14				0.01	0.10	0.56	2.07	5.64	12.13	21.82	47.45	72.05	88.17	95.95	98.86
	15					0.03	0.23	0.97	3.01	7.26	14.40	36.66	62.30	82.06	93.10	97.82
	16					0.01	0.09	0.43	1.51	4.10	8.99	26.99	51.88	74.42	89.00	96.12
	17						0.03	0.18	0.72	2.18	5.31	18.93	41.50	65.52	83.49	93.53
	18						0.01	0.07	0.32	1.10	2.97	12.64	31.84	55.82	76.57	89.81
150	0	77.86	95.17	98.96	99.78	99.95	99.99	100.00	100.00	100.00	100.00	100.00	100.00	100.00	100.00	100.00
	1	44.30	80.39	94.15	98.41	99.60	99.90	99.98	100.00	100.00	100.00	100.00	100.00	100.00	100.00	100.00
	2	19.05	57.91	83.07	94.16	98.19	99.48	99.86	99.96	99.99	100.00	100.00	100.00	100.00	100.00	100.00
	3	6.47	35.28	66.16	85.42	94.52	98.14	99.42	99.83	99.95	99.99	100.00	100.00	100.00	100.00	100.00
	4	1.80	18.30	46.93	72.04	87.44	95.01	98.20	99.40	99.81	99.95	100.00	100.00	100.00	100.00	100.00
	5	0.42	8.19	29.57	55.76	76.56	89.17	95.52	98.31	99.41	99.81	99.98	100.00	100.00	100.00	100.00
	6	0.08	3.20	16.60	39.37	62.71	80.16	90.66	96.03	98.45	99.44	99.94	99.99	100.00	100.00	100.00
	7	0.02	1.11	8.34	25.32	47.72	68.34	8312	91.94	96.50	98.60	99.82	99.98	100.00	100.00	100.00
	8		0.34	3.78	14.85	33.62	54.84	72.98	85.58	93.04	96.93	99.52	99.94	99.99	100.00	100.00
	9		0.10	1.55	7.97	21.91	41.26	60.93	76.85	87.65	94.00	98.89	99.84	99.98	100.00	100.00
	10		0.02	0.58	3.93	13.22	29.03	48.15	66.16	80.13	89.40	97.66	99.61	99.95	99.99	100.00
	11		0.01	0.20	1.79	7.40	19.09	35.90	54.32	70.66	82.91	95.54	99.14	99.87	99.98	100.00
	12			0.06	0.75	3.85	11.74	25.23	42.40	59.82	74.55	92.19	98.25	99.70	99.96	100.00
	13			0.02	0.29	1.87	6.77	16.70	31.39	48.43	64.70	87.34	96.70	99.35	99.90	99.99
	14				0.11	0.85	3.66	10.42	22.03	37.41	53.98	80.86	94.25	98.70	99.77	99.97
	15				0.04	0.36	1.86	6.13	14.64	27.53	43.18	72.85	90.62	97.58	99.52	99.92

Page 15

PROBABILITY THAT ERROR RATE IN UNIVERSE SIZE OF OVER 1000 IS LESS THAN:

SIZE OF SAMPLE EXAMINED	NO. OF ERRORS FOUND	1%	2%	3%	4%	5%	6%	7%	8%	9%	10%	12%	14%	16%	18%	20%
150 (Cont)	16				0.01	0.14	0.89	3.40	9.22	19.28	33.06	63.64	85.63	95.78	99.06	99.83
	17					0.05	0.40	1.72	5.51	12.86	24.19	53.74	79.24	93.07	98.24	99.65
	18					0.02	0.17	0.89	3.13	8.16	16.92	43.76	71.54	89.26	96.92	99.31
	19					0.01	0.07	0.42	1.68	4.93	11.30	34.31	62.84	84.21	94.90	98.72
	20						0.03	0.19	0.86	2.84	7.21	25.87	53.56	77.92	92.01	97.76
	21						0.01	0.08	0.42	1.56	4.40	18.74	44.22	70.50	88.08	96.28
	22							0.03	0.20	0.82	2.56	13.04	35.29	62.22	83.02	94.10
	23							0.01	0.09	0.41	1.43	8.72	27.20	53.43	76.84	91.07
	24								0.04	0.20	0.76	5.60	20.24	44.58	69.66	87.06
180	0	83.62	97.37	99.58	99.94	99.99	100.00	100.00	100.00	100.00	100.00	100.00	100.00	100.00	100.00	100.00
	1	53.84	87.69	97.27	99.45	99.90	99.98	100.00	100.00	100.00	100.00	100.00	100.00	100.00	100.00	100.00
	2	26.91	70.01	90.86	97.65	99.46	99.89	99.98	100.00	100.00	100.00	100.00	100.00	100.00	100.00	100.00
	3	10.77	48.61	79.10	93.20	98.10	99.52	99.89	99.98	100.00	100.00	100.00	100.00	100.00	100.00	100.00
	4	3.56	29.28	63.01	84.99	94.93	98.50	99.60	99.90	99.98	100.00	100.00	100.00	100.00	100.00	100.00
	5	1.00	15.39	45.49	72.95	89.05	96.21	98.84	99.68	99.92	99.98	100.00	100.00	100.00	100.00	100.00
	6	0.24	7.13	29.69	58.32	80.02	91.98	97.16	99.11	99.75	99.93	100.00	100.00	100.00	100.00	100.00
	7	0.05	2.93	17.54	43.17	68.21	85.15	94.03	97.88	99.32	99.80	99.99	100.00	100.00	100.00	100.00
	8	0.01	1.08	9.41	29.51	54.77	75.79	88.92	95.57	98.42	99.49	99.96	100.00	100.00	100.00	100.00
	9		0.36	4.61	18.64	41.26	64.37	81.58	91.72	96.70	98.82	99.88	99.99	100.00	100.00	100.00
	10		0.11	2.07	10.89	29.09	51.90	72.13	86.00	93.81	97.55	99.71	99.98	100.00	100.00	100.00
	11		0.03	0.86	5.90	19.20	39.61	61.13	78.32	89.38	95.37	99.35	99.94	99.99	100.00	100.00
	12		0.01	0.33	2.98	11.86	28.56	49.48	68.91	83.21	91.96	98.65	99.84	99.99	100.00	100.00
	13			0.12	1.40	6.87	19.44	38.14	58.33	75.33	87.06	97.42	99.65	99.96	100.00	100.00
	14			0.04	0.62	3.74	12.50	27.96	47.36	66.03	80.57	95.42	99.27	99.92	99.99	100.00
	15			0.01	0.25	1.92	7.59	19.48	36.81	55.86	72.59	92.41	98.58	99.81	99.98	100.00
	16				0.10	0.92	4.36	12.90	27.34	45.48	63.44	88.17	97.43	99.60	99.96	100.00
	17				0.04	0.42	2.38	8.12	19.40	35.57	53.63	82.59	95.62	99.23	99.90	99.99
	18				0.01	0.18	1.23	4.87	13.15	26.70	43.77	75.69	92.95	98.57	99.79	99.98
	19					0.08	0.60	2.78	8.51	19.22	34.42	67.68	89.25	97.52	99.59	99.95
	20					0.03	0.28	1.51	5.26	13.27	26.06	58.89	84.40	95.89	99.22	99.89
	21					0.01	0.13	0.78	3.11	8.78	18.98	49.75	78.39	93.54	98.62	99.78
	22						0.05	0.39	1.76	5.57	13.30	40.74	71.31	90.29	97.66	99.59
	23						0.02	0.18	0.96	3.40	8.96	32.31	63.40	86.05	96.21	99.25
	24						0.01	0.08	0.50	1.99	5.81	24.78	54.97	80.76	94.13	98.69
	25							0.04	0.25	1.12	3.62	18.38	46.41	74.47	91.27	97.83
	26							0.02	0.12	0.60	2.18	13.17	38.10	67.84	87.54	96.54

PROBABILITY THAT ERROR RATE IN UNIVERSE

Page 16 **SIZE OF** OVER 1000 IS LESS THAN:

SIZE OF SAMPLE EXAMINED	NO. OF ERRORS FOUND	1%	2%	3%	4%	5%	6%	7%	8%	9%	10%	12%	14%	16%	18%	20%
180 (Cont)	27							0.01	0.06	0.31	1.26	9.12	30.39	59.58	82.87	94.70
	28								0.02	0.16	0.70	6.11	23.52	51.51	77.27	92.19
	29								0.01	0.08	0.38	3.95	17.67	43.45	70.82	88.90
220	0	89.04	98.83	99.88	99.99	100.00	100.00	100.00	100.00	100.00	100.00	100.00	100.00	100.00	100.00	100.00
	1	64.69	98.55	99.04	99.87	99.98	100.00	100.00	100.00	100.00	100.00	100.00	100.00	100.00	100.00	100.00
	2	37.76	81.77	96.21	99.35	99.90	99.99	100.00	100.00	100.00	100.00	100.00	100.00	100.00	100.00	100.00
	3	17.99	64.30	89.84	97.75	99.58	99.93	99.99	100.00	100.00	100.00	100.00	100.00	100.00	100.00	100.00
	4	7.15	44.96	79.15	94.15	98.66	99.74	99.95	99.99	100.00	100.00	100.00	100.00	100.00	100.00	100.00
	5	2.42	27.91	64.88	87.67	96.58	99.21	99.84	99.97	99.99	100.00	100.00	100.00	100.00	100.00	100.00
	6	0.71	15.43	49.06	77.99	92.66	97.99	99.53	99.90	99.98	100.00	100.00	100.00	100.00	100.00	100.00
	7	0.19	7.65	34.10	65.67	86.34	95.61	98.81	99.72	99.94	99.99	100.00	100.00	100.00	100.00	100.00
	8	0.04	3.43	21.78	51.99	77.48	91.57	97.38	99.30	99.83	99.67	100.00	100.00	100.00	100.00	100.00
	9	0.01	1.39	12.81	88.57	66.51	85.50	94.83	98.43	99.59	99.90	100.00	100.00	100.00	100.00	100.00
	10		0.52	6.95	26.77	54.32	77.32	90.79	96.85	99.07	99.76	99.99	100.00	100.00	100.00	100.00
	11		0.18	3.50	17.38	42.07	67.35	84.98	94.21	98.09	99.45	99.98	100.00	100.00	100.00	100.00
	12		0.06	1.63	10.57	30.84	56.27	77.36	90.22	96.40	98.85	99.92	100.00	100.00	100.00	100.00
	13		0.02	0.71	6.03	21.39	44.96	68.19	84.67	93.72	97.78	99.81	99.99	100.00	100.00	100.00
	14			0.29	3.23	14.03	34.28	57.98	77.53	89.82	96.02	99.59	99.97	100.00	100.00	100.00
	15			0.11	1.63	8.71	24.92	47.43	69.00	84.51	93.35	99.18	99.94	100.00	100.00	100.00
	16			0.04	0.78	5.13	17.26	37.25	59.50	77.78	89.54	98.46	99.86	99.99	100.00	100.00
	17			0.01	0.35	2.86	11.40	28.06	49.59	69.80	84.46	97.28	99.71	99.98	100.00	100.00
	18				0.15	1.52	7.18	20.26	39.87	60.89	78.09	95.47	99.43	99.95	100.00	100.00
	19				0.06	0.77	4.31	14.02	30.89	51.53	70.57	92.85	98.94	99.90	99.99	100.00
	20				0.02	0.37	2.48	9.29	23.04	42.22	62.17	89.25	98.15	99.80	99.98	100.00
	21				0.01	0.17	1.36	5.91	16.53	33.46	53.29	84.58	96.93	99.61	99.97	100.00
	22					0.07	0.71	3.60	11.42	25.61	44.36	78.82	95.12	99.23	99.98	100.00
	23					0.03	0.36	2.11	7.59	18.94	35.81	72.06	92.59	98.76	99.86	99.99
	24					0.01	0.17	1.19	4.85	13.52	28.02	64.49	89.21	97.93	99.74	99.98
	25						0.08	0.64	2.99	9.31	21.23	56.40	84.90	96.69	99.52	99.95
	26						0.04	0.33	1.78	6.20	15.58	48.12	79.63	94.93	99.17	99.91
	27						0.02	0.17	1.02	3.98	11.06	40.01	73.47	92.51	98.62	99.83
	28						0.01	0.08	0.56	2.47	7.60	32.39	66.56	89.84	97.78	99.69
	29							0.04	0.30	1.48	5.06	25.51	59.11	85.34	96.56	99.45
	30							0.02	0.15	0.86	3.26	19.54	51.39	80.49	94.85	99.09
	31							0.01	0.08	0.48	2.03	14.54	43.69	74.83	92.56	98.52

Page 17

PROBABILITY THAT ERROR RATE IN UNIVERSE SIZE OF OVER 1000 IS LESS THAN:

SIZE OF SAMPLE EXAMINED	NO. OF ERRORS FOUND	1%	2%	3%	4%	5%	6%	7%	8%	9%	10%	12%	14%	16%	18%	20%
220 (Cont)	32								0.04	0.26	1.23	10.52	36.28	68.46	89.58	97.69
	33								0.02	0.14	0.72	7.40	29.41	61.54	85.86	96.50
	34								0.01	0.07	0.41	5.05	23.26	54.30	81.87	94.86
	35									0.04	0.23	3.36	17.94	46.97	76.13	92.69
	36									0.02	0.12	2.17	13.49	39.80	70.21	89.90
240	0	91.04	99.22	99.93	99.99	100.00	100.00	100.00	100.00	100.00	100.00	100.00	100.00	100.00	100.00	100.00
	1	69.31	95.38	99.44	99.94	99.99	100.00	100.00	100.00	100.00	100.00	100.00	100.00	100.00	100.00	100.00
	2	43.08	86.01	97.60	99.66	99.66	99.99	100.00	100.00	100.00	100.00	100.00	100.00	100.00	100.00	100.00
	3	22.06	70.85	93.10	98.75	99.81	99.97	100.00	100.00	100.00	100.00	100.00	100.00	100.00	100.00	100.00
	4	9.49	52.52	84.85	96.49	99.34	99.90	99.99	100.00	100.00	100.00	100.00	100.00	100.00	100.00	100.00
	5	3.49	34.86	72.81	90.04	98.19	99.66	99.94	99.99	100.00	100.00	100.00	100.00	100.00	100.00	100.00
	6	1.12	20.75	58.23	84.79	95.80	99.06	99.82	99.97	99.99	100.00	100.00	100.00	100.00	100.00	100.00
	7	0.32	11.12	43.15	74.69	91.60	97.78	99.51	99.91	99.98	100.00	100.00	100.00	100.00	100.00	100.00
	8	0.08	5.40	29.57	62.44	85.16	95.40	98.83	99.74	99.95	99.99	100.00	100.00	100.00	100.00	100.00
	9	0.02	2.39	18.74	49.27	76.43	91.49	97.50	99.38	99.87	99.98	100.00	100.00	100.00	100.00	100.00
	10		0.97	11.00	36.60	65.82	85.73	95.20	98.65	99.68	99.93	100.00	100.00	100.00	100.00	100.00
	11		0.36	6.00	25.56	54.14	78.04	91.58	97.33	99.28	99.83	99.99	100.00	100.00	100.00	100.00
	12		0.12	3.05	16.78	42.40	68.67	86.38	95.14	98.53	99.62	99.98	100.00	100.00	100.00	100.00
	13		0.04	1.45	10.37	31.57	58.17	79.51	91.79	97.24	99.20	99.96	100.00	100.00	100.00	100.00
	14		0.01	0.64	6.04	22.33	47.32	71.13	87.07	95.16	98.46	99.90	100.00	100.00	100.00	100.00
	15			0.27	3.32	15.00	36.88	61.63	80.88	92.07	97.21	99.77	99.99	100.00	100.00	100.00
	16			0.11	1.72	9.57	27.50	51.57	73.32	87.76	95.25	99.54	99.97	100.00	100.00	100.00
	17			0.04	0.85	5.81	19.62	41.60	64.65	82.15	92.39	99.11	99.94	100.00	100.00	100.00
	18			0.01	0.40	3.36	13.39	32.30	55.32	75.28	88.45	98.40	99.87	99.99	100.00	100.00
	19				0.18	1.85	8.74	24.11	45.83	67.34	83.33	97.26	99.74	99.98	100.00	100.00
	20				0.07	0.97	5.46	17.31	36.72	58.66	77.06	95.55	99.50	99.96	100.00	100.00
	21				0.03	0.49	3.27	11.94	28.42	49.66	69.75	93.10	99.09	99.93	100.00	100.00
	22				0.01	0.23	1.87	7.92	21.23	40.81	61.66	89.77	98.43	99.85	99.99	100.00
	23					0.11	1.03	5.06	15.31	32.51	53.15	85.47	97.42	99.72	99.98	100.00
	24					0.05	0.55	3.10	10.65	25.09	44.59	80.17	95.92	99.49	99.96	100.00
	25					0.02	0.28	1.84	7.15	18.74	36.38	73.92	93.82	99.12	99.92	100.00
	26					0.01	0.14	1.05	4.64	13.56	28.84	66.88	90.99	98.52	99.85	99.99
	27						0.06	0.57	2.90	9.49	22.19	59.27	87.34	97.63	99.72	99.98

Page 18

PROBABILITY THAT ERROR RATE IN UNIVERSE
SIZE OF OVER 1000 IS LESS THAN:

SIZE OF SAMPLE EXAMINED	NO. OF ERRORS FOUND	1%	2%	3%	4%	5%	6%	7%	8%	9%	10%	12%	14%	16%	18%	20%
240 (Cont)	28						0.03	0.30	1.76	6.43	16.58	51.37	82.82	96.33	99.51	99.96
	29						0.01	0.16	1.03	4.22	12.01	43.50	77.44	94.53	99.17	99.92
	30							0.08	0.58	2.68	8.45	35.95	71.28	92.11	98.64	99.85
	31							0.04	0.32	1.65	5.77	28.98	64.48	88.99	97.86	99.73
	32							0.02	0.17	0.98	3.82	22.76	57.26	85.11	96.74	99.54
	33							0.01	0.09	0.57	2.45	17.43	49.85	80.45	95.19	99.24
	34								0.04	0.32	1.53	12.99	42.50	75.05	93.12	98.78
	35								0.02	0.17	0.93	9.44	35.46	68.99	90.45	98.10
	36								0.01	0.09	0.55	6.67	28.94	62.42	87.11	97.14
	37									0.05	0.31	4.60	23.08	55.52	83.06	95.81
	38									0.02	0.17	3.08	17.99	48.50	78.32	94.04
	39									0.01	0.09	2.02	13.70	41.57	72.92	91.74
	40										0.05	1.28	10.18	34.94	66.97	88.86
260	0	92.67	99.48	99.96	100.00	100.00	100.00	100.00	100.00	100.00	100.00	100.00	100.00	100.00	100.00	100.00
	1	73.42	96.70	99.67	99.97	100.00	100.00	100.00	100.00	100.00	100.00	100.00	100.00	100.00	100.00	100.00
	2	48.23	89.36	98.50	99.83	99.93	100.00	100.00	100.00	100.00	100.00	100.00	100.00	100.00	100.00	100.00
	3	26.36	76.48	95.39	99.31	99.91	99.99	100.00	100.00	100.00	100.00	100.00	100.00	100.00	100.00	100.00
	4	12.16	59.59	89.20	97.93	99.68	99.96	99.99	100.00	100.00	100.00	100.00	100.00	100.00	100.00	100.00
	5	4.82	41.94	79.39	94.99	99.06	99.86	99.98	100.00	100.00	100.00	100.00	100.00	100.00	100.00	100.00
	6	1.66	26.64	66.51	89.79	97.67	99.57	99.93	99.99	100.00	100.00	100.00	100.00	100.00	100.00	100.00
	7	0.51	15.30	52.05	81.92	95.02	98.92	99.81	99.97	100.00	100.00	100.00	100.00	100.00	100.00	100.00
	8	0.14	7.99	37.91	71.54	90.60	97.60	99.50	99.91	99.99	100.00	100.00	100.00	100.00	100.00	100.00
	9	0.03	3.81	25.67	59.44	84.05	95.24	98.86	99.77	99.96	99.99	100.00	100.00	100.00	100.00	100.00
	10	0.01	1.67	16.16	46.79	75.49	91.46	97.64	99.46	99.89	99.98	100.00	100.00	100.00	100.00	100.00
	11		0.67	9.48	34.80	65.21	85.98	95.55	98.85	99.75	99.95	100.00	100.00	100.00	100.00	100.00
	12		0.25	5.19	24.44	53.97	78.72	92.30	97.74	99.44	99.88	99.99	100.00	100.00	100.00	100.00
	13		0.09	2.66	16.21	42.69	69.88	87.63	95.90	98.87	99.74	99.99	100.00	100.00	100.00	100.00
	14		0.03	1.28	10.15	32.22	59.92	81.42	93.09	97.88	99.45	99.98	100.00	100.00	100.00	100.00
	15		0.01	0.58	6.01	23.18	49.50	73.76	89.07	96.26	98.93	99.94	100.00	100.00	100.00	100.00
	16			0.25	3.38	15.89	39.31	64.94	83.73	93.82	98.04	99.88	99.99	100.00	100.00	100.00
	17			0.10	1.80	10.39	29.98	55.40	77.05	90.35	96.62	99.74	99.99	100.00	100.00	100.00
	18			0.04	0.91	6.48	21.94	45.71	69.22	85.71	94.49	99.49	99.97	100.00	100.00	100.00
	19			0.01	0.44	3.86	15.40	36.42	60.54	79.87	91.47	99.07	99.94	100.00	100.00	100.00

PROBABILITY THAT ERROR RATE IN UNIVERSE

Page 19 **SIZE OF** OVER 1000 IS LESS THAN:

SIZE OF SAMPLE EXAMINED	NO. OF ERRORS FOUND	1%	2%	3%	4%	5%	6%	7%	8%	9%	10%	12%	14%	16%	18%	20%
260 (Cont)	20				0.20	2.19	10.37	28.00	51.45	72.91	87.44	98.36	99.88	99.99	100.00	100.00
	21				0.09	1.19	6.71	20.75	42.42	65.05	82.31	97.27	99.77	99.99	100.00	100.00
	22				0.04	0.62	4.16	14.82	33.88	56.60	76.12	95.64	99.56	99.97	100.00	100.00
	23				0.01	0.31	2.48	10.21	26.21	47.95	69.01	93.35	99.22	99.95	100.00	100.00
	24					0.15	1.42	6.78	19.61	39.50	61.21	90.27	98.67	99.89	99.99	100.00
	25					0.07	0.78	4.34	14.20	31.62	53.02	86.29	97.83	99.80	99.99	100.00
	26					0.03	0.42	2.68	9.94	24.57	44.80	81.40	96.59	99.64	99.98	100.00
	27					0.01	0.21	1.60	6.74	18.53	36.89	75.61	94.83	99.37	99.95	199.00
	28						0.11	0.92	4.42	13.55	29.57	69.05	92.46	98.94	99.91	99.99
	29						0.05	0.51	2.80	9.62	23.06	61.88	89.37	98.30	99.83	99.99
	30						0.02	0.28	1.72	6.62	17.50	54.37	85.49	97.35	99.71	99.98
	31						0.01	0.14	1.03	4.42	12.91	46.76	80.81	96.00	99.50	99.96
	32							0.07	0.59	2.87	9.26	39.33	75.36	94.17	99.17	99.93
	33							0.04	0.33	1.81	6.46	32.34	69.23	91.77	98.68	99.87
	34							0.02	0.18	1.10	4.38	25.97	62.56	85.70	97.96	99.77
	35							0.01	0.09	0.66	2.89	20.36	55.56	84.94	96.93	99.61
	36								0.05	0.38	1.85	15.58	48.43	80.46	95.52	99.36
	37								0.02	0.21	1.16	11.64	41.40	75.28	93.65	98.99
	38								0.01	0.12	0.70	8.48	34.69	69.51	91.24	98.44
	39									0.06	0.42	6.03	28.47	63.24	88.22	97.66
	40									0.03	0.24	4.18	22.88	56.65	84.57	96.57
	41									0.02	0.13	2.83	17.99	49.91	80.26	95.12
	42									0.01	0.07	1.87	13.85	43.21	75.34	93.24
	43										0.04	1.20	10.42	36.75	69.85	90.84
300	0	95.10	99.77	99.99	100.00	100.00	100.00	100.00	100.00	100.00	100.00	100.00	100.00	100.00	100.00	100.00
	1	80.24	98.34	99.89	99.99	100.00	100.00	100.00	100.00	100.00	100.00	100.00	100.00	100.00	100.00	100.00
	2	57.79	93.98	99.43	99.96	100.00	100.00	100.00	100.00	100.00	100.00	100.00	100.00	100.00	100.00	100.00
	3	35.28	85.15	98.01	99.80	99.98	100.00	100.00	100.00	100.00	100.00	100.00	100.00	100.00	100.00	100.00
	4	18.39	71.77	94.76	99.32	99.93	99.99	100.00	100.00	100.00	100.00	100.00	100.00	100.00	100.00	100.00
	5	8.29	55.59	88.80	98.14	99.77	99.98	100.00	100.00	100.00	100.00	100.00	100.00	100.00	100.00	100.00
	6	3.28	39.37	79.74	95.72	99.34	99.92	99.99	100.00	100.00	100.00	100.00	100.00	100.00	100.00	100.00
	7	1.15	25.46	67.97	91.49	98.40	99.77	99.97	100.00	100.00	100.00	100.00	100.00	100.00	100.00	100.00
	8	0.36	15.07	54.64	85.03	96.59	99.41	99.92	99.99	100.00	100.00	100.00	100.00	100.00	100.00	100.00
	9	0.10	8.18	41.26	76.30	93.50	98.68	99.79	99.97	100.00	100.00	100.00	100.00	100.00	100.00	100.00
	10	0.03	4.10	29.22	65.71	88.77	97.32	99.50	99.93	99.99	100.00	100.00	100.00	100.00	100.00	100.00
	11	0.01	1.90	19.40	54.07	82.20	95.03	98.93	99.82	99.97	100.00	100.00	100.00	100.00	100.00	100.00

PROBABILITY THAT ERROR RATE IN UNIVERSE

Page 20 **SIZE OF** OVER 1000 IS LESS THAN:

SIZE OF SAMPLE EXAMINED	NO. OF ERRORS FOUND	1%	2%	3%	4%	5%	6%	7%	8%	9%	10%	12%	14%	16%	18%	20%
300 (Cont)	12		0.82	12.09	42.40	73.88	91.50	97.91	99.59	99.93	99.99	100.00	100.00	100.00	100.00	100.00
	13		0.33	7.08	31.63	64.17	86.52	96.19	99.15	99.85	99.98	100.00	100.00	100.00	100.00	100.00
	14		0.12	3.90	22.42	53.70	80.00	93.54	98.37	99.67	99.94	100.00	100.00	100.00	100.00	100.00
	15		0.04	2.03	15.11	43.19	72.07	89.74	97.08	99.33	99.87	100.00	100.00	100.00	100.00	100.00
	16		0.01	1.00	9.69	33.34	63.05	84.64	95.08	98.74	99.74	99.99	100.00	100.00	100.00	100.00
	17			0.46	5.91	24.67	53.43	78.23	92.17	97.76	99.48	99.98	100.00	100.00	100.00	100.00
	18			0.21	3.43	17.50	43.77	70.65	88.19	96.24	99.03	99.96	100.00	100.00	100.00	100.00
	19			0.09	1.90	11.91	34.63	62.18	83.05	94.01	98.29	99.92	100.00	100.00	100.00	100.00
	20			0.03	1.01	7.76	26.43	53.22	76.78	90.91	97.13	99.84	99.99	100.00	100.00	100.00
	21			0.01	0.51	4.86	19.45	44.23	69.51	86.82	95.42	99.69	99.99	100.00	100.00	100.00
	22				0.25	2.92	13.79	35.65	61.49	81.69	93.01	99.43	99.98	100.00	100.00	100.00
	23				0.12	1.68	9.44	27.84	53.06	75.56	89.77	99.01	99.95	100.00	100.00	100.00
	24				0.05	0.94	6.22	21.05	44.60	68.56	85.61	98.34	99.90	100.00	100.00	100.00
	25				0.02	0.50	3.96	15.42	36.47	60.92	80.51	97.33	99.82	99.99	100.00	100.00
	26				0.01	0.26	2.43	10.93	29.00	52.92	74.52	95.88	99.67	99.99	100.00	100.00
	27					0.13	1.44	7.50	22.41	44.90	67.76	93.87	99.43	99.97	100.00	100.00
	28					0.05	0.83	4.99	16.82	37.16	60.44	91.20	99.04	99.94	100.00	100.00
	29					0.03	0.46	3.21	12.26	29.99	52.81	87.78	98.46	99.90	100.00	100.00
	30					0.01	0.25	2.00	8.68	23.58	45.16	83.57	97.59	99.81	99.99	100.00
	31						0.13	1.21	5.97	18.05	37.75	78.57	96.36	99.68	99.89	100.00
	32						0.06	0.71	3.99	13.46	30.83	72.84	94.69	99.46	99.97	100.00
	33						0.03	0.40	2.59	9.77	24.58	66.49	92.47	99.12	99.94	100.00
	34						0.02	0.22	1.63	6.91	19.14	59.69	89.63	98.61	99.90	100.00
	35						0.01	0.12	1.00	4.75	14.53	52.65	86.12	97.87	99.82	99.99
	36							0.06	0.60	3.19	10.77	45.57	81.92	96.84	99.70	99.98
	37							0.03	0.35	2.08	7.79	38.69	77.03	95.44	99.50	99.97
	38							0.02	0.20	1.32	5.49	32.20	71.53	94.59	99.21	99.94
	39							0.01	0.11	0.82	3.78	26.25	65.51	91.22	98.77	99.90
	40								0.06	0.50	2.54	20.96	59.12	88.28	98.15	99.83
	41								0.03	0.29	1.66	16.38	52.52	84.73	97.29	99.72
	42								0.02	0.17	1.06	12.53	45.90	80.56	96.12	99.56
	43								0.01	0.09	0.67	9.33	39.43	75.79	94.57	99.81

PROBABILITY THAT ERROR RATE IN UNIVERSE

Page 21 SIZE OF OVER 1000 IS LESS THAN:

SIZE OF SAMPLE EXAMINED	NO. OF ERRORS FOUND	1%	2%	3%	4%	5%	6%	7%	8%	9%	10%	12%	14%	16%	18%	20%
300 (Cont)	44									0.05	0.41	6.88	33.27	70.49	92.60	98.94
	45									0.03	0.24	4.93	27.58	64.74	90.13	98.42
	46									0.01	0.14	3.46	22.44	58.68	87.12	97.70
	47										0.08	2.38	17.91	52.43	83.56	96.72
	48										0.04	1.60	14.03	46.16	79.43	95.43
400	0	98.20	99.97	100.00	100.00	100.00	100.00	100.00	100.00	100.00	100.00	100.00	100.00	100.00	100.00	100.00
	1	90.95	99.72	99.99	100.00	100.00	100.00	100.00	100.00	100.00	100.00	100.00	100.00	100.00	100.00	100.00
	2	76.34	98.69	99.95	100.00	100.00	100.00	100.00	100.00	100.00	100.00	100.00	100.00	100.00	100.00	100.00
	3	56.75	95.91	99.79	99.99	100.00	100.00	100.00	100.00	100.00	100.00	100.00	100.00	100.00	100.00	100.00
	4	37.12	90.27	99.30	99.97	100.00	100.00	100.00	100.00	100.00	100.00	100.00	100.00	100.00	100.00	100.00
	5	21.41	81.15	98.10	99.88	99.99	100.00	100.00	100.00	100.00	100.00	100.00	100.00	100.00	100.00	100.00
	6	10.96	68.91	95.65	99.65	99.98	100.00	100.00	100.00	100.00	100.00	100.00	100.00	100.00	100.00	100.00
	7	5.02	54.85	91.38	99.10	99.94	100.00	100.00	100.00	100.00	100.00	100.00	100.00	100.00	100.00	100.00
	8	2.08	40.74	84.90	97.99	99.83	99.99	100.00	100.00	100.00	100.00	100.00	100.00	100.00	100.00	100.00
	9	0.78	28.21	76.16	95.97	99.58	99.97	100.00	100.00	100.00	100.00	100.00	100.00	100.00	100.00	100.00
	10	0.27	18.21	65.60	92.67	99.06	99.92	99.99	100.00	100.00	100.00	100.00	100.00	100.00	100.00	100.00
	11	0.08	10.97	54.01	87.80	98.10	99.80	99.98	100.00	100.00	100.00	100.00	100.00	100.00	100.00	100.00
	12	0.02	6.19	42.40	81.23	96.45	99.56	99.96	100.00	100.00	100.00	100.00	100.00	100.00	100.00	100.00
	13	0.01	3.27	31.68	73.05	93.86	99.10	99.91	99.99	100.00	100.00	100.00	100.00	100.00	100.00	100.00
	14		1.62	22.52	63.63	90.10	98.28	99.79	99.98	100.00	100.00	100.00	100.00	100.00	100.00	100.00
	15		0.76	15.23	53.53	85.01	96.95	99.57	99.96	100.00	100.00	100.00	100.00	100.00	100.00	100.00
	16		0.34	9.80	43.40	78.56	94.90	99.18	99.90	99.99	100.00	100.00	100.00	100.00	100.00	100.00
	17		0.14	6.01	33.87	70.88	91.94	98.50	99.80	99.98	100.00	100.00	100.00	100.00	100.00	100.00
	18		0.06	3.51	25.42	62.29	87.93	97.42	99.61	99.96	100.00	100.00	100.00	100.00	100.00	100.00
	19		0.02	1.96	18.34	53.20	82.78	95.78	99.27	99.91	99.99	100.00	100.00	100.00	100.00	100.00
	20		0.01	1.04	12.72	44.09	76.51	93.43	98.71	99.81	99.98	100.00	100.00	100.00	100.00	100.00
	21			0.53	8.49	35.41	69.27	90.23	97.83	99.65	99.96	100.00	100.00	100.00	100.00	100.0
	22			0.26	5.45	27.54	61.32	86.08	96.52	99.37	99.91	100.00	100.00	100.00	100.00	100.00
	23			0.12	3.37	20.73	52.97	80.95	94.64	98.91	99.83	100.00	100.00	100.00	100.00	100.00
	24			0.06	2.00	15.10	44.60	74.88	92.07	98.20	99.69	100.00	100.00	100.00	100.00	100.00
	25			0.02	1.15	10.65	36.56	68.02	88.72	97.14	99.46	99.99	100.00	100.00	100.00	100.00
	26			0.01	0.64	7.27	29.16	60.56	84.50	95.63	99.08	99.98	100.00	100.00	100.00	100.00
	27				0.34	4.80	22.62	52.79	79.43	93.56	98.51	99.96	100.00	100.00	100.00	100.00

PROBABILITY THAT ERROR RATE IN UNIVERSE

Page 22 SIZE OF OVER 1000 IS LESS THAN:

SIZE OF SAMPLE EXAMINED	NO. OF ERRORS FOUND	1%	2%	3%	4%	5%	6%	7%	8%	9%	10%	12%	14%	16%	18%	20%
400 (Cont)	28				0.18	3.07	17.06	44.99	78.56	90.83	97.65	99.93	100.00	100.00	100.00	100.00
	29				0.09	1.90	12.51	37.47	67.01	87.87	96.43	99.87	100.00	100.00	100.00	100.00
	30				0.04	1.14	8.92	30.46	59.96	83.14	94.76	99.78	100.00	100.00	100.00	100.00
	31				0.02	0.67	6.18	24.17	52.65	78.15	92.54	99.62	99.99	100.00	100.00	100.00
	32				0.01	0.38	4.16	18.71	45.31	72.45	89.70	99.37	99.98	100.00	100.00	100.00
	33					0.21	2.73	14.12	38.20	66.17	86.18	98.99	99.97	100.00	100.00	100.00
	34					0.11	1.74	10.40	31.53	59.47	81.95	98.43	99.95	100.00	100.00	100.00
	35					0.06	1.08	7.47	25.46	52.53	77.04	97.64	99.91	100.00	100.00	100.00
	36					0.03	0.65	5.23	20.11	45.58	71.51	96.54	99.84	100.00	100.00	100.00
	37					0.01	0.38	3.57	15.53	38.81	65.47	95.07	99.74	99.99	100.00	100.00
	38						0.22	2.38	11.73	32.42	59.05	93.15	99.58	99.99	100.00	100.00
	39						0.12	1.55	8.66	26.55	52.44	90.72	99.33	99.98	100.00	100.00
	40						0.07	0.98	6.25	21.31	45.80	87.73	98.97	99.96	100.00	100.00
	41						0.04	0.61	4.41	16.77	39.33	84.15	98.45	99.94	100.00	100.00
	42						0.02	0.37	3.04	12.92	33.18	79.98	97.73	99.89	100.00	100.00
	43						0.01	0.22	2.05	9.76	27.49	75.24	96.75	99.83	100.00	100.00
	44							0.13	1.36	7.21	22.37	70.00	95.46	99.72	99.99	100.00
	45							0.07	0.88	5.23	17.86	64.35	93.80	99.56	99.99	100.00
	46							0.04	0.55	3.71	14.00	58.40	91.71	99.33	99.98	100.00
	47							0.02	0.34	2.58	10.76	52.29	89.15	98.99	99.96	100.00
	48							0.01	0.21	1.76	8.12	46.16	86.09	98.52	99.93	100.00
	49								0.12	1.17	6.01	40.16	82.51	97.88	99.89	100.00
	50								0.07	0.77	4.36	34.42	78.41	97.01	99.82	99.99
	51								0.04	0.49	3.11	29.04	73.84	95.89	99.72	99.99
	52								0.02	0.31	2.17	24.12	68.84	94.45	99.57	99.98
	53								0.01	0.19	1.49	19.71	63.50	92.65	99.35	99.97
	54									0.12	1.00	15.85	57.91	90.45	99.05	99.96
	55									0.07	0.66	12.54	52.18	87.81	98.62	99.93
	56									0.04	0.43	9.75	46.44	84.71	98.06	99.89
	57									0.02	0.27	7.46	40.80	81.15	97.30	99.82
	58									0.01	0.17	5.62	35.38	77.14	96.32	99.73
	59										0.11	4.16	30.25	72.71	95.08	99.59
	60										0.06	3.03	25.51	67.91	93.52	99.39
	61										0.04	2.17	21.21	62.82	91.62	99.12
	62										0.02	1.52	17.38	57.52	89.34	98.75
	63										0.01	1.06	14.04	52.10	86.65	98.25

Index

R

S

T

U